D1796908

Community structure and co-operation in biofilms

The study of biofilm considers the close association of micro-organisms with each other at interfaces and the consequent physiological adaptation to the proximity of other cells and surfaces. As such it is relevant to a variety of disciplines, including medicine, dentistry, bioremediation, biofouling, water technology, engineering and food science. Although the habitats studied, and their associated cellular communities, differ widely, some common elements exist such as method of attachment, coadhesion, regulation of biofilm phenotype and biofilm architecture. This book aims to distil the common principles of biofilm physiology and growth for all interested disciplines. It will appeal to the specialist biofilm researcher as well as to students wishing to introduce themselves to the topic.

David Allison is a Senior Lecturer in Microbiology in the School of Pharmacy and Pharmaceutical Sciences at the University of Manchester, UK.
Peter Gilbert is a Reader in Microbiology in the School of Pharmacy and Pharmaceutical Sciences at the University of Manchester, UK.
Hilary Lappin-Scott is Professor of Environmental Microbiology in the School of Biological Sciences at the University of Exeter, UK.
Michael Wilson is Professor of Microbiology in the Eastman Dental Institute at University College London, UK.

Symposia of the Society for General Microbiology

Managing Editor: Dr Melanie Scourfield, SGM, Reading, UK
Volumes currently available:

FIFTY-NINTH SYMPOSIUM OF THE
SOCIETY FOR GENERAL MICROBIOLOGY
HELD AT THE UNIVERSITY OF EXETER SEPTEMBER 2000

Edited by
D. G. Allison, P. Gilbert, H. M. Lappin-Scott and M. Wilson

Community structure and co-operation in biofilms

Published for the Society for General Microbiology

PUBLISHED BY THE PRESS SYNDICATE OF THE UNIVERSITY OF CAMBRIDGE
The Pitt Building, Trumpington Street, Cambridge, United Kingdom

CAMBRIDGE UNIVERSITY PRESS
The Edinburgh Building, Cambridge CB2 2RU, UK   http://www.cup.cam.ac.uk
40 West 20th Street, New York, NY 10011–4211, USA   http://www.cup.org
10 Stamford Road, Oakleigh, Melbourne 3166, Australia
Ruiz de Alarcón 13, 28014 Madrid, Spain

First published 2000

Printed in Great Britain at the University Press, Cambridge

Typeface Sabon (Adobe) 10/13.5pt   *System* QuarkXPress™   [SE]

A catalogue record for this book is available from the British Library

ISBN 0 521 79302 5 hardback

Front cover illustration: *Pseudomonas aeruginosa* PAN067 biofilm streamers growing in turbulent flow. The streamers were attached to the glass surface by the upstream 'head' while the downstream 'tails' oscillated rapidly in the flow. The scanning electron photomicrographs were provided by Paul Stoodley [Center for Biofilm Engineering (CBE), Bozeman, MT, USA, and Exeter University, UK], Frieda Jørgensen and Hilary Lappin-Scott (Exeter University, UK). Digital enhancement by Kathy Lange, CBE Graphic Design. For experimental details see Stoodley *et al.* this volume.

CONTENTS

CONTRIBUTORS

Allison, D. G.
School of Pharmacy and Pharmaceutical Sciences, University of Manchester, Oxford Road, Manchester M13 9PL, UK

Andersen, R. N.
Oral Infection and Immunity Branch, National Institute of Dental and Craniofacial Research, National Institutes of Health, Bethesda, MD 20892, USA

Barken, K. B.
Molecular Microbial Ecology Group, Department of Microbiology, Technical University of Denmark, DK-2800 Lyngby, Denmark

Bayston, R.
Biomaterials-Related Infection Group, University of Nottingham, Division of Microbiology, Clinical Sciences Building, City Hospital, Nottingham NG5 1PB, UK

Borchard, W.
Institute for Physical and Theoretical Chemistry, University of Duisburg, Germany

Bowden, G. H. W.
Department of Oral Biology, Faculty of Dentistry, The University of Manitoba, Winnipeg, Canada

Boyle, J. D.
School of Engineering, Exeter University, Exeter, UK

Busscher, H. J.
Department of Biomedical Engineering, University of Groningen, Antonius Deusinglaan 1, 9713 AV Groningen, The Netherlands

Caldwell, D. E.
Department of Applied Microbiology & Food Science, University of Saskatchewan, Saskatoon, SK, Canada S7N 5A8

Costerton, J. W.
Center for Biofilm Engineering, Montana State University, Bozeman, MT 59717, USA

Davies, D. G.
Department of Biological Sciences, Binghamton University, Binghamton, NY 13902, USA

Dunsmore, B. C.
Environmental Microbiology Research Group, Exeter University, Exeter, UK

Ehlers, L. J.
National Research Council, Washington, DC, USA

Flemming, H.-C.
Department of Aquatic Microbiology, University of Duisburg, Germany

Gilbert, P.
School of Pharmacy and Pharmaceutical Sciences, University of Manchester, Oxford Road, Manchester M13 9PL, UK

Haagensen, J. A. J.
Molecular Microbial Ecology Group, Department of Microbiology, Technical University of Denmark, DK-2800 Lyngby, Denmark

Hall-Stoodley, L.
Center for Biofilm Engineering, Montana State University, Bozeman, MT, USA

Heijnen, J. J.
Kluyver Institute for Biotechnology, Delft University of Technology, Julianalaan 67, 2628 BC Delft, The Netherlands

Jass, J.
Department of Microbiology, Umeå University, Umeå, Sweden

Jones, S. M.
Environmental Microbiology Research Group, Exeter University, Exeter, UK

Jørgensen, F.
Public Health Laboratory Service, Exeter, UK

Karthikeyan, S.
Department of Applied Microbiology & Food Science, University of Saskatchewan, Saskatoon, SK, Canada S7N 5A8

Kazmerzak, K. M.
Oral Infection and Immunity Branch, National Institute of Dental and Craniofacial Research, National Institutes of Health, Bethesda, MD 20892, USA

Kolenbrander, P. E.
Oral Infection and Immunity Branch, National Institute of Dental and Craniofacial Research, National Institutes of Health, Bethesda, MD 20892, USA

Korber, D. R.
Department of Applied Microbiology & Food Science, University of Saskatchewan, Saskatoon, SK, Canada S7N 5A8

Körstgens, V.
Institute for Physical and Theoretical Chemistry, University of Duisburg, Germany

Lappin-Scott, H. M.
Environmental Microbiology Research Group, Exeter University, Exeter, UK

McBain, A. J.
School of Pharmacy and Pharmaceutical Sciences, University of Manchester, Oxford Road, Manchester M13 9PL, UK

Marsh, P. D.
Research Division, CAMR, Salisbury SP4 0JG, UK, and Leeds Dental Institute, University of Leeds, LS2 9LU, UK

Mayer, C.
Institute for Physical and Theoretical Chemistry, University of Duisburg, Germany

Molin, S.
Molecular Microbial Ecology Group, Department of Microbiology, Technical University of Denmark, DK-2800 Lyngby, Denmark

Moore, G. F.
Environmental Microbiology Research Group, Exeter University, Exeter, UK

Palmer, R. J., Jr
Oral Infection and Immunity Branch, National Institute of Dental and Craniofacial Research, National Institutes of Health, Bethesda, MD 20892, USA

Picioreanu, C.
Kluyver Institute for Biotechnology, Delft University of Technology, Julianalaan 67, 2628 BC Delft, The Netherlands

Smejkal, C. W.
Environmental Microbiology Research Group, Exeter University, Exeter, UK

Sternberg, C.
Molecular Microbial Ecology Group, Department of Microbiology, Technical University of Denmark, DK-2800 Lyngby, Denmark

Stoodley, P.
Center for Biofilm Engineering, Montana State University, Bozeman, MT, USA

van der Mei, H. C.
Department of Biomedical Engineering, University of Groningen, Antonius Deusinglaan 1, 9713 AV Groningen, The Netherlands

van Loosdrecht, M. C. M.
Kluyver Institute for Biotechnology, Delft University of Technology, Julianalaan 67, 2628 BC Delft, The Netherlands

Wimpenny, J.
Cardiff School of Biosciences, Cardiff University, Cathays Park, Cardiff CF1 3TL, UK

Wingender, J.
Department of Aquatic Microbiology, University of Duisburg, Germany

Wolfaardt, G. M.
Department of Microbiology, University of Stellenbosch, Private Bag X1, 7602 Stellenbosch, South Africa

EDITORS' PREFACE

The study of biofilm has been embraced by the microbiological community as it recognizes the profound effect that attachment of cells and cell populations to surfaces has upon their physiology and combined metabolic potential. Particularly, growth of microbial cells as communities, associated with interfaces, has been found to more directly address the many problems and opportunities associated with micro-organisms than do planktonic mono-culture studies. Fifteen years ago, the term 'biofilm' was mentioned in the abstracts and titles of approximately one scientific publication per week. Today, such citations occur every few hours and the wealth of literature captured by this umbrella term has burgeoned. The term biofilm is no longer definitive; rather it is an epithet indicative of an organism's or community's relationship to its natural habitat. Biofilm research, particularly at the community level, does not lend itself to reductionist experiments. Rather, the more one approaches the perfect experiment then the less flexible and informative it sometimes becomes! Inevitably, as the complexity of the system is increased then the range of outcomes and their interpretation broaden. In selecting the contributions to this symposium volume, we have tried not only to reflect the dynamic nature of microbial communities but also to represent the wide range of diverse disciplines that have been brought to bear on this topic. We particularly hope that the book and symposium will kindle the 'biofilm' spirit in the young researcher.

We would like to thank all of the contributors for their input to both the meeting and to the book, and express our sincere gratitude to Melanie Scourfield of the Society for her efficient and gentle handling of the Editors in the production of this volume.

David Allison
Peter Gilbert
Hilary Lappin-Scott
Michael Wilson

An overview of biofilms as functional communities

Julian Wimpenny

Cardiff School of Biosciences, Cardiff University, Cathays Park, Cardiff CF1 3TL, UK

INTRODUCTION

A vast number of microbial aggregates fall into this 'catch-all' name – biofilm. Whether this unifying term does a great service or the opposite to a core branch of microbiology is open to some doubt – for biofilm is found in almost every environment graced with surfaces, sufficient nutrient and some water. A gentle digital examination of the waste outlet of the average kitchen sink will reveal a certain sliminess which embraces the quintessential soul of a biofilm! That 'dirt' which can block car windscreen washer jets is from the same stable. It does not seem necessary to list all possible examples of such structures. They range from growth on the leads of cardiac pacemakers, through biofilm attached to the inner surfaces of water distribution pipes, to the epilithon of rocks in streams and accumulated plaque on the surface of teeth.

There are almost as many definitions of biofilm as there are scientists working in the field or types of the structure itself. Any reasonable definition needs to incorporate the idea of a surface or interface on or at which microbes proliferate; it should also invoke the unifying effect of extracellular polymers which can envelop and probably protect the microbial colonies forming. It might also embrace a sense of community with the implication of emergent properties.

It is worth trying to classify a number of microbial systems that seem to be related to biofilm since they share many of its properties. In fact, members of the whole family of microbial aggregates have more in common than separates them (Table 1). The term microbial aggregate is chosen to mean those associations of micro-organisms that are

SGM symposium 59: Community structure and co-operation in biofilms. Editors D. Allison, P. Gilbert, H. Lappin-Scott, M. Wilson. Cambridge University Press. ISBN 0 521 79302 5 ©SGM 2000.

Table 1. Some different types of microbial aggregate

Type of microbial aggregate	Description
Biofilm	Community forming at a phase boundary generally, but not always, at a liquid : solid interface. Spatially and temporally heterogeneous. May have specific mechanisms for attachment to surface. Generates EPS for adhesion, protection and to facilitate community interactions.
Bacterial colony	A group of organisms growing on a surface, often fed with nutrient from below and incorporating gas exchange from above. May be a clone formed from a single cell. Shows recognizable pattern, limited morphogenesis and spatial and temporal heterogeneity.
Effluent treatment floc	A loosely associated mixed community showing irregular radial symmetry and temporal and spatial heterogeneity
Anaerobic digester granules	A reasonably symmetrical radially organized microbial community showing spatial differentiation and metabolic co-operation, often leading to the oxidation of organic substrates, leading in the end to methane
Food-associated systems; e.g. Kefir grains, the ginger beer plant	Irregular radially organized communities often of EPS-producing lactic acid bacteria and yeast. Used in the production of low-alcohol beverages.
Marine snow	Loose associations of microbes with organic detritus
Mycelial balls	Tightly intertwined mycelia generated in fungal fermentations by carefully controlling growth conditions. Radially symmetrical, often spatially heterogeneous as conditions can become anoxic at the centre.
'Wolf-packs'	Associations, generally motile, of swarming bacteria which interact through the transmission of density-dependent signals and feed through engulfing and digesting organic detritus and other microbes
Pellicles	Predominantly two-dimensional structures forming on the surface of liquids; e.g. neuston, pellicles of bacteria including *Acetobacter* and of fungi such as *Penicillium* and *Aspergillus niger* used in the fermentation industry. Oxidant and reductants from opposite sides of the structure.
Algal mat communities	Variably dense, often layered systems whose biology is predominantly driven by sunlight

Table 2. Biofilms forming at different phase interfaces

Interface	Biofilm
Solid : liquid	Most biofilms: epilithon, medical prostheses, the inner surface of water conduits, ship hulls, marine installations, tooth and epithelial surfaces, etc., etc.
Gas : solid – though often exposed to liquids	Bacterial colonies, myxobacterial swarms, lichen, acetic acid production by the 'Quick' vinegar process, trickling filters, surface biofilms using gas phase nutrients
Gas : liquid	Neuston, vinegar production by the Orleans process, penicillin and citric acid production by traditional fungal fermentations
Liquid : liquid	Hydrocarbon oxidizing biofilm at oil : water interfaces; growth in some food emulsions
Solid : solid – though exposed to liquid phase from time to time	Endolithotrophic communities

largely microbial biomass plus varying amounts of extracellular polymeric materials produced by the microbes themselves. This definition excludes communities which are associated with significant amounts of inanimate or other materials, for example soil and many sediments.

Aggregates all represent communities in which the microbial population is concentrated so that there is the possibility of significant interactions (exchange of substrates, products, inhibitors, deployment of signal molecules, etc.) between them. Cell density, the size and geometry of the aggregate and its metabolic activity will almost certainly lead to diffusion barriers, which may be minimal or large enough to cause significant changes in its biology. Most commonly, because of its low solubility in water and high rate of utilization by bacteria, a steep gradient in oxygen tension develops which can lead to anoxic regions and the proliferation of anaerobic species.

Biofilm itself is distinguished from other microbial aggregates since, by definition, it forms at a phase boundary (Table 2). Although most phase boundaries can be colonized, the commonest type of biofilm appears at a liquid : solid interface.

INVESTIGATIVE METHODS

As in so much scientific endeavour, research on biofilms has depended largely on the development of powerful new techniques to investigate their structure and function. These have been reviewed in the past by different authors, for example Costerton *et al.* (1994, 1995) and Caldwell *et al.* (1992, 1997). A few of these are discussed below.

Growth systems

The type of equipment needed to investigate biofilm formation is very much dependent on the type of questions asked. If *in situ* or *in vivo* investigations are required, the experimenter is obviously restricted in his choice since the biofilm itself proliferates in its natural growth system. The subject has been reviewed by Wimpenny (1996, 1999). Table 3 indicates a selection of growth systems with their main attributes.

The wide range of systems available means that there is almost certainly one that will apply to the majority of problems associated with biofilm biology. It is sensible to distinguish between experimental models and microcosms when discussing the investigation of biological systems. A model system represents the cultivation of a completely defined community (one or more species!) in whatever growth device is selected. In contrast, a microcosm is a collection of microbes from a natural community and may include some species that have not yet been isolated. More representative perhaps, but less well understood! To be quite clear, the growth system can be *either* a model or a microcosm according to the above criteria.

Microscopy

Microscopy embraces a wide range of traditional and modern techniques. The electron microscope has been a tremendous source of structural information, always remembering that on the microbial scale, preparation techniques can generate artefacts. Sutton *et al.* (1994) have compared conventional scanning electron microscopy (SEM), low temperature SEM and electroscan wet mount SEM in *Streptococcus crista* to reveal wide differences in the final image. For example, under 'environmental' conditions, one sees mostly a blanket of extracellular polymeric substances (EPS) giving no hint of the structure of the biofilm below.

The application of the confocal scanning laser optical microscope (CSLM) to biological samples (White *et al.*, 1987; Shotton, 1989) provided a powerful tool, especially in conjunction with fluorescent probe techniques. The optical geometry of the CSLM meant that coherent light beams had a very narrow depth of focus whilst all out-of-focus information was rejected. This allowed a series of narrow focal planes to be recorded at different depths throughout a sample. These images can be assembled using image-processing techniques to generate a three-dimensional digitized image. Given the latter, it becomes possible to reconstruct vertical sections through the array, generating an in-depth profile of a biofilm sample. These techniques have helped reveal the highly heterogeneous structure of microbial biofilm.

The use of fluorescent probes with the CSLM added discrimination to what was already a powerful technique. Thus it became possible to distinguish not only between classes,

Table 3. Some of the many biofilm growth systems used in the laboratory

Growth system	Attributes	Reference
Glass slide	Transparent surface allows optical microscopy. Good for attachment and early biofilm growth	Caldwell & Lawrence (1988); Bos et al. (1994)
Chemostat-based systems	Surfaces exposed to steady state cultures, though biofilm not steady state	Keevil et al. (1987); Keevil (1989); Marsh (1995)
Channel reactor	Models flow: channel can be open or closed as a tube. Sample ports at intervals along the channel. Example: the Robbins device.	McCoy et al. (1981); Ruseska et al. (1982)
Solid particle support	Downflow systems Trickling filter, film grows on a solid substratum with air spaces irrigated with nutrient solutions	Diz & Novak (1999)
	Upflow systems Airlift and related systems. Constant motion leads to attrition and the removal of excess biofilm from the support. System tends towards a steady state.	Gjaltema et al. (1994); van Loosdrecht et al. (1995)
Constant shear device	The rototorque: Two concentric cylinders, the outer stationary, the inner rotating. Removable glass slides in outer wall, constant flow of medium through the system. Growth on slides at known shear	Trulear & Characklis (1982); Bakke et al. (1984)
	The Gilbert rotator: Four chambers formed by four sets of intercalating cylinders, each set having fluid inputs and outputs. At a constant rotation speed, four different shear rates applied.	Allison et al. (1999)
	The Fowler Cell Adhesion Monitor: A stationary flat disc is aligned near a rotating disc. Cell and nutrient feed pass in near the centre. The cells attach. At a constant rotation rate a shear gradient develops across the plate and cells detach at a critical value.	Fowler & Mackay (1980)
Membrane reactors	A permeable membrane separates oxidant from reductant (e.g. air and growth medium). Biofilm grows on the membrane receiving essential nutrients from each side.	Rothemund et al. (1994); Wilderer (1995); Watanabe et al. (1997)
Rotating drums	As in the rototorque, though not designed specifically to apply reproducible shear fields. Growth on the inside of the outer cylinder and on the outside of the inner cylinder.	Arcangeli & Arvin (1995, 1997)

Table 3 (cont.)

Growth system	Attributes	Reference
Steady state systems	*The Constant Depth Film Fermenter:* Generates steady state in recessed film pans by passing support ring under scraper blades. Operates aseptically under well-controlled conditions	Coombe *et al.* (1982); Peters & Wimpenny (1988)
	The Gilbert 'Baby factory': System based on the Cooper–Helmstetter device for synchronizing cell populations by attaching them to cellulose acetate membranes, inverting and irrigating with sterile warm growth media. Mother cells attach and form a thin biofilm. Newborn cells are released as soon as the mother cell has completed a division. As in a chemostat growth rate controlled by dilution rate.	Gilbert *et al.* (1989); Gander & Gilbert (1997)

genera and species, but between the viability and even the Gram reaction of individual organisms, as well as monitoring some chemical properties within the biofilm. Probes include negative stains such as fluorescein, which provides a fluorescent background upon which the bacteria can be viewed as non-fluorescent objects. Agents such as resazurin can distinguish between 'live' and 'dead' cells. The former reduce the dye to a colourless non-fluorescent form, in contrast to non-living cells, which retain the fluorescent dye. Living and dead are contentious terms. In microbiology, a living cell can only truly be determined as one that can grow and reproduce, in the end developing a colony. The assumption here is that organisms capable of catalysing oxidation/reduction reactions are alive. Obviously this is an oversimplification. Tetrazolium and related agents operate in a similar fashion. Viability, this time based on membrane integrity, is the basis of a commercial agent, the Baclite viability probe. Here, live cells with intact membranes fluoresce green whilst cells whose membrane structure is compromised (supposed dead) fluoresce red.

Attachment of fluorophores to other agents can increase their value. For example, linked to dextrans they can be used to determine diffusion coefficients and cell distribution, and with polyanionic dextrans charge distribution. Conjugated lectins can help to reveal the distribution of oligosaccharides. Attached to polyclonal antibodies, fluorophores can be used to determine the position of species within a biofilm. A most powerful technique is to attach fluorophores to 16S rRNA probes, allowing the identification of microbes at almost any taxonomic level (see later).

Carboxyfluorescein is a probe whose fluorescence is modulated by pH and has been used to determine spatial differences in pH in biofilms of *Vibrio parahaemolyticus*

(Caldwell *et al.*, 1992). Recently, there has been a major advance in confocal microscopy with the multiphoton laser confocal microscope. The fluorescent agent is excited by two or more photons simultaneously, generally in the infrared. A powerful laser is employed in ultrashort pulses. Because of the longer wavelength used, there are fewer problems with photo-bleaching and the penetration of the laser beam is much deeper than with normal CSLM. In addition, the use of a pulsed beam means that the rate of fluorescence decay can give information which is *not* dependent on the actual concentration of the fluorophore. Using carboxyfluorescein, it is possible to determine the pH around groups of cells further into a biofilm than ever before. This has been demonstrated clearly by Vroom *et al.* (1999) using a defined 10-membered community grown in the CDFF. pH measurements were determined up to 140 µm into the biofilm. pH gradients around cell clusters revealed that values as low as 3.0 were possible.

CSLM has also been employed to determine flow rates in heterogeneous biofilm. Here, fluorescently labelled latex beads were tracked at intervals as they moved through the voids and interstices of a biofilm (Stoodley *et al.*, 1994).

Microelectrode experiments

Once it is accepted that there is heterogeneity on the microscale it is clear that the application of sensors with appropriate geometries are needed to map changes in the physico-chemical environment. Many of these now exist thanks amongst others to Bø Barker Jorgensen and Nils Peter Revsbech and more recently to Dirk deBeer and Zbigniew Lewandowski. The most commonly used are dissolved oxygen and pH electrodes, though others are available for measuring nitrogen oxides and sulphide and there are even enzyme electrodes capable of measuring glucose. Amongst the seminal work on oxygen distribution in biofilms was that of deBeer & Stoodley (1994), who mapped oxygen partial pressure in model biofilms and showed that within cell clusters pO_2 fell to zero whilst around and beneath clusters in void spaces there was always measurable oxygen. More recently, Rasmussen & Lewandowski (1998) have used oxygen probes to determine mass transfer rates in heterogeneous biofilms.

Molecular methods

Fluorescence *in situ* hybridization (FISH). The ability to identify individual bacterial cells represents an important advance in the techniques needed to understand the organization of microbial ecosystems, including biofilm. Oligonucleotide probes are made to recognize specific regions of 16S rRNA; these are then labelled with different fluorescent dyes. Whole families of probes can be generated: for example, one that recognizes prokaryotes, another for the *Archaea* and successively more specific probes for particular groups of bacteria right down to individual species. A particularly good example of the use of 16S rRNA to map the diversity of a microbial population from earliest colonization to maturity used a river water community grown in a rotating annular reactor (Manz *et al.*, 1999).

Green fluorescent protein (GFP). The jellyfish *Aequorea victoria* synthesizes a fluorescent green pigment. The latter has proved to be a most useful probe since it can be inserted with no obvious ill effects into the genomes of many different types of organism, including animals, plants, yeasts and bacteria (Chalfie *et al.*, 1994; Anderson *et al.*, 1998). If incorporated constitutively GFP expression can be used to recognize particular species right down to the individual level. When located adjacent to specific promoters, GFP fluorescence can indicate which genes are turned on. Since its original discovery, GFP has been altered to provide a range of markers with enhanced fluorescence and/or different spectral characteristics. One or two examples of the use of GFP markers will indicate the tremendous power of the technique. Andersen *et al.* (1998) engineered the GFP protein by attaching a polypeptide sequence to its carboxy-terminus. This sequence, AANDENYALAA, is recognized in *Escherichia coli* by the tail-specific *tsp* protease, which then degrades the whole protein so that the fluorescence disappears. The authors realized that slight alterations to the AANDENYALAA tail sequence could alter the rate at which the polypeptide and hence fluorescence was degraded. They created a family of such tagged proteins which they could then use to examine time-dependent gene expression. These constructs were used to monitor growth rates in terms of the rate of synthesis of rRNA in individual cells or groups of cells in a heterogeneous community (Sternberg *et al.*, 1999).

GFP was used to examine community interactions by the same research group (Møller *et al.*, 1998). Here, two species of bacteria were involved: *Pseudomonas putida* and an *Acinetobacter* sp. The pathway investigated was toluene and related aromatic degradation by the *pu* and the *pm* pathways. Promoters from each pathway were labelled with GFP in the pseudomonad. In pure or mixed cultures of the two species, the *pu* promoter was expressed in the presence of benzyl alcohol whilst the *pm* promoter was only expressed in the pseudomonad when both species were present. There was clearly an important interaction between the two species. The latter were individually tagged with fluorescently labelled 16S rRNA probes as well as GFP so that the identity of the bacteria as well as expression of the aromatic degradation could be monitored *independently* at the level of single cells.

Other methods

Many other investigative methods have been developed. For example, nuclear magnetic resonance imaging has been used to monitor flow regimes in biofilm communities (Lewandowski *et al.*, 1993), and Fourier transform infrared has been used to examine attachment and growth of microbes on different surfaces. I do not plan to discuss these further.

ADHESION AND EARLY EVENTS IN BIOFILM DEVELOPMENT

It is generally accepted that a very clean surface is quickly covered with a conditioning film of organic molecules, and that this precedes attachment of bacteria to the clean surface. In a liquid:solid system, bacteria penetrate the viscous sublayer by eddy diffusion and attach to the surface through long-range, weak interactions with low specificity, namely electrostatic or van der Waals forces. Irreversible attachment follows through short-range, generally highly specific, interactions. These can be dipole, ionic, hydrogen bonding or hydrophobic interactions (see Denyer *et al.*, 1993). Some of the latter are expressed by the secretion of EPS and by the deployment of a range of fibrillar structures, including fimbriae or fibrils. Many of these are equipped with specific adhesins that can attach to elements of the conditioning film or to other bacteria, especially in complex locations such as the oral environment (Handley *et al.*, 1999).

Once the cells are attached they start to grow and to produce more EPS. At the same time, they often develop strategies for capturing space by moving from where the first few cell divisions have taken place (Caldwell *et al.*, 1992; Korber *et al.*, 1995).

The biofilm develops, generating an architecture which may be more or less porous depending on the physico-chemical characteristics of the environment in which it grows. During its formation, a succession of different species will flourish influenced by changes in the local environment. Species will be imported and exported and other organic and inorganic matter may be incorporated into the structure. At some point, due to shear forces, the development of anaerobic zones forming gas pockets, etc., pieces of the biofilm may slough off. This will be followed by recolonization and regeneration of the structure. What determines the actual three-dimensional structure of a biofilm?

BIOFILM STRUCTURE

The role of physico-chemical factors

It is simply not good enough to ignore the part that physico-chemical factors play in regulating biofilm architecture and function. Unfortunately, the biofilm world seems divided into those who posit that complete control of all aspects of biofilm development, structure, morphology and physiology is due to genetic mechanisms. On the other side are some who believe that structural determinants are completely regulated by local physico-chemical factors. Of course, the truth lies somewhere between these extremes.

A good paradigm for the role of environment in biofilm pattern formation comes from the increasingly sophisticated research into bacterial colony morphogenesis. I have described the interplay between genotype and environment regarding colonies as follows:

> 'Formation of the detailed structure of a bacterial colony is a combination of two separate factors intrinsic and extrinsic. Intrinsic factors are products of the genetics of the cell itself. They determine the morphology of the individual cell, the mode of cell reproduction, the possession of extracellular appendages (flagella, fimbriae, pili etc.) production of extracellular products (exopolysaccharides, proteins etc.) motility, energy metabolism, pigment formation and so on. Extrinsic factors include the prevailing physico-chemical environment which influences the physiology of the cell plus the transport of solutes into and out of the growing colony and the inevitable formation of solute diffusion gradients within the colony and the surrounding medium' (Wimpenny, 1992).

Whilst all of this seems still to be true, we do need to add intercellular signalling to the list of genetically controlled attributes which respond to environmental factors, including propinquity (Thomas *et al.*, 1997).

One of the earliest examples of pattern formation due to diffusion was the work of Cooper *et al.* (1968), who reported that the 'snowflake' pattern of 2-week-old colonies of *Aerobacter aerogenes* could be explained as acute substrate-limited growth. The appellation 'snowflake' indicates not only its morphological resemblance to the snow flake but a similar mechanism for its formation, since the unique and beautiful patterns of a snowflake are due to the restricted diffusion of water molecules to the developing structure. Much more recently there has been a critical examination of colonies of the Gram-positive *Bacillus subtilis* and related species and strains (Matsushita & Fujikawa, 1990; Ohgiwara *et al.*, 1992). Schindler & Rovensky (1994) created a simple computer model of bacterial colony growth. They compared different models and pointed out the resemblance of diffusion-limited aggregation (DLA) models to bacterial colony growth. Ben-Jacob *et al.* (1994) have concentrated in this and a group of related papers on the growth of *B. subtilis* as a function of substrate and agar concentration. They monitored and photographed actual bacterial colonies and compared the latter with a cellular automaton model. Through much of the range of substrate concentrations, the latter (fundamentally a DLA model) reflected the colony structure accurately. At very low substrate concentrations, the actual colony began to differ from the simulation. The difference was interpreted as a physiological response of the bacteria, via a signalling system, to low substrate concentrations.

The biofilm world was in some disagreement as to what constituted a 'typical' biofilm structure. Some argued that biofilms were simple stalked or irregular branching structures well separated from their neighbours; others that biofilms were mushroom- or tulip-shaped structures penetrated by large and small pores. Yet others considered biofilm to be a more or less flat, homogeneous structure. A simple cellular automaton model was used by Wimpenny & Colasanti (1997) to suggest that all three models were actually correct since the final structure was largely dependent on resource concentration. Thus the first type appeared in water distribution systems where the substrate concentration was very low (Keevil, 1989; Walker & Keevil, 1994). The second type was generated in the laboratory using media containing significant nutrient concentrations (see Costerton *et al.*, 1994, 1995). The third was dense relatively uniform biofilm (Nyvad & Fejerskov, 1989) found in habitats (for example, the human mouth) where nutrient levels are generally high, or periodically extremely high. Recent work by Wood *et al.* (2000) has indicated the presence of channels in dental plaque biofilm. Huang *et al.* (1998) describe a dense flat biofilm formed under conditions of phosphate starvation.

van Loosdrecht *et al.* (1995) proposed that resource concentration and shear rate were both determinants of biofilm morphology. Picioreanu *et al.* (1998, 1999) working in the same laboratory in Delft, Holland, extended this work with a beautiful series of hybrid computer models that agreed substantially with our own proposals. These workers made models of growth in gel beads used in the biotechnology industry and showed that they could reproduce the structures seen very accurately. They went on to investigate biofilm structure using realistic variables and parameters in the simulation. Two- and three-dimensional representations of the resulting predictions clearly showed the range from stacked tower-like configurations at low resource levels to dense confluent growth when substrate availability exceeded its utilization rate. In addition, Picioreanu included the effects of shear rate and indicated the way in which biofilm could erode and slough off as this parameter was increased. Additional support using CA models came from the USA (Hermanowicz, 1998), who presented very similar results from a simpler model.

Use of the CSLM gave interesting evidence that porous structures *were* related to substrate type as well as concentration. Wolfaardt *et al.* (1994) showed that a degradative community degrading diclofop methyl, a commercial herbicide, led to the formation of mushroom-shaped stacks of multicellular aggregates. Growth on tryptone soya broth (TSB) at low concentrations led to groups of cells separated by pores or spaces, whilst high concentrations of TSB gave a denser confluent appearance to the biofilm.

Cellular control systems

Molecular techniques are starting to throw light on some of the processes associated with biofilm formation. Pratt & Kolter (1998) have isolated mutants of *E. coli* that can no longer attach to surfaces. They were either non-motile or could not make type 1 pili. The latter are mannose-sensitive adhesins. Both motility and the adhesin seem therefore to be associated with attachment.

Similarly, surface attachment defective (*sad*) mutants of *Pseudomonas aeruginosa* were isolated, one of which had defective flagella and hence was non-motile whilst a second could not make the type IV pilus. This last group formed a flat layer on the substratum rather than compact microcolonies (O'Toole & Kolter, 1998).

Extracellular cues seem to be involved in the expression of different regulatory systems. Davies *et al.* (1993) reported that polysaccharide production was switched on when *P. aeruginosa* contacted a surface. This organism seems to express *algC*, *algD* and *algU*:: *lacZ*, all associated with alginate production (Davies & Geesey, 1995).

It has become quite clear in recent years that other cues include signalling molecules, often associated with density-dependent phenomena. Two such systems were reported in *P. aeruginosa*. LasR–LasI regulates virulence in the organisms; however, it also controls the formation of a second system, Rh1R–Rh1I, which is a regulator for a number of secondary metabolites. Both systems encode signal molecules: *rh1I*, butyryl homoserine lactone; *lasI*, 3-oxododecanoyl-homoserine lactone. Both mutant strains as well as the wild-type organism can attach to surfaces; however, LasI cannot make the step from microcolonies to a differentiated thick biofilm. Addition of the signal molecule 3-oxododecanoyl-homoserine lactone reverses the effect. Davies *et al.* (1998) conclude that signal molecules are involved in the control of biofilm formation, at least in some species.

Costerton *et al.* (1999) discuss these processes and present a scheme for possible regulation of the development of a biofilm.

Recently, Loo *et al.* (2000) isolated 18 mutants of *Streptococcus gordonii* which were unable to generate biofilm. Whilst nine of these were associated with quorum sensing, signal transduction and osmoadaptation, the remainder were of unknown function, suggesting that much remains to be learned about the processes regulating biofilm formation. Another interesting recent finding is that biofilm formation in *P. aeruginosa* is also controlled by catabolite repression, more usually associated with regulation of carbon metabolism. O'Toole *et al.* (2000) reported that Crc mutants of this organism made only a simple monolayer film instead of the denser punctuated structure that the wild-type could generate.

COMMUNITY INTERACTIONS
Another area which has excited interest in recent years is the concept of 'community'. By community I think we mean an interacting group of living organisms in the same geographical area. The word can indicate that the sum of its activities is greater than the sum of all the activities of its constituent members. In other words, a community might have *emergent* properties.

Model communities
A productive strand in the investigation of biofilm communities has been the use of model systems. These have the advantage that the system is completely understood and interactions between species can be deduced in an unequivocal fashion. The main disadvantage is the uncomfortable feeling that one cannot properly extrapolate to the natural system since the latter may have additional unrecognized components. It is only possible to mention a few examples of model communities. Bradshaw *et al.* (1989) described a nine-member oral community which was investigated in a chemostat system containing enamel discs. This system was used later by Bradshaw *et al.* (1997) and by Kinniment *et al.* (1996), who grew the community under steady state conditions in a constant depth biofilm fermenter. Most recently, Shu *et al.* (2000) used a four-membered community of cariogenic bacteria as a model to investigate enamel and tooth root caries. Stoodley *et al.* (1999) investigated responses to shear levels on a community consisting of *Klebsiella pneumoniae*, *P. aeruginosa* and *Pseudomonas fluorescens*. In low shear fields, the community grew as roughly spherical colonies, whilst on increasing shear so that flow became turbulent, colonies elongated to generate filamentous streamers. There are many other examples of the use of model communities. It is quite clear that these have an important part to play in biofilm research; one that complements investigations into natural communities. Both approaches are necessary.

Community structure has been the subject of considerable discussion. Caldwell and colleagues (e.g. Caldwell *et al.*, 1997) have developed a most interesting *proliferation* theory. Caldwell points out that success at any level amongst living things comes as a result of the ability to grow and reproduce. Success results not from Darwinian concepts of competition, survival of the fittest and so on, but rather just by the ability to grow and survive with inevitably those that grow fastest in a particular set of physico-chemical conditions surviving best. Proliferation is an iterative process: it can apply at the molecular level, at the level of subcellular structures, at the cellular, organ and species level. What is more, proliferation is not, as Darwin would have it, restricted to 'species'. A group of species can form a community, and the latter may have emergent properties which allow it greater success by increasing its genetic potential, allowing it to use a wider range of resources, modify its own environment possibly in a favourable direction and in the end to proliferate faster. The Caldwell reference is an interesting, detailed, thought-provoking article, which discusses microbial communities in all their disparate forms, including homogeneous chemostat and gradostat cultures as well as gel-stabilized gradient systems. In addition, the physiology of anaerobic digester granules as communities is discussed. The concept of 'proliferation' seems perfectly acceptable and raises interesting suggestions that a community can become, under steady state conditions, an entity with some kind of deterministic stoichiometry. It might be true under rigidly controlled conditions though even here, as Caldwell points out, the genotypic flexibility of constituent species will ensure that continual changes occur. For example, the Dalapon community grown in a chemostat over prolonged periods consisted of primary Dalapon degraders and secondary organisms. In the end, a secondary organism mutated to be able to use the herbicide (Senior *et al.*, 1976).

There are a number of other interesting observations, both structural and functional, concerning biofilm communities. Oral bacteria were first shown to coaggregate by Gibbons & Nygaard (1970). Kolenbrander (1988) and most recently Kolenbrander *et al.* (1999) have carried the analysis of coaggregation to a highly sophisticated level, demonstrating complex patterns of possible structures that have been deduced first from simple test-tube coaggregation studies, and secondly by a careful analysis of specific receptor sites and adhesins on oral species. Here we should recognize the idea of a *structural* community where the members are arranged in a specific spatial order. Events in oral communities have progressed further, perhaps in an understanding of these complex interactions, than for most other biofilm systems. In the same environment, there is good evidence for a *functional* community. Bradshaw *et al.* (1994) demonstrated metabolic co-operation between a selected model community of oral bacteria, which together could degrade hog mucin better than the sum of activities of all the individual species. This was due to the collaborative efforts of community members, each of which was able to degrade one step in the breakdown of the substrate. This work has most recently been reviewed by Marsh & Bradshaw (1999), who discuss both the sequential cleavage of sugars of the oligosaccharide side chain and the concerted attack on host proteins and glycoproteins using sulphatases, glycosidases and proteinases.

Community structure has been elegantly investigated by Molin and his group using GFP plus fluorescent rRNA probes to visualize the manner in which species in microbial aggregates in a biofilm are organized. This group has recently shown that two bacterial species, a *Pseudomonas* and a *Burkholderia*, growing on citrate would develop as separate individual microcolonies. However, only the *Burkholderia* was capable of growing on the aromatic compound chlorobiphenyl, which was oxidized to chlorobenzoate, which the pseudomonad could use. Under these conditions, the pseudomonad moved, presumably chemotactically, towards the *Burkholderia*, leading to a mixed colony functioning as a 'working' community at the microcolony level (Nielsen *et al.*, 2000). A beautiful example of ecology in action!

Stochastic or deterministic factors?

Molin (1999) has pointed out that there are two quite different approaches to biofilm community structure. The first view is that such communities are merely random accretions of bacteria which can, if spatial constraints permit, associate and interact in a manner which benefits the community. A second point of view is that all communities are evolving structures associated in a more or less deterministic fashion as a specific answer to environmental problems.

Perhaps the truth is somewhere between these two extremes. Stochastic processes precede deterministic ones. Thus the colonization of clean surfaces is essentially random, consisting of a subset of all possible microbes that have the ability first to attach loosely and then firmly to a surface. These cells may start to grow or may be washed off the surface due to shear forces. Organisms capable of interaction *may* be located close enough to each other that interactions are possible. This is the *propinquity* factor (Thomas *et al.*, 1997). Now deterministic factors come into play. Interactions may lead to a hybrid association forming, where the two organisms co-operate. Chemoattractants, including pheromone molecules, may lead to the movement or even the growth of microbes towards one another. The emerging community might recruit additional contributing members by a combination of chance and necessity. So under reasonably stable environmental conditions, a community is formed whose properties *may* represent the best solution to immediate environmental problems/opportunities. One should not really be under the illusion that such a community could be compared to a *tissue* in the sense that a multicellular organism has a precise structure with only relevant cell types present. No, our microbial community will consist of a melange of types. These will include primary resource converters; secondary and subsequent species relying on products of a food chain; scavengers that do not contribute to the efficiency of the community or may even detract from it; parasites, predators and competitors, none of which represent added value for the association. What is more, as time goes by, other species will be imported or exported so that the community will change in ways that may or may not be energetically favourable. Some of these processes are indicated in Fig. 1.

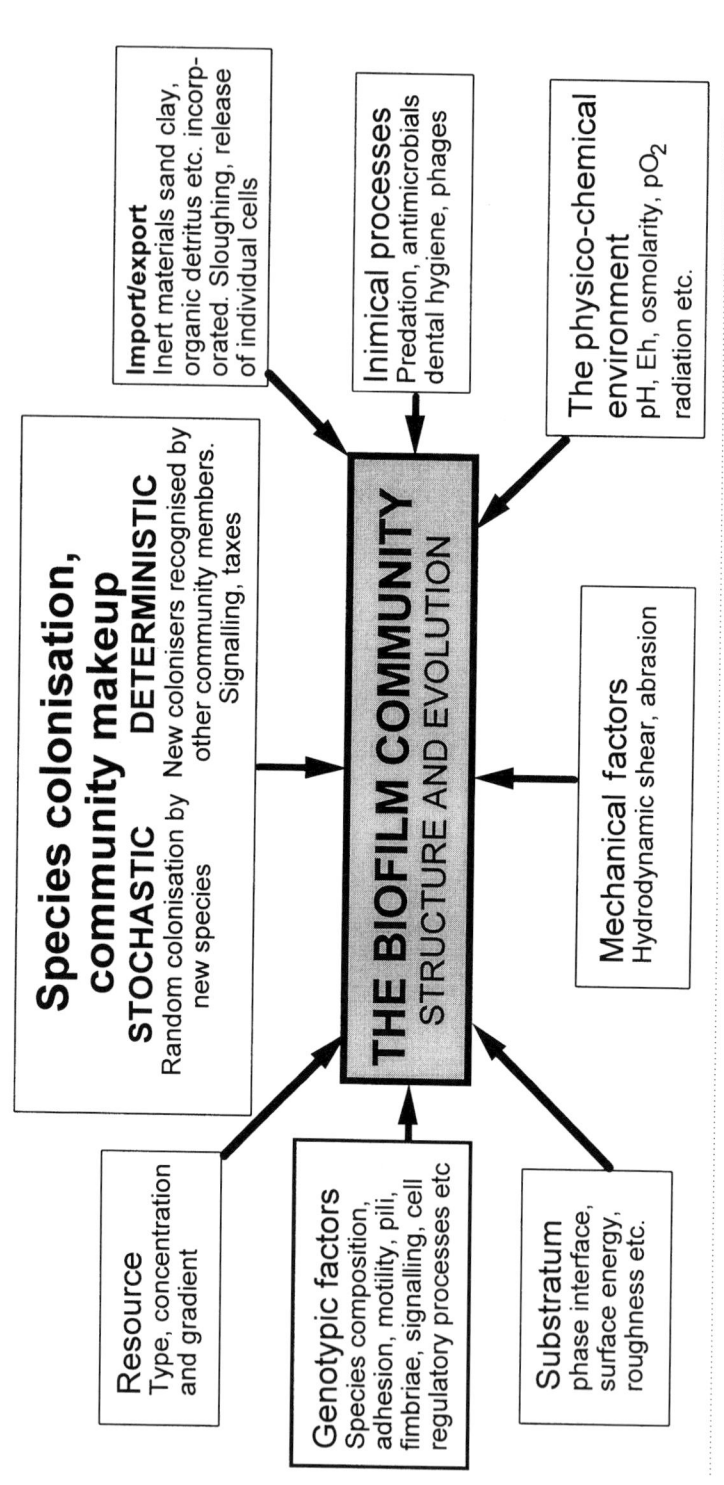

Species colonisation, community makeup
STOCHASTIC   DETERMINISTIC
Random colonisation by New colonisers recognised by
new species   other community members.
        Signalling, taxes

Import/export
Inert materials sand clay,
organic detritus etc. incorporated. Sloughing, release
of individual cells

Inimical processes
Predation, antimicrobials
dental hygiene, phages

The physico-chemical
environment
pH, Eh, osmolarity, pO_2
radiation etc.

THE BIOFILM COMMUNITY
STRUCTURE AND EVOLUTION

Resource
Type, concentration
and gradient

Genotypic factors
Species composition,
adhesion, motility, pili,
fimbriae, signalling, cell
regulatory processes etc

Substratum
phase interface,
surface energy,
roughness etc.

Mechanical factors
Hydrodynamic shear, abrasion

Fig. 1. Some of the many factors that influence the development and biology of a biofilm.

Factors affecting possible levels of community interaction

Even a cursory examination of microbial systems reveals that there are many different levels of association between microbial communities. At one end of the scale are obligate endosymbiotic associations like mitochondria or chloroplasts; at the other, very loose temporary associations in dynamically changing temporally and spatially heterogeneous systems. There seem to be two dominant themes that govern the process of organization into a community. The first of these is the benefit of the association; the second is stability in terms of space and time. The energetic value of possessing prokaryote-derived organelles transcends most other considerations as far as eukaryotic microbes and metazoans are concerned. From this extreme form of co-operation one can discern a spectrum of levels of association. This is illustrated in Fig. 2. Below organelle development come endosymbionts, some of which can exist apart from their hosts. These might be green algae associated with *Hydra*, or possibly some sulphur-oxidizing species allied to marine animals found around deep submarine black-smokers. Tight associations between cellular types epitomize lichens as well as the two bacteria that make up '*Methanobacillus omelianski*'. Other syntrophic associations can be found in anaerobic communities, where one organism donates hydrogen and a second scavenges it to generate methane. Some of these lead to discrete organized structures like anaerobic digester granules. At the other end of the scale are very loose associations, such as those found in many biofilm communities, though even here there is a wide range of levels of interdependence. For instance, high in the interdependency stakes are dental plaque communities.

There may be some lessons for us in this spectrum of dependency amongst microbial communities. For many microbial habitats there is not much evidence that recognizable tightly organized consortia exist, suggesting that we might look for other explanations for the loose associations which seem to be the *sine qua non* of microbial ecology. One explanation could be based on two main factors: micro-scale heterogeneity on the one hand and response time on the other.

At the micro-scale, the physico-chemical universe is very different from the way we perceive it at the macro-scale. Diffusion is the main transport process and is extraordinarily fast when cell–cell distances are short. At the one or a few microbes level, diffusion gradients around the cell are meaningless, since mass transfer rates exceed the cell's ability to use a substrate or generate a product. However, larger multicellular associations *can* generate meaningful solute gradients, leading to structural and metabolic differentiation or the accumulation of molecules (for example, digestive enzymes or quorum-sensing products). Again it follows that the development of a structure means that processes of chemotaxis and of intercellular signalling can operate successfully.

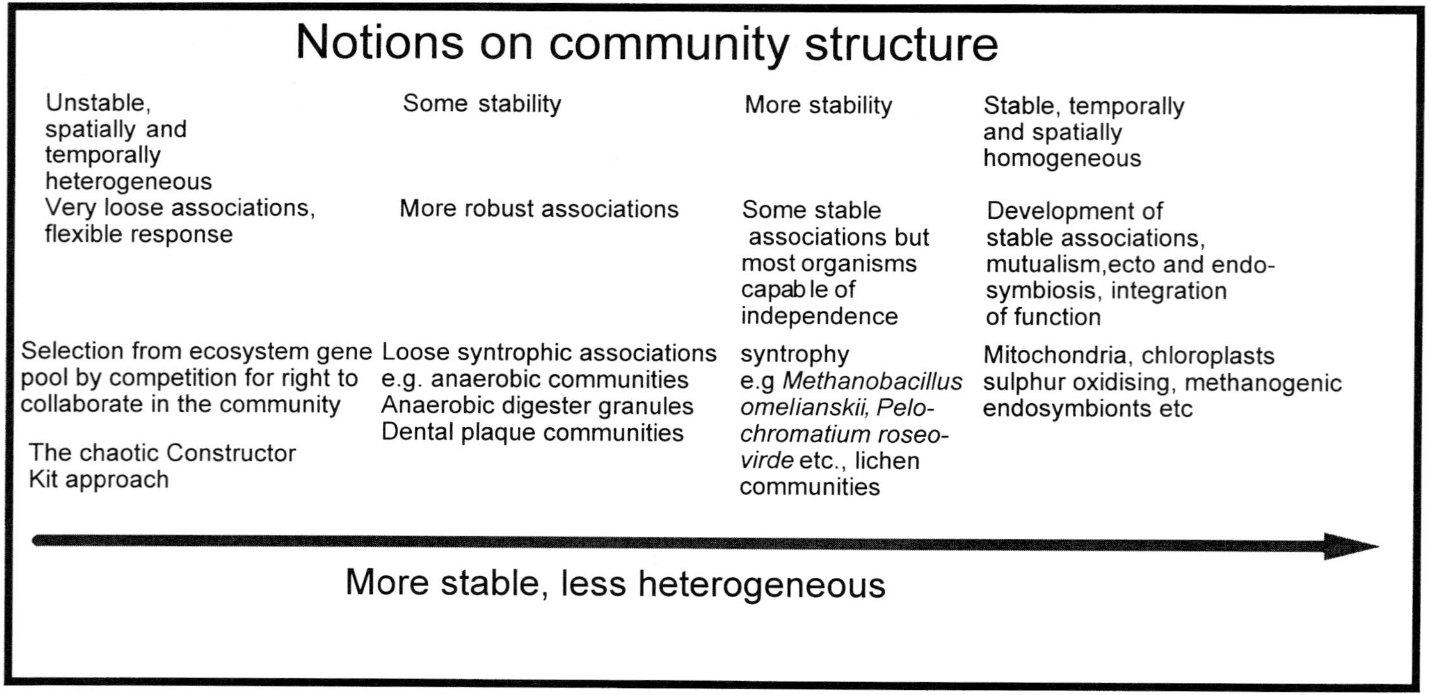

Notions on community structure

Unstable, spatially and temporally heterogeneous	Some stability	More stability	Stable, temporally and spatially homogeneous
Very loose associations, flexible response	More robust associations	Some stable associations but most organisms capable of independence	Development of stable associations, mutualism, ecto and endo-symbiosis, integration of function
Selection from ecosystem gene pool by competition for right to collaborate in the community	Loose syntrophic associations e.g. anaerobic communities Anaerobic digester granules Dental plaque communities	syntrophy e.g Methanobacillus omelianskii, Pelo-chromatium roseo-virde etc., lichen communities	Mitochondria, chloroplasts sulphur oxidising, methanogenic endosymbionts etc
The chaotic Constructor Kit approach			

More stable, less heterogeneous

Fig. 2. One interpretation of the range of microbial communities found in nature.

In most microbial habitats, spatial heterogeneity on the micro-scale can lead to a multiplicity of microscopic niches, allowing the proliferation of different species. In a soil system, there are steep pH gradients around colloidal clay and humic materials. Organic debris means that localized 'hot spots' containing good substrate resources or conversely inhibitors may be present. Water potential, temperature, light, substrate and product gradients can alter quickly, sometimes in an irregular manner with very short time constants, sometimes as part of daily or yearly cycles.

For all these reasons, living cells at this scale must be capable of a rapid response. There is therefore a premium on very small units of metabolism (cells) which have high surface area to volume ratios, a simple structure and the minimum of diffusion restriction in their internal transport structures.

Biological constructor kits. The strategic consideration at this level is that of the Meccano (constructor) set! In the latter, it is possible to generate a crane from amongst the store of basic components (nuts, bolts, wheels, axles, frames, etc., etc.), or if later a truck is needed, the crane can be taken to pieces and the new vehicle generated from the same building blocks. Similarly, local physico-chemical and biotic factors determine the kind of biological 'machine' that is needed and the pieces are selected to put it together. As conditions change, so different 'machines' are needed and they duly form. It might be interesting to collect together some of the parts of the constructor kit. Naturally, it is tempting to regard each species in the microbial world as one such building block and in a way this is almost certainly true. The divisions of labour seen in microbiology are endlessly intriguing. Examples include the separation of ammonia oxidation from nitrite oxidation amongst nitrifying species. The dental plaque community seems to be quite a good example of a construction kit having emergent properties, as does the community found in anaerobic digester granules.

Of course the concept of a construction kit is too simplistic, since, as discussed earlier, our machines are much more chaotic that the mechanical variety just mentioned. Imagine the crane or the truck being *approximately* right for the job, however the mad creativity of Nature has led to the bolting on of all sorts of different bits, some with function, others rather hindering its ability to do its professed job, yet others hell bent on destroying it! Now *that's* a microbial community!

Biofilms are just one, albeit the most important, of a group of microbial aggregates that can operate as integrated communities. Community is a relatively novel idea in microbial ecology, and now we have the tools to investigate this properly, the future looks tremendously exciting!

REFERENCES

Allison, D. G., Heys, S. J. D., Willcock, L., Holah, J. & Gilbert, P. (1999). Cellular detachment and dispersal from bacterial biofilms: a role for quorum sensing. In *Biofilms: the Good, the Bad and the Ugly*, pp. 279–286. Edited by J. Wimpenny, P. Gilbert, J. Walker, M. Brading & R. Bayston. Cardiff: BioLine.

Andersen, J. B., Sternberg, C., Poulsen, L. K., Bjørn, S. P., Givskov, M. & Molin, S. (1998). New unstable variants of green fluorescent protein for studies of transient gene expression in bacteria. *Appl Environ Microbiol* **64**, 2240–2246.

Arcangeli, J. P. & Arvin, E. (1995). Growth of an aerobic and an anoxic toluene degrading biofilm – a comparative study. *Water Sci Technol* **32**, 125–132.

Arcangeli, J. P. & Arvin, E. (1997). Modelling of the growth of a methanotrophic biofilm. *Water Sci Technol* **36**, 199–204.

Bakke, R., Trulear, M. G., Robinson, J. A. & Characklis, W. G. (1984). Activity of *Pseudomonas aeruginosa* in biofilms: steady state. *Biotechnol Bioeng* **26**, 1418–1424.

Ben-Jacob, E., Schochet, O., Tenenbaum, A., Cohen, I., Czirok, A. & Tamas, V. (1994). Generic modelling of cooperative growth patterns in bacterial colonies. *Nature* **368**, 46–49.

Bos, R., van der Mei, H. C., Meinders, J. M. & Busscher, H. J. (1994). Quantitative method to study co-adhesion of microorganisms in a parallel plate-flow chamber – basic principles of the analysis. *J Microbiol Methods* **20**, 289–305.

Bradshaw, D. J., McKee, A. S. & Marsh, P. D. (1989). The use of defined inocula stored in liquid nitrogen for mixed culture chemostat studies. *J Microbiol Methods* **9**, 123–128.

Bradshaw, D. J., Homer, K. A., Marsh, P. D. & Beighton, D. (1994). Metabolic cooperation in oral microbial communities during growth on mucin. *Microbiology* **140**, 3407–3412.

Bradshaw, D. J., Marsh, P. D., Watson, G. K. & Allison, C. (1997). Evidence for multi-species interactions in oral biofilm development. *J Dent Res* **76**, 1058.

Caldwell, D. E. & Lawrence, J. R. (1988). Study of attached cells in continuous flow slide culture. In *CRC Handbook of Laboratory Model Systems for Microbial Ecosystems*, pp. 117–138. Edited by J. W. T. Wimpenny. Boca Raton, FL: CRC Press.

Caldwell, D. E., Korber, D. R. & Lawrence, J. R. (1992). Confocal laser microscopy and digital image analysis in microbial ecology. *Adv Microb Ecol* **12**, 1–67.

Caldwell, D. E., Wolfaardt, G. M., Korber, D. R. & Lawrence, J. R. (1997). Do bacterial communities transcend Darwinism? *Adv Microb Ecol* **15**, 105–191.

Chalfie, M., Tu, Y., Euskirchen, G., Ward, W. W. & Prasher, D. C. (1994). Green fluorescent proteins as a marker for gene expression. *Science* **263**, 802–805.

Coombe, R. A., Tatevossian, A. & Wimpenny, J. W. T. (1982). An *in vitro* model for dental plaque. *Paper given at the 30th Meeting of the International Association for Dental Research*, British Division, University of Edinburgh, Scotland.

Cooper, A. L., Dean, A. C. R. & Hinshelwood, C. (1968). Factors affecting the growth of bacterial colonies on agar plates. *Proc R Soc Lond Ser B* **171**, 175–199.

Costerton, J. W., Lewandowski, Z., deBeer, D., Caldwell, D., Korber, D. & James, G. (1994). Biofilms, the customized microniche. *J Bacteriol* **176**, 2137–2142.

Costerton, J. W., Lewandowski, D. E., Caldwell, D. E., Korber, D. R. & Lappin-Scott, H. M. (1995). Microbial biofilms. *Annu Rev Microbiol* **49**, 711–745.

Costerton, J. W., Stewart, P. S. & Greenberg, E. P. (1999). Bacterial biofilms: a common cause of persistent infections. *Science* **284**, 1318–1322.

Davies, D. G. & Geesey, G. G. (1995). Regulation of the alginate biosynthesis gene algC in

Pseudomonas aeruginosa during biofilm development in continuous culture. *Appl Environ Microbiol* **61**, 860–867.

Davies, D. G., Chakrabarty, A. M. & Geesey, G. G. (1993). Exopolysaccharide production in biofilms – substratum activation of alginate gene expression by *Pseudomonas aeruginosa*. *Appl Environ Microbiol* **59**, 1181–1186.

Davies, D. G., Parsek, M. R., Pearson, J. P., Iglewski, B. H., Costerton, J. W. & Greenberg, E. P. (1998). The involvement of cell-to-cell signals in the development of a bacterial biofilm. *Science* **280**, 295–298.

deBeer, D. & Stoodley, P. (1994). Effects of biofilm structure on oxygen distribution and mass transport. *Biotechnol Bioeng* **43**, 1131–1138.

Denyer, S. P., Hanlon, G. W. & Davies, M. C. (1993). Mechanisms of microbial adherence. In *Microbial Biofilms: Formation and Control. Society for Applied Bacteriology Technical Series 30*, pp. 13–27. Edited by S. P. Denyer, S. P. Gorman & M. Sussman. Oxford: Blackwell Scientific.

Diz, H. R. & Novak, J. T. (1999). Modelling biooxidation of iron in a packed-bed reactor. *J Environ Eng – ASCE* **125**, 109–116.

Fowler, H. W. & Mackay, A. J. (1980). The measurement of microbial adhesion. In *Microbial Adhesion to Surfaces*, pp. 143–161. Edited by R. C. W. Berkeley, J. M. Lynch, J. Melling, P. R. Rutter & B. Vincent. Chichester: Ellis Horwood.

Gander, S. & Gilbert, P. (1997). The development of a small-scale biofilm model suitable for studying the effects of antibiotics on biofilm of Gram negative bacteria. *J Antimicrob Chemother* **40**, 329–334.

Gibbons, R. J. & Nygaard, M. (1970). Interbacterial aggregation of plaque bacteria. *Arch Oral Biol* **15**, 1397–1400.

Gilbert, P., Allison, D. G., Evans, D. J., Handley, P. S. & Brown, M. R. W. (1989). Growth rate control of adherent bacterial populations. *Appl Environ Microbiol* **55**, 1308–1311.

Gjaltema, A., Arts, P. A. M., van Loosdrecht, M. C. M., Kuenen, J. G. & Heijnen, J. J. (1994). Heterogeneity of biofilms in rotating annular reactors. *Biotechnol Bioeng* **44**, 194–204.

Handley, P. S., McNab, R. & Jenkinson, H. F. (1999). Adhesive surface structures on oral bacteria. In *Dental Plaque Revisited*, pp. 145–170. Edited by H. N. Newman & M. Wilson. Cardiff: BioLine.

Hermanowicz, S. W. (1998). A model of two dimensional biofilm morphology. *Water Sci Technol* **37**, 219–222.

Huang, C.-T., Xu, K. D., McFeters, G. & Stewart, P. S. (1998). Spatial patterns of alkaline phosphatase expression within bacterial colonies and biofilms in response to phosphate starvation. *Appl Environ Microbiol* **64**, 1526–1531.

Keevil, C. W. (1989). Chemostat models of human and aquatic corrosive biofilms. In *Recent Advances in Microbial Ecology*, pp. 151–156. Edited by T. Hattori. Tokyo: Japan Scientific Society.

Keevil, C. W., Bradshaw, D. J., Dowsett, A. B. & Feary, T. W. (1987). Microbial film formation: dental plaque deposition on acrylic tiles using continuous culture techniques. *J Appl Bacteriol* **62**, 129–138.

Kinniment, S. L., Wimpenny, J. W. T., Adams, D. & Marsh, P. D. (1996). Development of a steady state oral microbial community using the constant-depth film fermenter. *Microbiology* **142**, 631–638.

Kolenbrander, P. E. (1988). Intergeneric coaggregation among human oral bacteria and ecology of dental plaque. *Annu Rev Microbiol* **42**, 627–656.

Kolenbrander, P. E., Anderson, R. N., Clemans, D. L., Whittaker, C. J. & Klier, C. M. (1999). Potential role of functionally similar coaggregation mediators in bacterial succession. In *Dental Plaque Revisited*, pp. 171–176. Edited by H. N. Newman & M. Wilson. Cardiff: BioLine.

Korber, D. R., Lawrence, J. R., Lappin-Scott, H. M. & Costerton, J. W. (1995). The formation of microcolonies and functional consortia within biofilms. In *Bacterial Biofilms*, pp. 15–45. Edited by H. M. Lappin-Scott & J. W. Costerton. Cambridge: Cambridge University Press.

Lewandowski, Z., Altobelli, S. A. & Fukushima, E. (1993). Nmr and microelectrode studies of hydrodynamics and kinetics in biofilms. *Biotechnol Prog* **9**, 40–45.

Loo, C. Y., Corliss, D. A. & Ganeshkumar, N. (2000). *Streptococcus gordonii* biofilm formation: identification of genes that code for biofilm phenotypes. *J Bacteriol* **182**, 1374–1382.

McCoy, W. F., Bryers, J. D., Robbins, J. & Costerton, J. W. (1981). Observations of fouling biofilm formation. *Can J Microbiol* **29**, 910–917.

Manz, W., Wendt-Potthoff, K., Neu, T. R., Szewzyk, U. & Lawrence, J. R. (1999). Phylogenetic composition, spatial structure and dynamics of lotic bacterial biofilms investigated by fluorescent in situ hybridization and confocal laser scanning microscopy. *Microb Ecol* **37**, 225–237.

Marsh, P. D. (1995). Dental plaque. In *Microbial Biofilms*, pp. 282–300. Edited by H. M. Lappin-Scott & J. W. Costerton. Cambridge: Cambridge University Press.

Marsh, P. D. & Bradshaw, D. J. (1999). Microbial community aspects of dental plaque. In *Dental Plaque Revisited*, pp. 237–253. Edited by H. N. Newman & M. Wilson. Cardiff: BioLine.

Matsushita, M. & Fujikawa, H. (1990). Diffusion limited growth in bacterial colony formation. *Physica A* **168**, 498–506.

Molin, S. (1999). Microbial activities in biofilm communities. In *Dental Plaque Revisited*, pp. 73–78. Edited by H. N. Newman & M. Wilson. Cardiff: BioLine.

Møller, S., Sternberg, C., Andersen, J. B., Christensen, B. B., Ramos, J. L., Givskov, M. & Molin, S. (1998). *In situ* gene expression in mixed culture biofilms. *Appl Environ Microbiol* **64**, 721–732.

Nielsen, A. T., Tolker-Nielsen, T., Barken, K. B. & Molin, S. (2000). Role of commensal relationships on the spatial structure of a surface-attached microbial consortium. *Environ Microbiol* **2**, 59–68.

Nyvad, B. & Fejerskov, O. (1989). Structures of dental plaque and the plaque-enamel interface in human experimental caries. *Caries Res* **23**, 151–158.

Ohgiwari, M., Matsushita, M. & Matsuyama, T. (1992). Morphological changes in growth phenomena of bacterial colony patterns. *J Phys Soc Jpn* **61**, 816–822.

O'Toole, G. A. & Kolter, R. (1998). Flagellar and twitching motility are necessary for *Pseudomonas aeruginosa* biofilm development. *Mol Microbiol* **30**, 295–305.

O'Toole, G. A., Gibbs, K. A., Hager, P. W., Phibbs, P. V. & Kolter, R. (2000). The global carbon metabolism regulator Crc is a component of a signal transduction pathway required for biofilm development by *Pseudomonas aeruginosa*. *J Bacteriol* **182**, 425–431.

Peters, A. C. & Wimpenny, J. W. T. (1988). A constant depth laboratory model film fermentor. *Biotechnol Bioeng* **32**, 263–270.

Picioreanu, C., van Loosdrecht, M. C. M. & Heijnen, J. J. (1998). Mathematical modeling of biofilm structure with a hybrid differential-discrete cellular automaton approach. *Biotechnol Bioeng* **58**, 101–116.

Picioreanu, C., van Loosdrecht, M. C. M. & Heijnen, J. J. (1999). Discrete-differential modeling of biofilm structure. *Water Sci Technol* **39**, 115–122.

Pratt, L. A. & Kolter, R. (1998). Genetic analysis of *Escherichia coli* biofilm formation: roles of flagella, motility, chemotaxis and type 1 pili. *Mol Microbiol* **30**, 285–293.

Rasmussen, K. & Lewandowski, Z. (1998). Microelectrode measurements of local mass transport rates in heterogeneous biofilms. *Water Res* **59**, 302–309.

Rothemund, C., Camper, A. & Wilderer, P. A. (1994). Biofilms growing on gas permeable membranes. *Water Sci Technol* **29**, 447–454.

Ruseska, L., Robbins, J., Costerton, J. W. & Lashen, E. (1982). Biocide testing against corrosion causing oil-field bacteria helps control plugging. *Oil Gas J* March, 153–164.

Schindler, J. & Rovensky, L. (1994). A model of intrinsic growth of a *Bacillus* colony. *Binary* **6**, 105–108.

Senior, E., Slater, A. T. & Bull, J. H. (1976). Enzyme evolution in a microbial community growing on the herbicide, Dalapon. *Nature* **263**, 476.

Shotton, D. M. (1989). Confocal scanning optical microscopy and its applications for biological specimens. *J Cell Sci* **94**, 175–206.

Shu, M., Wong, L., Miller, J. H. & Sissons, C. H. (2000). Development of multispecies consortia biofilms of oral bacteria as an enamel and root caries model system. *Arch Oral Biol* **45**, 27–40.

Sternberg, C., Christensen, B. B., Johansen, T., Nielsen, A. T., Andersen, J. B., Givskov, M. & Molin, S. (1999). Distribution of bacterial growth activity in flow chamber biofilms. *Appl Environ Microbiol* **65**, 4108–4117.

Stoodley, P., deBeer, D. & Lewandowski, Z. (1994). Liquid flow in biofilm systems. *Appl Environ Microbiol* **60**, 2711–2716.

Stoodley, P., Lewandowski, Z., Boyle, J. D. & Lappin-Scott, H. M. (1999). The formation of migratory ripples in a mixed species bacterial biofilm growing in turbulent flow. *Environ Microbiol* **1**, 447–455.

Sutton, N. A., Hughes, N. & Handley, P. S. (1994). A comparison of conventional SEM techniques, low temperature SEM and the electroscan wet scanning electron microscope to study the structure of a biofilm of *Streptococcus crista* CR3. *J Appl Bacteriol* **76**, 448–454.

Thomas, L. V., Wimpenny, J. W. T. & Barker, G. C. (1997). Spatial interactions between subsurface bacterial colonies in a model system: a territory model describing the inhibition of *Listeria monocytogenes* by a nisin-producing lactic acid bacterium. *Microbiology* **143**, 2575–2582.

Trulear, M. G. & Characklis, W. G. (1982). Dynamics of biofilm formation. *J Water Pollut Control Fed* **54**, 1288–1301.

van Loosdrecht, M. C. M., Eikelboom, D., Gjaltema, A., Mulder, A., Tijhuis, L. & Heijnen, J. J. (1995). Biofilm structures. *Water Sci Technol* **32**, 35–43.

Vroom, J. M., DeGraw, K. J., Gerritsen, H. C., Bradshaw, D. J. & Marsh, P. D. (1999). Depth penetration and detection of pH gradients in biofilms by two-photon excitation microscopy. *Appl Environ Microbiol* **65**, 3502–3511.

Walker, J. T. & Keevil, C. W. (1994). A study of microbial biofilms using light microscope techniques. *Int Biodeterior Biodegrad* **34**, 223–236.

Watanabe, Y., Kimura, K., Okabe, S., Ozawa, G. & Ohkuma, N. (1997). A novel biofilm-membrane reactor for ammonia oxidation at low concentrations. *Water Sci Technol* **36**, 51–60.

White, J. G., Amos, W. B. & Fordham, M. (1987). An evaluation of confocal versus conventional imaging of biological structure by fluorescence light microscopy. *J Cell Biol* **105**, 41–48.

Wilderer, P. A. (1995). Technology of membrane biofilm reactors operated under periodically changing process conditions. *Water Sci Technol* **31**, 173–183.

Wimpenny, J. W. T. (1992). Microbial systems – patterns in time and space. *Adv Microb Ecol* **12**, 469–522.

Wimpenny, J. W. T. (1996). Laboratory growth systems in biofilm research. *Scanning Microsc Int* **6**, 221–232.

Wimpenny, J. W. T. (1999). Laboratory models of biofilm. In *Dental Plaque Revisited*, pp. 89–110. Edited by H. N. Newman & M. Wilson. Cardiff: BioLine.

Wimpenny, J. W. T. & Colasanti, R. (1997). A unifying hypothesis for the structure of microbial biofilms. *FEMS Microbiol Ecol* **22**, 1–16.

Wolfaardt, G. M., Lawrence, J. R., Robarts, R. D. & Caldwell, S. J. (1994). Multicellular organization in a degradative biofilm community. *Appl Environ Microbiol* **60**, 434–446.

Wood, S. R., Kirkham, J., Marsh, P. D., Shore, R. C., Nattress, B. & Robinson, C. (2000). Architecture of intact natural human plaque biofilms studied by confocal laser scanning microscopy. *J Dent Res* **79**, 21–27.

Initial microbial adhesion events: mechanisms and implications

Henk J. Busscher and Henny C. van der Mei

Department of Biomedical Engineering, University of Groningen, Antonius Deusinglaan 1, 9713 AV Groningen, The Netherlands

INTRODUCTION

In most textbooks, biofilm formation on surfaces has been depicted as a sequence of events. Depending on the particular interest of the author, these events are split up into four or more steps (Escher & Characklis, 1990; Van Loosdrecht et al., 1990). The sequence is presented in Fig. 1. Traditionally, biofilm formation is said to begin with mass transport of micro-organisms towards a substratum surface (Fig. 1b), but in almost all environments, the mass transport step is preceded by the adsorption of conditioning film components (Fig. 1a), such as: an adsorbed tear film on a contact lens (Baguet et al., 1995; Landa et al., 1998); the salivary pellicle on surfaces in the oral cavity (Busscher et al., 1989; Bradshaw et al., 1997); a film of adsorbed urinary components on urogenital surfaces (Reid et al., 1998); and adsorbed macromolecules on marine surfaces (Schneider & Marshall, 1994). Many more examples of the formation of a film conditioning a surface as the first step in biofilm formation can be given. Once transported to a substratum surface, organisms may or may not adhere, depending on the interaction forces (Rutter & Vincent, 1980). This initial adhesion (Fig. 1c) is generally reversible (Norde & Lyklema, 1989), but even in the absence of exopolymer production, becomes less reversible within minutes due to the progressive removal of water from in-between the interacting surfaces (Meinders et al., 1995). The unfolding of binding molecules and other non-metabolic mechanisms eventually lead to the strong anchoring of the initially adhering organisms (Fig. 1e). At this stage, there is often a neglected step which occurs almost simultaneously, that is, coadhesion (Fig. 1d). In natural habitats, more than one organism can colonize a surface, while within a microbial community only a limited number of strains act as primary colonizers. Once

SGM symposium 59: Community structure and co-operation in biofilms. Editors D. Allison, P. Gilbert, H. Lappin-Scott, M. Wilson. Cambridge University Press. ISBN 0 521 79302 5 ©SGM 2000.

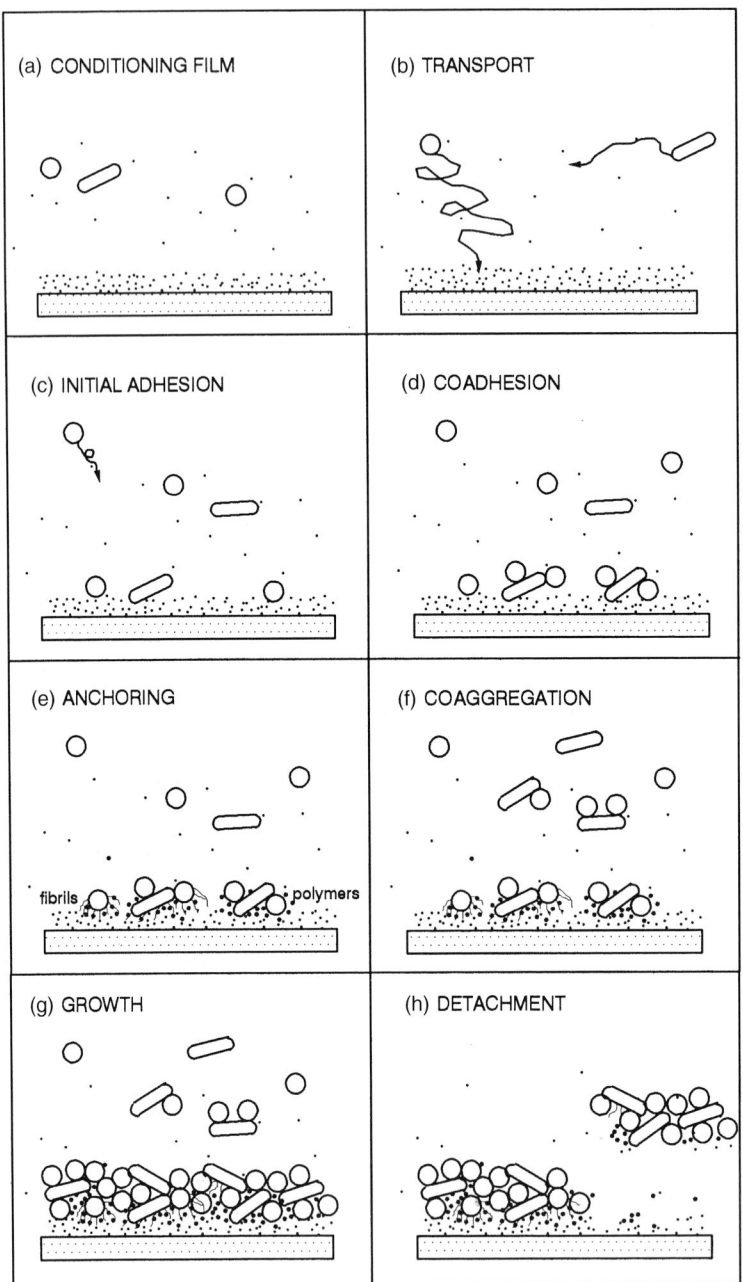

Fig. 1. Sequential steps in biofilm formation. (a) Conditioning film formation; (b) microbial mass transport; (c) initial adhesion; (d) coadhesion between different strains; (e) anchoring of adhering organisms; (f) coaggregation between organisms in suspension; (g) growth of adhering organisms; (h) detachment of parts of the biofilm.

primary colonizers have adhered, secondary colonizers coadhere with organisms already adhering to the surface as the onset of a multispecies biofilm (Kolenbrander, 1989). Coadhesion events have been described for the oral cavity (Bos *et al.*, 1995), the oropharynx (Neu *et al.*, 1992), the urogenital tract (Reid *et al.*, 1988) and water pipe lines (Buswell *et al.*, 1997). Although microbial densities in suspension are low in natural environments (Bos *et al.*, 1999a), coaggregates of organisms may form in suspension and adhere as coaggregates (Fig. 1f), which leads to a biofilm similar in structure to that arising from coadhesion. Mechanistically, the only difference described so far between coaggregation and coadhesion is the availability of interaction sites on the microbial cell surfaces. These may become unavailable for coadhesion when they possess a large affinity for the substratum surface and adsorb there (Bos *et al.*, 1996a). Subsequently, growth is the main mechanism of microbial accumulation in the biofilm (Fig. 1g), which continues until naturally occurring shear forces, such as the blinking of the eyelid over a contact lens surface, waves over a marine surface or the mechanical action of a toothbrush, detach the entire biofilm or parts of it. The initially adhering organisms that link the biofilm directly with the substratum surface play a pivotal role here (Busscher *et al.*, 1995). If they cannot withstand the shear forces and detach, then the entire biofilm detaches too (Fig. 1h). In certain industrial applications, it is important to maintain a biofilm on a surface (Diks & Ottengraf, 1991). Also, in medicine, the maintenance of a protective, indigenous microflora is associated with health benefits (Costerton *et al.*, 1987). These beneficial biofilms can be found in the digestive and urinary tracts (Reid *et al.*, 1990), and in dental plaque, where they may be essential for the balance between re- and demineralization of dental hard tissues. However, due to a variety of endogenous and exogenous factors, the microbial composition of such 'healthy' biofilms can become disturbed (Neu *et al.*, 1992) to produce a pathogenic biofilm that can lead to disease or deterioration of the substratum, as in dental caries (Marsh & Martin, 1992) or the biodeterioration of silicone rubber voice prostheses (Mahieu *et al.*, 1986; Neu *et al.*, 1994).

CONDITIONING FILM FORMATION

The generally accepted notion that organisms never adhere to a bare substratum surface in natural environments is based on the ubiquitous presence of macromolecules. Owing to their smaller dimensions, diffusional mass transport of macromolecules is many times greater than that of micro-organisms (Van der Mei *et al.*, 1994). The adsorption of macromolecules to a substratum occurs within seconds. This is illustrated in Fig. 2, where there is rapid formation of a salivary conditioning film on a polished enamel surface (Busscher *et al.*, 1989). Despite the fact that the adhering micro-organisms never 'meet' a substratum that is not covered with macromolecules, this does not imply that the properties of a substratum surface have no influence on further microbial adhesion events. The composition and conformation of adsorbed

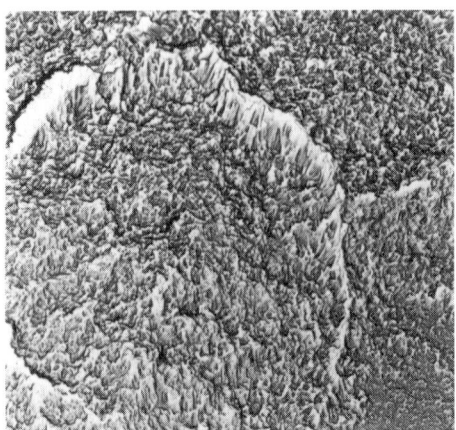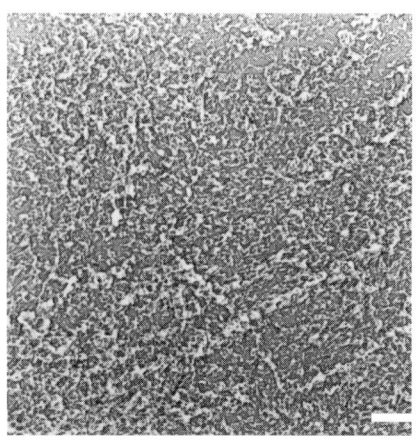

Fig. 2. Electron micrograph of a bare dental enamel surface after polishing, obtained by a shadow replica technique to enhance molecular details, showing the crystallite structure of the enamel prisms (left), which have already disappeared after seconds of exposure to saliva (right). Bar, 1 μm. Micrographs are reprinted from *Archives of Oral Biology*, 34, H. J. Busscher, H. M. W. Uyen, I. Stokroos & W. L. Jongebloed, A transmission electron microscopy study of the adsorption patterns of early developing artificial pellicles on human enamel, pp. 803–810, copyright 1989, with permission from Elsevier Science.

macromolecules critically depend on the substratum surface properties. It has been suggested that salivary mucins adsorb with their carbohydrate end to the hydrophilic substrata therewith exposing their protein end, while on hydrophobic substrata the mucins adsorb with their protein end. This 'flip-flop' behaviour caused adsorption of salivary mucins to hydrophobic Teflon to enhance adhesion of *Streptococcus mutans*, while adsorption to hydrophilic glass discourages *S. mutans* adhesion (Pratt-Terpstra & Busscher, 1991).

Other examples of conditioning film formation are numerous, and an extensive review with emphasis on the marine environment has been written by Schneider (1996).

MICROBIAL MASS TRANSPORT

It is of utmost importance to take the different modes of microbial mass transport into account (Elimelech *et al.*, 1995) for understanding microbial adhesion in natural environments and in experimental systems. Strong convective mass transport exists for the upstream side of a rock in the middle of a river, while at its other sides, a combination of convection and diffusion transports the organisms towards the rock. In cross-section, convective-diffusion and sedimentation constitute the modes of microbial mass transport towards the bottom of the river. Often, mass transport is neglected in the design of microbial adhesion assays (Sjollema *et al.*, 1989). As a consequence, erroneous conclusions can be drawn concerning microbial adhesion, as it can not be decided, in the absence of a verified sufficient mass transport, whether adhesion is low due to a lack of affinity of the organisms for a substratum or due to insufficient mass transport. Hydrodynamics explain mass transport conditions for microbial adhesion in experimental systems such as the parallel plate or stagnation point flow chamber (Yang et al., 1999a, b). It has now been demonstrated that when the convective mass transport component is towards a substratum surface, such as in the stagnation point flow chamber, this yields a higher deposition efficiency than when the convective mass transport is parallel to the substratum, such as in the parallel plate flow chamber (Yang et al., 1999b). This probably indicates that the convective mass transport towards a surface facilitates the close approach between the interacting surfaces needed for adhesion to occur.

INITIAL ADHESION AND ANCHORING

Initial microbial adhesion is a result of non-specific attractive Lifshitz–van der Waals forces operating over separation distances of several hundred nanometres (Rutter & Vincent, 1980). As most substratum and microbial cell surfaces possess a net overall negative charge (Jucker et al., 1996; Busscher et al., 2000), electrostatic interactions are repulsive and become operative, depending on ionic strength, from separation distances of around 10–20 nm. Only when approach is closer than about 5 nm do distinct features on the interacting surfaces, such as specific adhesion receptors, become effective and facilitate strong adhesion (Busscher & Weerkamp, 1987; Busscher et al., 1992), which eventually becomes irreversible. Specific adhesion receptors also operate through the long-range Lifshitz–van der Waals attraction, but in addition include localized attraction from acid–base interactions (Van Oss, 1995) and positively charged domains (Cowan et al., 1994).

The transition from reversible to irreversible adhesion has been measured for thermophilic streptococci adhering to hydrophobic and hydrophilic substrata. This showed that over a time span of 5–10 min, adhesion became more than 100 times less reversible (Meinders et al., 1995). When dealing with exopolymer-producing organisms, the transition between reversible and irreversible adhesion may not only be faster, but irreversibility may possibly become absolute. With regards to the reversibility of microbial adhesion, the ionic and shear conditions that exist and alterations in interactions due to change of ionic strength, pH and fluid flow may stimulate microbial detachment.

COAGGREGATION AND COADHESION

Coaggregation is defined as the adhesive interaction between two micro-organisms in suspension, while coadhesion is the interaction between a planktonic and an already adhering organism (Bos *et al.*, 1995; Ellen *et al.*, 1994). Kolenbrander and co-workers (Kolenbrander, 1989; Ganeshkumar *et al.*, 1998) extensively studied coaggregation of oral micro-organisms. Brownian motion dynamics have indicated (Bos *et al.*, 1998), however, that coaggregation may be an unlikely event because in most natural circumstances, the densities of planktonic organisms in a suspension, such as saliva, urine or potable water systems, are too low for frequent encounters between partner strains. A much more likely event is coadhesion.

Coaggregation, nevertheless, can generally be considered as a good model for the interactions that govern coadhesion (Bos *et al.*, 1995). Coaggregation and coadhesion are critical colloid-chemical phenomena, easily influenced by temperature variations, pH, ionic strength and the presence of adhesion-mediating cations or inhibitors like lactose (Bos *et al.*, 1996a, b, c). Both phenomena have hitherto only been described for strains interacting in the absence of strong electrostatic repulsion, that is, at least one of the interacting pairs had an almost zero zeta potential. Interactions between early colonizers of enamel surfaces, such as between streptococci and actinomyces, occur only in the presence of divalent Ca^{2+} cations and could not be explained by acid–base interactions. This is opposite to the acid–base interactions between streptococci and *Prevotella* (Cookson *et al.*, 1995) in the absence of divalent cations (Bos *et al.*, 1999b).

Coadhesion between microbial pairs can create niches in a biofilm for optimal growth of both pairs (Bradshaw *et al.*, 1994). Furthermore, recent work on the removal by brushing of coadhering oral streptococci and actinomyces from enamel surfaces has demonstrated that coadhering micro-organisms were less easily detached than streptococci and actinomyces that did not coadhere (Bos *et al.*, 1999a). Moreover, *de novo* adhesion of streptococci to brushed surfaces was higher for coadhering pairs than for non-coadhering pairs.

SURFACE GROWTH

Growth of adhering micro-organisms is not an initial adhesion event, except for the surface-associated growth of the initially adhering organisms. In this respect, it is interesting that Harkes *et al.* (1991) suggested that micro-organisms adhering under conditions of electrostatic attraction would not be able to multiply and grow. Recently, Gottenbos *et al.* (2000) demonstrated that generation times of pseudomonads adhering on substrata were longer when the association of the organisms with the surface was stronger. Recently, Barton *et al.* (1996) compared the initial surface-growth rate of *Staphylococcus epidermidis*, *Pseudomonas aeruginosa* and *Escherichia coli* on different orthopaedic implant materials in a parallel plate flow chamber in whole growth medium. They found a correlation between the generation time of *P. aeruginosa* and the free energy of adhesion of the organisms for the different biomaterials. These observations are most relevant in the design of non-fouling surfaces, which are generally based on low energy surfaces, to which few organisms can adhere.

DETACHMENT

Once a mature biofilm has formed, it has to be able, by its internal cohesivity and adhesivity to the substratum surface, to withstand detachment forces (Busscher *et al.*, 1995), which are numerous in natural and industrial environments. Sometimes, biofilms detach partly as a result of insufficient internal cohesivity (Rittmann, 1989). However, if the link between the biofilm and the substratum surface, as made by the initially adhering organisms, is weak, an entire biofilm may detach under environmental shear conditions.

A lack of adhesivity between oral biofilms and hydrophobic surfaces has been described, based on the observation that, under the well-controlled and constant shear conditions of a parallel plate flow chamber, oral streptococci adhered almost equally well to salivary conditioning films on hydrophobic and hydrophilic surfaces. After 9 d, *in vivo* hydrophobic surfaces harboured far less dental plaque than hydrophilic surfaces (Quirynen *et al.*, 1994). It was hypothesized that, under the dynamic shear conditions of the oral cavity, organisms would nevertheless adhere and grow to form a biofilm, but that due to a lack of adhesivity of the initially adhering bacteria, detachment would regularly occur yielding an *in toto* lower formation of dental plaque on hydrophobic surfaces (Busscher & Van der Mei, 1997).

On voice prostheses in laryngectomized patients, a similarly strong influence of the prosthesis surface hydrophobicity on oropharyngeal biofilm formation has been observed over a time period of 6 weeks (Everaert *et al.*, 1997). Both *in vivo* observations over extended periods of time indicate that under dynamic shear conditions, the link between a biofilm and a substratum surface is weakest for hydrophobic biomaterials.

ANTI-ADHESIVE INITIAL ADHESION EVENTS IN A BIOFILM

So far, only adhesive initial adhesion events have been described. However, especially in multispecies biofilms, defence mechanisms exist by which adhering organisms can prevent other colonizing organisms from adhering (Rosenberg, 1986). One of the most powerful mechanisms utilized to this end is the release of biosurfactants by adhering organisms that adsorb to the substratum surface to alter it in such a way as to discourage adhesion of other strains and species (Neu, 1996). Since the application of axisymmetric drop shape analysis by profile (Rotenberg *et al.*, 1983), the release of minute amounts of biosurfactants by micro-organisms in a suspension droplet has become detectable. This has enabled many more strains to be identified as

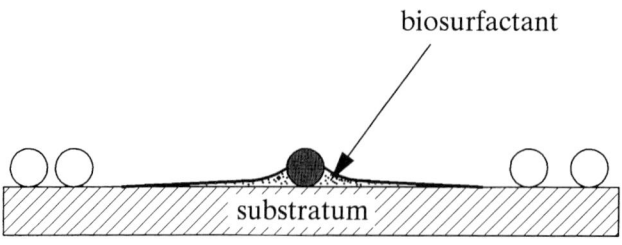

Fig. 3. Schematic presentation of biosurfactant release by an adhering organism and its adsorption to a substratum surface. The substratum surface properties have become unfavourable for other strains to adhere.

biosurfactant-releasing, including lactobacilli (Reid *et al.*, 1999), *Streptococcus mitis* (Van der Vegt *et al.*, 1991) and *Streptococcus thermophilus* (Busscher *et al.*, 1997) strains. Lactobacilli have been demonstrated to interfere with initial uropathogen adhesion (Velraeds *et al.*, 1996), *S. mitis* strains interfere with *S. mutans* adhesion (Van Hoogmoed *et al.*, 2000) and *S. thermophilus* strains may inhibit *Candida* adhesion (Busscher *et al.*, 1997).

Recently, we have grown mixed bacterial and fungal oropharyngeal biofilms on silicone rubber in a modified Robbins device (Leunisse *et al.*, 1999) and subsequently exposed the biofilms formed to suspensions of biosurfactant-releasing *S. thermophilus* B. In line with the *in vitro* results on initial adhesion inhibition of *Candida* by streptococci (Busscher *et al.*, 1997), this reduced the prevalence of *Candida* in the biofilm to 33% of the control (Van der Mei *et al.*, 2000).

Although the amounts released by adhering micro-organisms are small, the power of this defence mechanism must not be underestimated. A simple calculation shows that the amount of biosurfactant released by one adhering organism is sufficient to cover a substratum surface area several hundred times its own geometrical area (Busscher & Van der Mei, 1997), as schematically indicated in Fig. 3.

SUMMARY

We have reviewed the traditional sequence of events during biofilm formation on a substratum surface and associated the process with a number of events thought important from our perspective. This included conditioning film formation, coaggregation and coadhesion, as well as detachment. Evidence for biosurfactant release as a defence mechanism in mixed-species biofilms exists, but it does not appear to be sufficiently widespread to be included in a general sequence of events governing biofilm formation.

REFERENCES

Baguet, J., Sommer, F., Claudon-Eyl, V. & Duc, T. M. (1995). Characterization of lacrymal component accumulation on worn soft contact lens surfaces by atomic force microscopy. *Biomaterials* **16**, 3–9.

Barton, A. J., Sagers, R. D. & Pitt, W. D. (1996). Measurement of bacterial growth rates on polymers. *J Biomed Mater Res* **32**, 271–278.

Bos, R., Van der Mei, H. C. & Busscher, H. J. (1995). A quantitative method to study co-adhesion of microorganisms in a parallel plate flow chamber. II: Analysis of the kinetics of co-adhesion. *J Microbiol Methods* **23**, 169–182.

Bos, R., Van der Mei, H. C. & Busscher, H. J. (1996a). Influence of ionic strength and substratum hydrophobicity on the co-adhesion of oral microbial pairs. *Microbiology* **142**, 2355–2361.

Bos, R., Van der Mei, H. C. & Busscher, H. J. (1996b). Co-adhesion of oral microbial pairs under flow in the presence of saliva and lactose. *J Dent Res* **75**, 809–815.

Bos, R., Van der Mei, H. C. & Busscher, H. J. (1996c). Influence of temperature on the co-adhesion of oral microbial pairs in saliva. *Eur J Oral Sci* **104**, 372–377.

Bos, R., Van der Mei, H. C. & Busscher, H. J. (1998). On the role of co-aggregation and co-adhesion in dental plaque formation. In *Oral Biofilms and Plaque Control*, pp. 163–174. Edited by H. J. Busscher & L. V. Evans. India: Harwood Publishers.

Bos, R., Yang, J., Belder, G. F. & Busscher, H. J. (1999a). Interspecies binding as a design factor in oral biofilms. In *Biofilms: the Good, the Bad and the Ugly*, pp. 237–242. Edited by J. Wimpenny, P. Gilbert, M. Brading & R. Bayston. Wiltshire: BioLine.

Bos, R., Van der Mei, H. C. & Busscher, H. J. (1999b). Physico-chemistry of initial microbial adhesive interactions – its mechanisms and methods for study. *FEMS Microbiol Rev* **23**, 179–230.

Bradshaw, D. J., Homer, K. A., Marsh, P. D. & Beighton, D. (1994). Metabolic cooperation in oral microbial communities during growth on mucin. *Microbiology* **140**, 3407–3412.

Bradshaw, D. J., Marsh, P. D., Watson, G. K. & Allison, C. (1997). Effects of conditioning films on oral microbial biofilm development. *Biofouling* **11**, 217–226.

Busscher, H. J. & Van der Mei, H. C. (1997). Physico-chemical interactions in initial microbial adhesion and relevance for biofilm formation. *Adv Dent Res* **11**, 24–32.

Busscher, H. J. & Weerkamp, A. II. (1987). Specific and non-specific interactions in bacterial adhesion to solid substrata. *FEMS Microbiol Rev* **46**, 165–173.

Busscher, H. J., Uyen, H. M. W., Stokroos, I. & Jongebloed, W. L. (1989). A transmission electron microscopy study of the adsorption patterns of early developing artificial pellicles on human enamel. *Arch Oral Biol* **34**, 803–810.

Busscher, H. J., Cowan, M. M. & Van der Mei, H. C. (1992). On the relative importance of specific and non-specific approaches to oral microbial adhesion. *FEMS Microbiol Rev* **88**, 199–200.

Busscher, H. J., Bos, R. & Van der Mei, H. C. (1995). Initial microbial adhesion is a determinant for the strength of biofilm adhesion. *FEMS Microbiol Lett* **128**, 229–234.

Busscher, H. J., Van Hoogmoed, C. G., Geertsema-Doornbusch, G. I., Van der Kuijl-Booij, M. & Van der Mei, H. C. (1997). *Streptococcus thermophilus* and its biosurfactants inhibit adhesion by *Candida* spp. on silicone rubber. *Appl Environ Microbiol* **63**, 3810–3817.

Busscher, H. J., Bos, R., Van der Mei, H. C. & Handley, P. S. (2000). Physicochemistry of microbial adhesion from an overall approach to the limits. In *Physical Chemistry of Biological Surfaces*, pp. 431–458. Edited by A. Baszkin & W. Norde. New York: Marcel Dekker.

Buswell, C. M., Herlihy, Y. M., Marsh, P. D., Keevil, C. D. & Leach, S. A. (1997). Coaggregation amongst aquatic biofilm bacteria. *J Appl Microbiol* **83**, 477–484.

Cookson, A. L., Handley, P. S., Jacob, A. E., Watson, G. K. & Allison, C. (1995). Coaggregation between *Prevotella nigrescens* and *Prevotella intermedia* with *Actinomyces naeslundii* strains. *FEMS Microbiol Lett* **132**, 291–296.

Costerton, J. W., Cheng, K. J., Geesey, G. G., Ladd, T. I., Nickel, J. C., Dasgupta, M. & Marrie, T. J. (1987). Bacterial biofilms in nature and disease. *Annu Rev Microbiol* **41**, 435–464.

Cowan, M. M., Mikx, F. H. M. & Busscher, H. J. (1994). Electrophoretic mobility and hemagglutination of *Treponema denticola* ATCC33520. *Colloids Surf B: Biointerfaces* **2**, 407–410.

Diks, R. M. M. & Ottengraf, S. P. (1991). Verification studies of a simplified model for the removal of dichloromethane from waste gases using a biological trickling filter. *Bioprocess Eng* **6**, 93–99.

Elimelech, M., Gregory, J., Jia, X. & Williams, R. A. (1995). *Particle Deposition and Aggregation: Measurement, Modelling and Simulation*, 1st edn, pp. 68–159. Oxford: Butterworth-Heinemann.

Ellen, R. P., Veisman, H., Buivids, I. A. & Rosenberg, M. (1994). Kinetics of lactose-reversible coadhesion of *Actinomyces naeslundii* WVU 398A and *Streptococcus oralis* 34 on the surface of hexadecane droplets. *Oral Microbiol Immunol* **9**, 364–371.

Escher, A. & Characklis, W. G. (1990). Modeling the initial events in biofilm accumulation. In *Biofilms*, pp. 445–487. Edited by W. G. Characklis & K. C. Marshall. New York: Wiley.

Everaert, E. P. J. M., Mahieu, H. F., Wong Chung, R. P., Verkerke, G. J., Van der Mei, H. C. & Busscher, H. J. (1997). A new method for *in vivo* evaluation of biofilms on surface-modified silicone rubber voice prostheses. *Eur Arch Otorhinolaryngol* **254**, 261–263.

Ganeshkumar, N., Hughes, C. V. & Weiss, E. I. (1998). Co-aggregation in dental plaque formation. In *Oral Biofilms and Plaque Control*, pp. 163–174. Edited by H. J. Busscher & L. V. Evans. India: Harwood Publishers.

Gottenbos, B., Van der Mei, H. C. & Busscher, H. J. (2000). Initial adhesion and surface-growth of *Staphylococcus epidermidis* and *Pseudomonas aeruginosa* on biomedical polymers. *J Biomed Mater Res* **50**, 208–214.

Harkes, G., Feijen, J. & Dankert, J. (1991). Adhesion of *Escherichia coli* on to a series of poly(methacrylates) differing in charge and hydrophobicity. *Biomaterials* **12**, 853–860.

Jucker, B. A., Harms, H. & Zehnder, A. J. B. (1996). Adhesion of the positively charged bacterium *Stenotrophomona* (*Xanthomonas*) *maltophilia* 70401 to glass and teflon. *J Bacteriol* **178**, 5472–5479.

Kolenbrander, P. E. (1989). Surface recognition among oral bacteria: multi-generic coaggregations and their mediators. *Crit Rev Microbiol* **17**, 137–155.

Landa, A. S., Van der Mei, H. C., Van Rij, G. & Busscher, H. J. (1998). Efficacy of ophthalmic solutions to detach adhering *Pseudomonas aeruginosa* from contact lenses. *Cornea* **17**, 293–300.

Leunisse, C., Van Weissenbruch, R., Busscher, H. J., Van der Mei, H. C. & Albers, F. W. J.

(1999). The artificial throat: a new method for standardization of *in vitro* experiments with tracheo-oesophageal voice prostheses. *Acta Otolaryngol (Stockh)* **119**, 604–608.

Mahieu, H. F., Van Saene, J. J. M., Den Besten, J. & Van Saene, H. K. F. (1986). Oropharynx decontamination preventing *Candida* vegetation on voice prostheses. *Arch Otolaryngol Head Neck Surg* **112**, 321–325.

Marsh, P. D. & Martin, M. V. (1992). *Oral Microbiology*, 3rd edn. London: Chapman & Hall.

Meinders, J. M., Van der Mei, H. C. & Busscher, H. J. (1995). Deposition efficiency and reversibility of bacterial adhesion under flow. *J Colloid Interface Sci* **176**, 329–341.

Neu, T. R. (1996). Significance of bacterial-surface active compounds in interactions of bacteria with interfaces. *Microbiol Rev* **6**, 151–166.

Neu, T. R., Van der Mei, H. C. & Busscher, H. J. (1992). Biofilms associated with health. In *Biofilms – Science and Technology*, pp. 21–34. Edited by L. F. Melo, T. R. Bott, M. Fletcher & B. Capdeville. London: Kluwer.

Neu, T. R., De Boer, C. E., Verkerke, G. J., Schutte, H. K., Rakhorst, G., Van der Mei, H. C. & Busscher, H. J. (1994). Biofilm development in time on a silicone voice prosthesis – a case study. *Microb Ecol Health Dis* **7**, 27–33.

Norde, W. & Lyklema, J. (1989). Protein adsorption and bacterial adhesion to solid surfaces: a colloid-chemical approach. *Colloids Surf* **38**, 1–13.

Pratt-Terpstra, I. H. & Busscher, H. J. (1991). Adsorption of salivary mucins onto enamel and artificial solid substrata and its influence on oral streptococcal adhesion. *Biofouling* **3**, 199–207.

Quirynen, M., Van der Mei, H. C., Bollen, C. M. L., Geertsema-Doornbusch, G. I., Busscher, H. J. & Van Steenberghe, D. (1994). Clinical relevance of the influence of surface free energy and roughness on the supragingival and subgingival plaque formation in man. *Colloids Surf B: Biointerfaces* **2**, 25–31.

Reid, G., McGroarty, J. A., Angotti, R. & Cook, R. L. (1988). Lactobacillus inhibitor production against *Escherichia coli* and coaggregation ability with uropathogens. *Can J Microbiol* **34**, 344–351.

Reid, G., Bruce, A. W., McGroarty, J. A., Cheng, K. J. & Costerton, J. W. (1990). Is there a role for lactobacilli in prevention of urogenital and intestinal infection? *Clin Microbiol Rev* **3**, 335–344.

Reid, G., Van der Mei, H. C. & Busscher, H. J. (1998). Microbial biofilms and urinary tract infections. In *Urinary Tract Infections*, pp. 111–116. Edited by W. Brumfitt, J. M. T. Hamilton-Miller & R. R. Bailey. London: Chapman & Hall.

Reid, G., Heinemann, C., Velraeds, M., Van der Mei, H. C. & Busscher, H. J. (1999). Biosurfactants produced by *Lactobacillus*. *Methods Enzymol* **310**, 426–433.

Rittmann, B. E. (1989). Detachment from biofilms. In *Structure and Function of Biofilms*, pp. 49–58. Edited by W. G. Characklis & P. A. Wilderer. Chichester: Wiley.

Rosenberg, E. (1986). Microbial surfactants. *CRC Crit Rev Microbiol* **3**, 109–131.

Rotenberg, Y., Boruvka, L. & Neumann, W. (1983). Determination of the surface tension and contact angle from the shape of axisymmetric fluid interfaces. *J Colloid Interface Sci* **93**, 169–183.

Rutter, P. R. & Vincent, B. (1980). The adhesion of microorganisms to surfaces: physico-chemical aspects. In *Microbial Adhesion to Surfaces*, pp. 79–91. Edited by R. C. W. Berkeley, J. M. Lynch, J. Melling, P. R. Rutter & B. Vincent. London: Ellis Horwood.

Schneider, R. R. (1996). Conditioning film-induced modification of substratum physicochemistry – analysis by contact angles. *J Colloid Interface Sci* **182**, 204–213.

Schneider, R. P. & Marshall, K. C. (1994). Retention of the Gram-negative marine bacterium SW8 on surfaces – effects of microbial physiology, substratum nature and conditioning films. *Colloids Surf B: Biointerfaces* **2**, 387–396.

Sjollema, J., Busscher, H. J. & Weerkamp, A. H. (1989). Real-time enumeration of adhering microorganisms in a parallel plate flow cell using automated image analysis. *J Microbiol Methods* **9**, 73–78.

Van Hoogmoed, C. G., Van der Kuijl-Booij, M., Van der Mei, H. C. & Busscher, H. J. (2000). Inhibition of *Streptococcus mutans* NS adhesion to glass with and without a salivary conditioning film by biosurfactant releasing *Streptococcus mitis* strains. *Appl Environ Microbiol* **66**, 659–663.

Van Loosdrecht, M. C. M., Lyklema, J., Norde, W. & Zehnder, J. B. (1990). Influences of interfaces on microbial activity. *Microbiol Rev* **54**, 75–87.

Van der Mei, H. C., Meinders, J. M. & Busscher, H. J. (1994). The influence of ionic strength and pH on diffusion of microorganisms with different structural surface structures. *Microbiology* **140**, 3413–3419.

Van der Mei, H. C., Free, R. H., Elving, G. J., Van Weissenbruch, R., Albers, F. W. J. & Busscher, H. J. (2000). Effect of probiotic bacteria on yeast prevalence in oropharyngeal biofilms on silicone rubber voice prostheses *in vitro. J Med Microbiol* (in press).

Van Oss, C. J. (1995). Hydrophobicity of biosurfaces – origin, quantitative determination and interaction energies. *Colloids Surf B: Biointerfaces* **5**, 91–110.

Van der Vegt, W., Van der Mei, H. C., Noordmans, J. & Busscher, H. J. (1991). Assessment of bacterial biosurfactant production through axisymmetric drop shape analysis by profile. *Appl Microbiol Biotechnol* **35**, 766–770.

Velraeds, M. C. M., Van der Mei, H. C., Reid, G. & Busscher, H. J. (1996). Inhibition of initial adhesion of uropathogenic *Enterococcus faecalis* by biosurfactants from *Lactobacillus* isolates. *Appl Environ Microbiol* **62**, 1958–1963.

Yang, J., Bos, R., Poortinga, A. T., Wit, P. J., Belder, G. F. & Busscher, H. J (1999a). Comparison of particle deposition in a parallel plate and stagnation point flow chamber. *Langmuir* **15**, 4671–4677.

Yang, J., Bos, R., Belder, G. F., Engel, J. & Busscher, H. J (1999b). Deposition of oral bacteria and polystyrene particles to quartz and dental enamel in a parallel plate and stagnation point flow chamber. *J Colloid Interface Sci* **220**, 410–418.

Physiological events in biofilm formation

David G. Davies

Department of Biological Sciences, Binghamton University, Binghamton, NY 13902, USA

INTRODUCTION

Bacteria have been shown to exist predominantly in nature as sessile populations attached to surfaces in contact with water (Geesey *et al.*, 1977). The numbers of bacteria attached to surfaces have been estimated to be between 1000 and 10000 times greater than the numbers of planktonic bacteria in any given environment (Watkins & Costerton, 1984).

The advantages to the micro-organisms of being attached to a surface have been largely attributed to enhanced scavenging of nutrients from both the bulk water and from the substratum with which the bacteria are associated (Marshall, 1976). Micro-organisms at surfaces are considered to be exposed to higher nutrient conditions than their planktonic counterparts and, therefore, are at an advantage in the attached state. As a demonstration of this, Kjelleberg *et al.* (1982) exposed starved planktonic and attached marine bacteria to 2 mg l^{-1} of both yeast extract and tryptone. The attached population was able to grow under low nutrient conditions, while the planktonic population was not. Further investigations revealed that when adsorbed to surfaces, both low (Power & Marshall, 1988) and high (Samuelsson & Kirchman, 1990) molecular mass compounds were available for microbial growth. These and other studies strengthened the idea that differences in growth rates of surface-associated bacteria and planktonic bacteria were primarily due to differences in nutrient availability. Thus attached bacteria, while nutritionally at an advantage over their planktonic counterparts, have for many years been considered to be similar in other respects to non-attached bacteria.

SGM symposium 59: Community structure and co-operation in biofilms. Editors D. Allison, P. Gilbert, H. Lappin-Scott, M. Wilson. Cambridge University Press. ISBN 0 521 79302 5 ©SGM 2000.

Continued research began to expose differences between planktonic and attached bacteria that implied physiological alterations following attachment to a surface. Davies & McFeters (1988) demonstrated that attached *Klebsiella oxytoca*, when grown on particles of granular activated carbon, had doubling times 11 times faster than planktonic cells grown in the same medium. Additionally, it was shown that DNA synthesis and RNA turnover rates were five and 11 times higher, respectively, in attached bacteria compared to bacteria in the bulk liquid. Furthermore, the attached bacteria were significantly larger. McCoy & Costerton (1982) noticed that *Pseudomonas aeruginosa* grew as filamentous cells when attached to surfaces in a Robbins device, but were classical rods when grown in liquid medium. Morisaki (1983) found that *Escherichia coli* attached to a solid surface had higher rates of cellular respiration than when grown in the same medium as planktonic bacteria. In another study, the substrate affinity and the substrate conversion rates of a *Vibrio* species were discovered to be different for attached and planktonic cells, respectively (Jeffrey & Paul, 1986). These observations, along with others from a wide variety of laboratories, stimulated the development of a hypothesis which stated that attached bacteria were potentially physiologically distinct from planktonic bacteria. This would suggest that historical studies performed on planktonic bacteria might not be directly relevant to attached bacteria. If this is the case, much of our understanding of bacterial behaviour would apply to only free-living bacteria. The implications of such a hypothesis are further broadened if it is considered that planktonic bacteria represent only a small percentage of the total number of bacteria found in nature.

Evidence has continued to accumulate in support of the belief that surfaces have an influence on the physiology of bacteria. Unfortunately, the evidence lacks consistency, showing in some cases that attached bacteria have higher metabolic activity than planktonic bacteria, while in other cases the reverse has been demonstrated. As a result, no consensus can be established from the data and no generalized conclusions can be made concerning the influence of attachment on bacterial physiology.

In the past decade, the approach to studying this problem has shifted towards the investigation of specific gene regulation, yielding some interesting insights. One of the first investigations into gene regulation at a surface was carried out by Dagostino *et al.* (1991) using *Pseudomonas* S9 growing on polystyrene microtitre plates. In their study, transposon mutagenesis was used to insert promoterless *lacZ* genes into recipient organisms, giving some the ability to display β-galactosidase activity. Those organisms that expressed β-galactosidase at the solid surface were subcultured into liquid medium or onto agar plates of the same medium. Some of these subcultured populations were no longer able to express the β-galactosidase gene. When these organisms were grown again on the solid surface, β-galactosidase production was recovered, thus demonstrating activation of the gene only at a surface. The authors, however, did not identify the specific target genes controlling *lacZ* activation. In a separate study by Belas *et al.* (1986), the *laf* gene, which is responsible for production of lateral flagella, was activated in *Vibrio parahaemolyticus* when the organism was cultured on agar medium but not when it was cultured as planktonic cells in liquid medium. Further experimentation led the authors to conclude that *laf* gene activation was a consequence of elevated medium viscosity, a common situation at the liquid boundary with surfaces. In another study by Davies *et al.* (1993), it was shown that activation of a specific alginate gene was mediated by association with a solid surface. In that study, *P. aeruginosa* was grown on a Teflon substratum and in liquid medium and the activity of the *algC* gene was monitored at the single cell level using a plasmid-borne *algC::lacZ* reporter. The gene *algC* is required for the production of alginate, an extracellular polymeric substance (EPS) produced and secreted by *P. aeruginosa*. When the bacteria attached to the Teflon surface, β-galactosidase specific activity was shown to increase 85-fold over the activity of the planktonic bacteria. This result demonstrated that the surface environment was either directly or indirectly responsible for activation of a specific gene at a surface. Furthermore, the production of alginate is known to facilitate the development of *P. aeruginosa* as a biofilm.

These results indicated that the genetic regulation of surface-associated bacteria is distinct from planktonic bacteria for at least some species. A number of efforts have begun to examine the degree to which gene regulation differs in these populations. Recently, it has been shown that *E. coli*, when grown as a biofilm, has a massive change in genetic regulation, altering transcription of 38% of its genes following attachment to a surface (Prigent-Combaret *et al.*, 1999). Presumably, multiple effects result from the change in regulation of such a significant portion of the *E. coli* genome. This evidence is only the most recent indication that the development of bacterial biofilms has a strong physiological component.

Considerable activity is currently under way to investigate the developmental characteristics of bacterial biofilms. This activity is supported in large measure by the availability of new technologies and the development of new methods that allow the study of individual bacteria within biofilms, their gene regulation, RNA and protein synthesis, *in situ* activity measurements and species identification using specific probes and stains. An interesting consequence of the new approaches and questions regarding the physiology of biofilm bacteria is that they have generated a renaissance in the use of the microscope as a central tool in the field of microbiology.

INITIAL EVENTS IN BIOFILM FORMATION

The initial event in biofilm formation is contact made between the micro-organism and a surface. The mechanisms by which bacteria are transported to a surface include random contacts with the substratum due to Brownian motion, sedimentation due to differences in specific gravity between the bacteria and the bulk liquid, convective transport in which cells are brought to the surface by the movement of the bulk liquid, and active transport mediated by flagellar activity which may or may not include chemotaxis (van Loosdrecht *et al.*, 1990).

Chemotaxis has been considered as an important mechanism enabling bacteria to reach and/or interact with a surface. Chemotaxis is dependent upon the presence and positioning of flagella in the cell envelope. In a recent study, Pratt & Kolter (1998) used a molecular approach to determine the role of chemotaxis in the development of an *E. coli* biofilm. Insertion mutagenesis was used to disrupt the *che* operon, which is responsible for chemotaxis. When these mutant organisms were compared with wild-type *E. coli*, no differences were found in the ability to form biofilms, the authors concluding that chemotaxis is not essential for bacterial attachment. Further study needs to be done on the role of chemotaxis in other organisms.

Once contact with a surface has been made, bacteria must develop an interaction with that surface which will bond the two together in a process known as attachment. Several factors have been described which play a role in bacterial attachment. One such factor involves the overall surface properties of bacteria. The relative degree of hydrophobicity of bacteria in relation to both the bulk liquid and the substratum has been demonstrated to influence attachment. Absolom *et al.* (1983) concluded that cell surface hydrophobicity played a role in attachment for five different species of proteobacteria. In their study, it was observed that if the bacteria had a hydrophobicity greater than the medium in which they were suspended, they attached better to hydrophobic substrata than to hydrophilic substrata. Conversely, when the bacteria were more hydrophilic than the suspending medium, they attached better to hydrophilic substrata. Stenström (1989) reported that *Streptococcus faecalis*, *Salmonella typhimurium* and *E. coli* adhered differentially to mineral particles, depending on cell surface hydrophobicity. In another study, Martinez-Martinez *et al.* (1991) observed that rates of bacterial attachment were influenced by the hydrophobicity of the bacterial cell surface and by the net surface charge.

Differences in surface hydrophobicity are believed to result from the net surface properties conferred upon the cell by surface molecules such as proteins and lipids. The production of these molecules at the surface of the bacterial envelope is inducible in at least some instances. For example, synthesis of lipopolysaccharide (LPS) O-side chain in *P. aeruginosa* has been shown to require the activity of the inducible gene *algC* (Goldberg *et al.*, 1993). Attachment rates for *P. aeruginosa* deficient in O-side chain LPS were found to be significantly higher than rates observed for the same bacterium with a complete O-side chain (D. G. Davies & G. G. Geesey, unpublished results). These data suggest that bacteria may be able to alter their surface hydrophobicity in ways that would either enhance or diminish attachment to a substratum. While hydrophobicity is considered to play a significant role in attachment of bacteria to surfaces, other specialized structures are also known to play a role.

Dalton & March (1998) have stated that flagella, fimbriae and other protein receptors are essential for bacterial attachment to surfaces. Numerous studies have been performed in which it has been observed that flagella facilitate attachment to various surfaces. As early as the 1970s, it had been noted that attachment of bacteria to a surface sometimes resulted in rotary motion of the bacteria around a fixed axis, presumably the flagellum (Meadows, 1971). Santos *et al.* (1991) noted that both pili and flagella were necessary for the attachment of *Pseudomonas fluorescens* to surfaces in flowing environments. In fact, the absence of flagella has been found by Korber *et al.* (1993) to severely restrict the ability of *P. fluorescens* to colonize surfaces. In another study, O'Toole & Kolter (1998a) used mutagenesis to disrupt random genes in *P. fluorescens* and tested the bacteria for their ability to attach to PVC. It was found that three of the mutants that were unable to attach to the substratum were defective in flagellar synthesis. In a separate study, O'Toole & Kolter (1998b) observed that flagella enhanced the sticking ability of *P. aeruginosa* to PVC, but it was unclear whether loss of motility or loss of the flagellum was responsible for the observed effect. To determine the actual role of motility in attachment to a surface, Pratt & Kolter (1998) examined the sticking efficiency of *E. coli* to PVC for mutants defective in flagellar biosynthesis and compared these with *E. coli* in which the flagella were paralysed. Their results showed that both mutants had decreased sticking efficiency to the substratum when compared with the wild-type organism. The authors concluded that motility plays an important role in the development of *E. coli* as a biofilm, suggesting that motility is required to overcome the repulsive forces found at the surface. These studies also showed that pili (Pratt & Kolter, 1998; O'Toole & Kolter, 1998b) are necessary for formation of a biofilm. Presumably, pili attach directly to a surface and, like flagella, may be necessary to overcome surface repulsion.

IRREVERSIBLE ATTACHMENT

Once attachment to a surface has been effected, the bacteria must maintain contact with the substratum and grow in order to develop a mature biofilm. Several studies have indicated that the initial attachment process is reversible, if not followed by the production of a more substantial bond to the substratum. This change from reversible to irreversible attachment was noted by ZoBell (1943), and has been characterized by Characklis (1990) as the transition from a weak interaction of the cell with the substratum to a permanent bonding, frequently mediated by the presence of extracellular polymers. Early investigators did not appreciate the possibility of surface transduction as a mechanism for inducing irreversible attachment. However, recent investigations have gone a long way to suggest that profound physiological changes may accompany the transition to permanent attachment at a surface. In fact, the transition to irreversible attachment marks the transformation of attached planktonic cells to cells having a true biofilm physiology.

One means of transition from reversible to irreversible attachment is mediated by type IV pili. Twitching motility is a mode of locomotion used by *P. aeruginosa* in which type IV polar pili are believed to extend and retract, propelling bacteria across a surface. Twitching motility is speculated by O'Toole & Kolter (1998b) to be involved in the formation of microcolonies. The authors suggest that interactions of bacteria with one another at a surface, forming groups of cells, help to strengthen the degree of attachment to the surface. The role of cell–cell adhesion in promoting surface attachment is also proposed by Cramton *et al.* (1999) for *Staphylococcus aureus*. This bacterium has been shown to produce a substance called polysaccharide intercellular adhesin (PIA) which the authors claim bonds *S. aureus* cells to one another. This in turn results in the formation of microcolonies and facilitates the maturation of the biofilm.

The hallmark of bacterial biofilms which segregates them from bacteria that are simply attached to a substratum is that biofilms contain EPS that surround the resident bacteria. The presence of EPS helps maintain the integrity of the biofilm, allowing large numbers of bacteria to coexist under flowing conditions, often in complexes that are many tens of microns in thickness. In addition to the role played by EPS in helping to cement bacteria to one another and to the substratum, these substances are also important in creating biofilm architecture, and in protecting the resident micro-organisms from the activity of grazing organisms and from harmful chemicals. In addition, EPS provides a matrix which can trap nutrients, concentrate intercellular communication molecules and allow the development of high cell densities required for degradation of polymeric substances by extracellular enzymes and for horizontal gene transfer.

Microbial EPS are biosynthetic polymers that can be highly diverse in chemical composition. Typically, EPS includes substituted and unsubstituted polysaccharides, substituted and unsubstituted proteins, nucleic acids and phospholipids (Wingender *et al.*, 1999). The structure of EPS is known to differ from species to species and with the exception of a small number of extracellular polysaccharides of commercial significance is largely uncharacterized.

Among the best characterized of all EPS is the bacterial product alginate. Research on bacterial alginate has included investigations into the physico-chemical properties of the polymer, the biochemistry and genetics of its synthesis and the physiological roles it plays. For these reasons, more information is available concerning the role of alginate in biofilm development than is the case for any other type of EPS.

Alginate has been shown to be involved in biofilm formation by *P. aeruginosa* in pulmonary lung infections (Doggett, 1969), in aquatic biofilms (Grobe *et al.*, 1995) and in industrial water systems (Wallace *et al.*, 1994). The production of alginate by *P. aeruginosa* has been shown to be physiologically regulated in response to a variety of environmental factors. The activation of a critical alginate promoter, *algD*, has been shown to take place during nitrogen limitation, membrane perturbation induced by ethanol and when cells were exposed to media of high osmolarity (Berry *et al.*, 1989; DeVault *et al.*, 1989, 1990). Similar to the *algD* promoter, the *algC* promoter has been shown to be activated by environmental signals such as high osmolarity, and this activation is dependent on the presence of the response regulator protein AlgR1, which has been shown to bind a consensus AlgR1-binding site (CCGTTCGTCN5) centred at -87 of the *algC* promoter (Zielinski *et al.*, 1991). These experiments have been performed in liquid medium and on agar-based medium and have not been duplicated for biofilm bacteria. However, these results hint that environmental activation of alginate genes may take place in growing bacterial biofilms.

The activities of the alginate biosynthesis enzymes are extremely low, even in mucoid strains of *P. aeruginosa*, and are either greatly reduced or absent in non-mucoid strains (Padgett & Phibbs, 1986; Piggot *et al.*, 1981; Sá-Correia *et al.*, 1987). Due to the low activity of these enzymes, reporter genes have been used to detect the activity of the alginate promoters, rather than measure the enzymes directly. The gene *algC*, encoding phosphomannomutase, was examined during biofilm development as a key regulation point in the alginate biosynthesis pathway. It was discovered that transcriptional activity of the *algC* gene was initiated within 15 min of attachment to a glass substratum; this activation was correlated to alginate biosynthesis by chemical assay (Davies & Geesey, 1995). This result indicates that the production of alginate is an early event in the formation of a biofilm by *P. aeruginosa*. It is presumed that the early onset of alginate synthesis during biofilm formation helps to cement the bacteria to a surface and to each other, assisting in the development of irreversible attachment and biofilm maturation.

Interestingly, the gene *algC* is also involved in synthesis of the O-side chain of LPS (see above). The switch in function of O-side chain synthesis to alginate synthesis by AlgC results in the production of a 'stickier' LPS as well as the production of alginate, further cementing the bacteria to the substratum.

BIOFILM MATURATION

As alginate synthesis continues, the next phase of biofilm development, maturation, is initiated. Maturation of microbial biofilms results in the development of complex architecture, channels, pores, and a redistribution of bacteria away from the substratum (Davies *et al.*, 1998). The complex architecture seen in microbial biofilms is presumed to be controlled, at least in part, by the resident micro-organisms. Differential growth rates, variation in EPS production and microbial dispersion are all believed to occur during biofilm maturation.

One of the mechanisms by which this regulation has been shown to occur is by intercellular communication. In the early 1970s, it was demonstrated that bacterial bioluminescence is controlled by intercellular communication in the marine bacteria *Vibrio fischeri* (= *Photobacterium fischeri*) and the related species *Vibrio harveyi* (Eberhard, 1972; Nealson *et al.*, 1970). As planktonic bacteria in the marine environment, these cells are found at densities of 10^2 ml^{-1} and are non-luminescent. As symbionts in the light organs of certain marine fishes and squids, these bacteria reach cell densities in excess of 10^{10}–10^{11} ml^{-1} and produce light via a chemical reaction using luciferin and luciferase. The trigger for chemiluminescence in these bacteria has been shown to be an acyl homoserine lactone (AHL) signal molecule which is secreted by the cells into their surroundings.

In all cases, AHL autoinducers are known to associate with a DNA-binding protein homologous to LuxR in *P. fischeri*, causing a conformational change in the protein, facilitating DNA polymerase binding, initiating transcription of the target genes. This process couples the transcription of specific genes to bacterial cell density (Latifi *et al.*, 1996). Regulation of this type has been referred to as 'quorum sensing' because it suggests the requirement for a 'quorate' population of bacterial cells necessary for activation of AHL responsive genes (Fuqua *et al.*, 1994).

The occurrence of AHLs in cells other than marine luminescent bacteria was revealed in 1992, when the terrestrial bacterium *Erwinia carotovora* was shown to use an AHL autoinducer system to regulate the production of the β-lactam antibiotic carbapenem (Bainton *et al.*, 1992b). This finding led to a general search for AHLs in a wide range of bacteria. To effect the search, bioluminescence sensor systems were developed and used to screen for AHL production in the spent supernatant liquids of a number of bacterial cultures. Many different organisms were shown by the screening to produce AHLs. These included: *P. aeruginosa*, *Serratia marcescens*, *Erwinia herbicola*, *Citrobacter freundii*, *Enterobacter agglomerans* and *Proteus mirabilis* (Bainton *et al.*, 1992a; Swift *et al.*, 1993). More recently, the list has grown to include *Erwinia stewartii* (Beck von Bodman & Farrand, 1995), *Yersinia enterocolitica* (Throup *et al.*, 1995),

Agrobacterium tumefaciens (Zhang *et al.*, 1993), *Chromobacterium violaceum* (Winson *et al.*, 1995), *Rhizobium leguminosarum* (Schripsema *et al.*, 1996) and others. It is now generally considered that all enteric bacteria (with the possible exception of *E. coli*), and the Gram-negative bacteria generally, are capable of cell density regulation using AHL autoinducers.

A number of pseudomonad virulence factors have been shown to be controlled by AHL compounds produced by the *lasI* and *rhlI* regulatory systems (Gambello *et al.*, 1993; Ochsner & Reiser, 1995; Winson *et al.*, 1995; Latifi *et al.*, 1995), in a manner reminiscent of the *lux* system. Latifi *et al.* (1996) have also shown that many stationary phase properties of *P. aeruginosa* are under the hierarchical control of the *lasI* and *rhlI* cell–cell signalling systems. Brown & Williams (1985) have suggested that many of the properties of biofilm bacteria, including their remarkable resistance to antibiotics (Nickel *et al.*, 1985), may derive from the fact that some of their component cells exhibit characteristics of stationary phase planktonic cells.

It is during stationary phase that Gram-negative bacteria have been shown to develop stress response resistance that is co-ordinately regulated through the induction of a stationary phase sigma factor known as RpoS (Hengge-Aronis, 1993). Biofilm bacteria are generally considered to show physiological similarity to stationary phase bacteria in batch cultures. Thus it is presumed that the synthesis and export of stationary phase autoinducer-mediated exoproducts occur generally within biofilms. The stationary phase behaviour of biofilm bacteria may be explained by the activity of accumulated AHL within cell clusters. The mechanism causing biofilm bacteria to demonstrate stationary phase behaviour is hinted at by the discovery that RpoS is produced in response to accumulation of the *rhlI* gene product in *P. aeruginosa* cultures (Latifi *et al.*, 1996). In a more recent study, Suh *et al.* (1999) have shown that in *P. aeruginosa* PAO1, RpoS is responsible for a decrease in the production of exotoxin A, elastase A, LasA protease and twitching motility. Additionally, it was found that *P. aeruginosa* FRD1 demonstrated a 70% loss in alginate synthesis when *rpoS* was inactivated.

McLean *et al.* (1997) have shown that AHL autoinducers are detectable in naturally occurring biofilms. In their study, the authors sampled a stream in Texas, USA, and demonstrated that the autoinduction system of *A. tumefaciens* could be activated on one side of an agar plate when biofilm-containing rocks were placed on the other side of the plate. No activation was observed when the biofilm-coated rocks had been previously autoclaved or if the rocks did not contain biofilm.

Recently, it has been shown that *P. aeruginosa* PAO1 requires the *lasI* gene product $3OC_{12}$-AHL in order to develop a normal differentiated biofilm (Davies *et al.*, 1998). In this study, it was observed that bacteria knocked out in *lasI* produced biofilm cell clusters that were 20% the thickness of the wild-type organism. In addition, these mutants grew as continuous sheets on the substratum, lacking differentiation and not demonstrating evidence of matrix polymer. By contrast, the wild-type organism formed characteristic microcolonies composed of groups of cells separated by intervening matrix polymer and separated from one another by water channels. When the autoinducer $3OC_{12}$-AHL was added to the medium of growing *lasI* mutant bacteria, these cells developed biofilms that were indistinguishable from the wild-type organism. These results indicated that $3OC_{12}$-AHL was responsible for the complex architecture observed in mature biofilms produced by *P. aeruginosa*. Since biofilm architecture is presumed to be influenced by matrix polymer, we can hypothesize that autoinduction is at least partly responsible for regulation of *P. aeruginosa* EPS. In support of this hypothesis, it has been demonstrated that the alginate genes *algC* and *algD* are induced in *P. aeruginosa* strain 8830 by $3OC_{12}$-AHL and by $3OC_4$-AHL (D. G. Davies & J. W. Costerton, unpublished data). These observations indicate the possible role of quorum sensing as a signal transduction system by which *P. aeruginosa* may initiate the production of alginate and possibly other types of EPS. In a separate study, Olvera *et al.* (1999) have observed that *algC* is necessary for the production of rhamnolipid. The transcription of rhamnolipid has been shown to be dependent on activation by *rhlI* (Passador *et al.*, 1993), further indicating that *algC* transcription is regulated at some level by bacterial quorum sensing.

In addition to activation of multiple factors by autoinduction, it has recently been shown that *P. aeruginosa* is able to repress the transcription of *lasI* in response to RsaL, an 11 kDa protein whose gene lies downstream from *lasR*. This protein has been shown to suppress LasB production, and presumably all other factors regulated by the *lasI* quorum-sensing system. This finding demonstrates a further level of control of quorum sensing in *P. aeruginosa* and suggests that cell–cell communication systems are highly complex.

Altogether, *P. aeruginosa* is known to have 39 genes that are under the regulation of the *lasI/rhlI* quorum-sensing systems (Whitely *et al.*, 1999). In addition to cell–cell communication via AHL-mediated quorum sensing, it has recently come to light that other communication systems are used by *P. aeruginosa*. Holden *et al.* (1999) have discovered that cell-free extracts of cultures in which *P. aeruginosa* were grown contain diketopiperazines (DKPs), cyclic dipeptides known to participate in cell–cell communication mostly in Gram-positive bacteria. This compound has been shown to be capable of activating a LuxR biosensing system and swarming motility in *S. marcescens*. In addition, a third extracellular sensing system has been discovered in *P. aeruginosa* by Pesci *et al.* (1999). In this system, *P. aeruginosa* produces a quinolone signalling molecule (2-heptyl-3-hydroxy-4-quinolone) which has been shown to induce LasB production via the *lasI/rhI* quorum-sensing system. This observation hints at a super-regulatory function for quinolones in bacterial communication.

The role played by quorum sensing and other intercellular signalling systems of bacteria during the development of biofilms is still unclear. Undoubtedly, these systems are complex and are under intracellular as well as extracellular control. The existence of these systems further suggests that interspecific communication is possible, as is evident from studies in which bioassays have been used to detect the presence of autoinducers from multiple species (see above).

Cell-density-dependent regulation may also be responsible for the release of enzymes which can degrade biofilm matrix polymers, allowing bacteria to disperse from a biofilm. It has been observed that when certain bacteria (including *P. aeruginosa*) reach high cell densities in biofilm cell clusters, the bacteria often undergo a dispersion event (D. G. Davies & J. W. Costerton; Stoodley *et al.*, 1999). Furthermore, biofilm dispersion has been shown to occur reproducibly under conditions of medium stagnation. Under both of these circumstances, AHL concentrations can build up to levels necessary for activation of cell-density-dependent genes.

The role of autoinduction in biofilm bacteria has been only superficially investigated. However, cell–cell communication as a biofilm phenomenon appears to be firmly established. Research on the degree to which quorum sensing and other forms of intercellular communication operate is needed to more perfectly understand the role of cell signalling in regulating biofilm development and persistence.

CONCLUSIONS

The complex regulation of surface attachment, irreversible surface binding, biofilm maturation and ultimately biofilm detachment has been shown in the past decade to be mediated to at least some degree by the physiology of the bacteria involved. This conclusion has resulted in a vigorous interest in the investigation of the physiological aspects of biofilm development. This interest is causing a shift of focus in the manner in which research on biofilms is conducted. In the 1970s and early 1980s, the majority of biofilm research was done using an engineering-based approach in which the physico-chemical parameters of the abiotic environment were considered to define the manner in which biofilm development would proceed. While the physical and chemical environment must have a significant impact on the way in which biofilms form and persist, it is now becoming evident that the bacteria play a significant role as well. The past 10 years have only seen a hint at the types of regulation bacteria use to orchestrate the progression of biofilm formation. In the next several years, it is probable that exact mechanisms of biological control of biofilm formation will be elucidated, at least for certain axenic biofilms and some of the more simple mixed biofilm communities. These mechanisms of regulation will open up new avenues for biofilm control, in the environment, in industry and in medicine. Research over the next 10 years promises to be exciting and fruitful as increased attention is given to investigating the physiology of the development of biofilms.

REFERENCES

Absolom, D. R., Lamberti, F. V., Policova, Z., Zingg, W., van Oss, C. J. & Neumann, A. W. (1983). Surface thermodynamics of bacterial adhesion. *Appl Environ Microbiol* **46**, 90–97.

Bainton, N. J., Bycroft, B. W., Chhabra, S. R. & 8 other authors (1992a). A general role for the *lux* autoinducer in bacterial cell signaling: control of antibiotic synthesis in *Erwinia. Gene* **116**, 87–91.

Bainton, N. J., Stead, P., Chhabra, S. R., Bycroft, B. W., Salmond, G. P. C., Stewart, G. S. A. B. & Williams, P. (1992b). N-(3-oxohexanoyl)-L-homoserine lactone regulates carbapenem antibiotic production in *Erwinia carotovora. Biochem J* **288**, 997–1004.

Beck von Bodman, S. & Farrand, S. K. (1995). Capsular polysaccharide biosynthesis and pathogenicity in *Erwinia stewartii* require induction by an N-acylhomoserine lactone autoinducer. *J Bacteriol* **177**, 5000–5008.

Belas, R., Simon, M. & Silverman, M. (1986). Regulation of lateral flagella gene transcription in *Vibrio parahaemolyticus. J Bacteriol* **167**, 210–218.

Berry, A., DeVault, J. D. & Chakrabarty, A. M. (1989). High osmolarity is a signal for enhanced *algD* transcription in mucoid and nonmucoid *Pseudomonas aeruginosa* strains. *J Bacteriol* **171**, 2312–2317.

Brown, M. R. W. & Williams, P. (1985). The influence of the environment on envelope properties affecting survival of bacteria in infections. *Annu Rev Microbiol* **39**, 527–556.

Characklis, W. G. (1990). Biofilm processes. In *Biofilms*, pp. 195–231. Edited by W. G. Characklis & K. C. Marshall. New York: Wiley.

Cramton, S. E., Gerke, C., Schnell, N. F., Nichols, W. W. & Gotz, F. (1999). The intercellular adhesion (*ica*) locus is present in *Staphylococcus aureus* and is required for biofilm formation. *Infect Immun* **67**, 5427–5433.

Dagostino, L., Goodman, A. E. & Marshall, K. C. (1991). Physiological responses induced in bacteria adhering to surfaces. *Biofouling* **4**, 113–119.

Dalton, H. M. & March, P. E. (1998). Molecular genetics of bacterial attachment and biofouling. *Curr Opin Biotechnol* **9**, 252–255.

Davies, D. G. & Geesey, G. G. (1995). Regulation of the alginate biosynthesis gene *algC* in *Pseudomonas aeruginosa* during biofilm development in continuous culture. *Appl Environ Microbiol* **61**, 860–867.

Davies, D. G. & McFeters, G. A. (1988). Growth and comparative physiology of *Klebsiella oxytoca* attached to granular activated carbon particles and in liquid media. *Microb Ecol* **15**, 165–175.

Davies, D. G., Chakrabarty, A. M. & Geesey, G. G. (1993). Exopolysaccharide production in biofilms: substratum activation of alginate gene expression by *Pseudomonas aeruginosa. Appl Environ Microbiol* **59**, 1181–1186.

Davies, D. G., Parsek, M. R., Pearson, J. P., Iglewski, B. H., Costerton, J. W. & Greenberg, E. P. (1998). The involvement of cell-to-cell signals in the development of a bacterial biofilm. *Science* **280**, 295–298.

DeVault, J. D., Berry, A., Misra, T. K. & Chakrabarty, A. M. (1989). Environmental sensory signals and microbial pathogenesis: *Pseudomonas aeruginosa* infection in cystic fibrosis. *Bio/Technology* **7**, 352–357.

DeVault, J. D., Kimbara, K. & Chakrabarty, A. M. (1990). Pulmonary dehydration and infection in cystic fibrosis: evidence that ethanol activates alginate gene expression and induction of mucoidy in *Pseudomonas aeruginosa*. *Mol Microbiol* **4**, 737–745.

Doggett, R. G. (1969). Incidence of mucoid *Pseudomonas aeruginosa* from clinical sources. *Appl Microbiol* **18**, 936–937.

Eberhard, A. (1972). Inhibition and activation of bacterial luciferase synthesis. *J Bacteriol* **109**, 1101–1105.

Fuqua, W. C., Winans, S. C. & Greenberg, E. P. (1994). Quorum sensing in bacteria: the LuxR-LuxI family of cell density-responsive transcriptional regulators. *J Bacteriol* **176**, 269–275.

Gambello, M. J., Kaye, S. & Iglewski, B. H. (1993). LasR of *Pseudomonas aeruginosa* is a transcriptional activator of the alkaline protease gene (*apr*) and an enhancer of exotoxin A expression. *Infect Immun* **61**, 1180–1184.

Geesey, G. G., Richardson, W. T., Yeomans, H. G., Irvin, R. T. & Costerton, J. W. (1977). Microscopic examination of natural sessile bacterial populations from Alpine streams. *Can J Microbiol* **23**, 1733–1736.

Goldberg, J. D., Hatano, K. & Pier, G. B. (1993). Synthesis of lipopolysaccharide O-side chains by *Pseudomonas aeruginosa* PAO1 requires the enzyme phosphomannomutase. *J Bacteriol* **175**, 1605–1611.

Grobe, W., Wingender, J. & Trüper, H. G. (1995). Characterization of mucoid *Pseudomonas aeruginosa* strains isolated from technical water systems. *J Appl Bacteriol* **79**, 94–102.

Hengge-Aronis, R. (1993). Survival of hunger and stress: the role of *rpoS* in early stationary phase regulation in *Escherichia coli*. *Cell* **72**, 165–168.

Holden, H. T., Ram Chhabra, S., deNys, R. & 14 other authors (1999). Quorum-sensing cross talk: isolation and chemical characterization of cyclic dipeptides from *Pseudomonas aeruginosa* and other gram negative bacteria. *Mol Microbiol* **33**, 1254–1266.

Jeffrey, W. H. & Paul, J. H. (1986). Activity of an attached and free-living *Vibrio* sp. as measured by thymidine incorporation, p-iodonitrotetrazolium reduction, and ATP/ADP ratios. *Appl Environ Microbiol* **51**, 150–156.

Kjelleberg, S., Humphrey, B. A. & Marshall, K. C. (1982). Effect of interfaces on small, starved marine bacteria. *Appl Environ Microbiol* **43**, 1166–1172.

Korber, D. R., Lawrence, J. R., Hendry, M. J. & Caldwell, D. E. (1993). Analysis of spatial variability within mot⁺ and mot⁻ *Pseudomonas fluorescens* biofilms using representative elements. *Biofouling* **7**, 339–358.

Latifi, A., Winson, K. M., Foglino, M., Bycroft, B. S., Stewart, G. S. A. B., Lazdunski, A. & Williams, P. (1995). Multiple homologues of LuxR and LuxI control expression of virulence determinants and secondary metabolites through quorum sensing in *Pseudomonas aeruginosa* PAO1. *Mol Microbiol* **17**, 333–344.

Latifi, A., Foglino, M., Tanaka, K., Williams, P. & Lazdunski, A. (1996). A hierarchical quorum-sensing cascade in *Pseudomonas aeruginosa* links the transcriptional activators LasR and RhlR (VsmR) to expression of the stationary-phase sigma factor RpoS. *Mol Microbiol* **21**, 1137–1146.

McCoy, W. F. & Costerton, W. J. (1982). Fouling biofilm development in tubular flow systems. *Dev Ind Microbiol* **23**, 551–558.

McLean, R. J., Whiteley, M., Stickler, D. J. & Fuqua, W. C. (1997). Evidence of autoinducer activity in naturally occurring biofilms. *FEMS Microbiol Lett* **154**, 259–263.

Marshall, K. C. (1976). *Interfaces in Microbial Ecology*. Cambridge, MA: Harvard University Press.

Martinez-Martinez, L., Pascual, A. & Perea, E. J. (1991). Effect of preincubation of *Pseudomonas aeruginosa* in subinhibitory concentrations of amikacin, ceftazidime and ciprofloxacin on adherence to plastic catheters. *Chemotherapy* **37**, 62–65.

Meadows, P. S. (1971). The attachment of bacteria to solid surfaces. *Arch Mikrobiol* **75**, 374–381.

Morisaki, H. (1983). Effect of solid-liquid interface on metabolic activity of *E. coli*. *J Gen Appl Microbiol* **29**, 195–204.

Nealson, K. H., Platt, T. & Hastings, J. W. (1970). Cellular control of the synthesis and activity of the bacterial luminescent system. *J Bacteriol* **104**, 313–322.

Nickel, J., Ruseska, C. K., Wright, J. B. & Costerton, J. W. (1985). Tobramycin resistance of cells of *Pseudomonas aeruginosa* growing as a biofilm on urinary catheter material. *Antimicrob Agents Chemother* **27**, 619–624.

Ochsner, U. A. & Reiser, J. (1995). Autoinducer-mediated regulation of rhamnolipid biosurfactant synthesis in *Pseudomonas aeruginosa*. *Proc Natl Acad Sci USA* **92**, 6424–6428.

Olvera, C., Goldberg, J. B., Sanchez, R. & Soeron-Chavez, G. (1999). The *Pseudomonas aeruginosa algC* gene product participates in rhamnolipid biosynthesis. *FEMS Microbiol Lett* **179**, 85–90.

O'Toole, G. A. & Kolter, R. (1998a). Initiation of biofilm formation in *Pseudomonas fluorescens* WCS365 proceeds via multiple, convergent signalling pathways: a genetic analysis. *Mol Microbiol* **28**, 449–461.

O'Toole, G. A. & Kolter, R. (1998b). Flagellar and twitching motility are necessary for *Pseudomonas aeruginosa* biofilm development. *Mol Microbiol* **30**, 295–304.

Padgett, P. J. & Phibbs, P. V., Jr (1986). Phosphomannomutase activity in wild-type and alginate-producing strains of *Pseudomonas aeruginosa*. *Curr Microbiol* **14**, 187–192.

Passador, L., Cook, J. M., Gambello, M. J., Rust, L. & Iglewski, B. H. (1993). Expression of *Pseudomonas aeruginosa* virulence genes requires cell-to-cell communication. *Science* **260**, 1127–1130.

Pesci, E. C., Milbank, J. B., Pearson, J. P., McKnight, S., Kende, A. S., Greenberg, E. P. & Iglewski, B. H. (1999). Quinolone signaling in the cell-to-cell communication system of *Pseudomonas aeruginosa*. *Proc Natl Acad Sci USA* **96**, 11229–11234.

Piggott, N. H., Sutherland, I. W. & Jarman, T. R. (1981). Enzymes involved in the biosynthesis of alginate by *Pseudomonas aeruginosa*. *Eur J Appl Microbiol Biotechnol* **13**, 179–183.

Power, K. & Marshall, K. C. (1988). Cellular growth and reproduction of marine bacteria on surface-bound substrate. *Biofouling* **1**, 163–174.

Pratt, L. A. & Kolter, R. (1998). Genetic analysis of *Escherichia coli* biofilm formation: roles of flagella, motility, chemotaxis and type I pili. *Mol Microbiol* **30**, 285–293.

Prigent-Combaret, C., Vidal, O., Dorel, C. & Lejeune, P. (1999). Abiotic surface sensing and biofilm-dependent regulation of gene expression in *E. coli*. *J Bacteriol* **181**, 5993–6002.

Sá-Correia, I., Darzins, A., Wang, S. K., Berry, A. & Chakrabarty, A. M. (1987). Alginate biosynthetic enzymes in mucoid and nonmucoid *Pseudomonas aeruginosa*: overproduction of phosphomannose isomerase, phosphomannomutase, and GDP-mannose pyrophosphorylase by overexpression of the phosphomannose isomerase (*pmi*) gene. *J Bacteriol* **169**, 3224–3231.

Samuelsson, M. O. & Kirchman, D. L. (1990). Degradation of adsorbed protein by attached bacteria in relationship to surface hydrophobicity. *Appl Environ Microbiol* **56**, 3643–3648.

Santos, R., Callow, M. E. & Bott, T. R. (1991). The structure of *Pseudomonas fluorescens* biofilms in contact with flowing systems. *Biofouling* **4**, 319–336.

Schripsema, J., de Rudder, K. E. E., van Vleit, T. G., Lankhorst, P. P., de Vroom, E., Kijne, J. W. & van Brussel, A. A. N. (1996). Bacteriocin *small* of *Rhizobium leguminosarum* belongs to the class of N-acyl-L-homoserine lactone molecules, known as autoinducers and as quorum sensing co-transcription factors. *J Bacteriol* **178**, 366–371.

Stenström, T. A. (1989). Bacterial hydrophobicity, an overall parameter for the measurement of adhesion potential to soil particles. *Appl Environ Microbiol* **55**, 142–147.

Stoodley, P., Jørgensen, F., Williams, P. & Lappin-Scott, H. M. (1999). The role of hydrodynamics and AHL signalling molecules as determinants of the structure of *Pseudomonas aeruginosa* biofilms. In *Biofilms: the Good, the Bad, and the Ugly*, pp. 223–230. Edited by J. Wimpenny, P. Gilbert, J. Walker, M. Brading & R. Bayston. Cardiff: BioLine.

Suh, S. J., Silo-Suh, L., Woods, D. E., Hasset, D. J., West, S. E. & Ohman, E. E. (1999). Effect of *rpoS* mutation on the stress response and expression of virulence factors in *Pseudomonas aeruqinosa*. *J Bacteriol* **181**, 3890–3897.

Swift, S., Winson, M. K., Chan, P. F., Bainton, N. J., Birstall, M., Reeves, P. J., Rees, C. E. C., Chhabra, S. R., Hill, P. J. & Stewart, G. S. A. B. (1993). A novel strategy for the isolation of *luxl* homologues: evidence for the widespread distribution of a LuxR: LuxI superfamily in enteric bacteria. *Mol Microbiol* **10**, 511–520.

Throup, J., Camara, M., Briggs, G., Winson, M. K., Chhabra, S. R., Bycroft, B. W., Williams, P. & Stewart, G. S. A. B. (1995). Characterization of the *yenl/yenR* locus from *Yersinia enterocolitica* mediating the synthesis of the N-acylhomoserine lactone signal molecules. *Mol Microbiol* **17**, 345–356.

van Loosdrecht, M. C. W., Lyklema, J., Jorde, W. & Zehnder, A. J. B. (1990). Influence of interfaces on microbial activity. *Microbiol Rev* **54**, 75–87.

Wallace, W. H., Fleming, J. T., White, D. C. & Sayler, G. S. (1994). An *algD*-bioluminescent reporter plasmid to monitor alginate production in biofilms. *Microb Ecol* **27**, 225–239.

Watkins, L. & Costerton, J. W. (1984). Growth and biocide resistance of bacterial biofilms in industrial systems. *Chem Times Trends* (October), 35–40.

Whitely, M., Lee, K. M. & Greenberg, E. P. (1999). Identification of genes controlled by quorum sensing in *Pseudomonas aeruginosa*. *Proc Natl Acad Sci USA* **96**, 13904–13909.

Wingender, J., Neu, T. R. & Flemming, H.-C. (1999). What are bacterial extracellular polymeric substances? In *Microbial Extracellular Polymeric Substances*, pp. 1–19. Edited by J. Wingender, T. R. Neu & H.-C. Flemming. Berlin: Springer.

Winson, M. K., Camara, M., Latifi, A. & 10 other authors (1995). Multiple N-acyl-L-homoserine lactone signal molecules regulate production of virulence determinants and secondary metabolites in *Pseudomonas aeruginosa*. *Proc Natl Acad Sci USA* **92**, 9427–9431.

Zhang, L., Murphy, P. J. & Max, I. T. (1993). Agrobacterium conjugation and gene regulation by N-acyl-L-homoserine lactones. *Nature* **362**, 446–448.

Zielinski, N. A., Chakrabarty, A. M. & Berry, A. (1991). Characterization and regulation of the *Pseudomonas aeruginosa algC* gene encoding phosphomannomutase. *J Biol Chem* **266**, 9754–9763.

ZoBell, C. E. (1943). The effect of solid surfaces upon bacterial activity. *J Bacteriol* **46**, 39–56.

Environmental and genetic factors influencing biofilm structure

Paul Stoodley,[1] Luanne Hall-Stoodley,[1] John D. Boyle,[2] Frieda Jørgensen[3] and Hilary M. Lappin-Scott[4]

[1] Center for Biofilm Engineering, Montana State University, Bozeman, MT, USA
[2] School of Engineering, Exeter University, Exeter, UK
[3] Public Health Laboratory Service, Exeter, UK
[4] Environmental Microbiology Research Group, Exeter University, Exeter, UK

INTRODUCTION

It is increasingly evident that biofilms growing in a diverse range of medical, industrial and natural environments form a similarly diverse range of complex structures (Stoodley *et al.*, 1999a). These structures often contain water channels which can increase the supply of nutrients to cells in the biofilm (deBeer & Stoodley, 1995) and prompted Costerton *et al.* (1995) to propose that the water channels may serve as a rudimentary circulatory system of benefit to the biofilm as a whole. This concept suggests that biofilm structure may be controlled, to some extent, by the organisms themselves and may be optimized for a certain set of environmental conditions. To date, most of the research on biofilm structure has been focused on the influence of external environmental factors such as surface chemistry and roughness, physical forces (that is, hydrodynamic shear) or nutrient conditions and the chemistry of the aqueous environment. However, there has been a recent increase in the number of researchers using molecular techniques to study the genetic regulation of biofilm formation and development. Davies *et al.* (1998) demonstrated that the structure of a *Pseudomonas aeruginosa* biofilm could be controlled through production of the cell signal (or pheromone) N-(3-oxododecanoyl)-L-homoserine lactone (OdDHL). In this paper, we will examine some of the research that has been conducted in our laboratories and those of others on the relative contribution of hydrodynamics, nutrients and cell signalling to the structure and behaviour of bacterial biofilms.

HYDRODYNAMICS

The hydrodynamic conditions of an aquatic environment will determine the transport rate of nutrients and planktonic cells to a surface, the shear stress acting on the biofilm

SGM symposium 59: Community structure and co-operation in biofilms. Editors D. Allison, P. Gilbert, H. Lappin-Scott, M. Wilson. Cambridge University Press. ISBN 0 521 79302 5 ©SGM 2000.

and the rate of erosion of cells from the biofilm. The morphology and physical properties of biofilms appear to be strongly influenced by the magnitude of the shear stresses under which the biofilm developed. At low laminar flows, individual biofilm microcolonies, although irregular in shape, commonly form isotropic patterns with no obvious directional component to the pattern (Møller *et al.*, 1998; Stoodley *et al.*, 1999b, c; Wolfaardt *et al.*, 1994) (Fig. 1a). However, biofilms grown at higher shear are commonly filamentous with the microcolonies being elongated in the downstream direction (Bryers & Characklis, 1981; McCoy *et al.*, 1981; Stoodley *et al.*, 1999c) (Fig. 1b). The length of the filaments or 'streamers' appears to be greatest in turbulent flows with Reynolds numbers (Re) between transition and 17000. At higher Re, the biofilm filaments are reduced in length, presumably because of continual shearing off of biofilm material at the tip (Bryers & Characklis, 1981). Other structures such as ripples and dunes have also been reported in pure and defined mixed-culture laboratory biofilms that were grown in turbulent flow (Gjaltema *et al.*, 1994; Stoodley *et al.*, 1999d).

Fluid-like flow of biofilm microcolonies over the substratum

In addition to the influence that hydrodynamics have on biofilm morphology, we have used digital time-lapse microscopy (DTLM) to demonstrate that hydrodynamics can also influence dynamic behaviour in bacterial biofilms (Stoodley *et al.*, 1999d). In this work, ripple-shaped and round microcolonies in mixed-culture biofilms, grown under turbulent flow (Re 3600), were transported downstream across the upper and lower surfaces of a square glass flow cell (Fig. 2). Some of the structures appeared to roll across the surface while others appeared to slide. The travel velocity of the microcolonies across the surface varied with short-term variations in the velocity of the bulk liquid. A maximum migration velocity of approximately 1 mm h^{-1} occurred in the transition region between laminar and turbulent flow. The ripple-shaped microcolonies were also observed to continually detach from the glass surface. These observations support the hypothesis made by Inglis (1993) that ventilator-associated pneumonia may be related to the detachment of biofilm fragments from the walls of tracheal tubes. The biofilms that he observed had distinct wave patterns which led Inglis to hypothesize that the biofilm had been flowing and that this dynamic phenomenon may be related to biofilm detachment and dissemination into the lungs. Time-lapse movies of biofilms in turbulent flow taken at frame intervals of 0.5–1 h over time periods of up to 24 h suggest that biofilms behave like viscous fluids flowing along channel walls (Stoodley *et al.*, 1999d). In addition to flow along channel walls, we have also observed similar flow phenomena around glass beads in a porous media flow cell (unpublished data). These observations are supported by several studies which show that biofilms can behave like viscoelastic liquids (Christensen & Characklis, 1990; Ohashi & Harada, 1994; Stoodley *et al.*, 1999e). Flowing biofilms have important consequences for the dissemination of bacterial infection or contamination since this

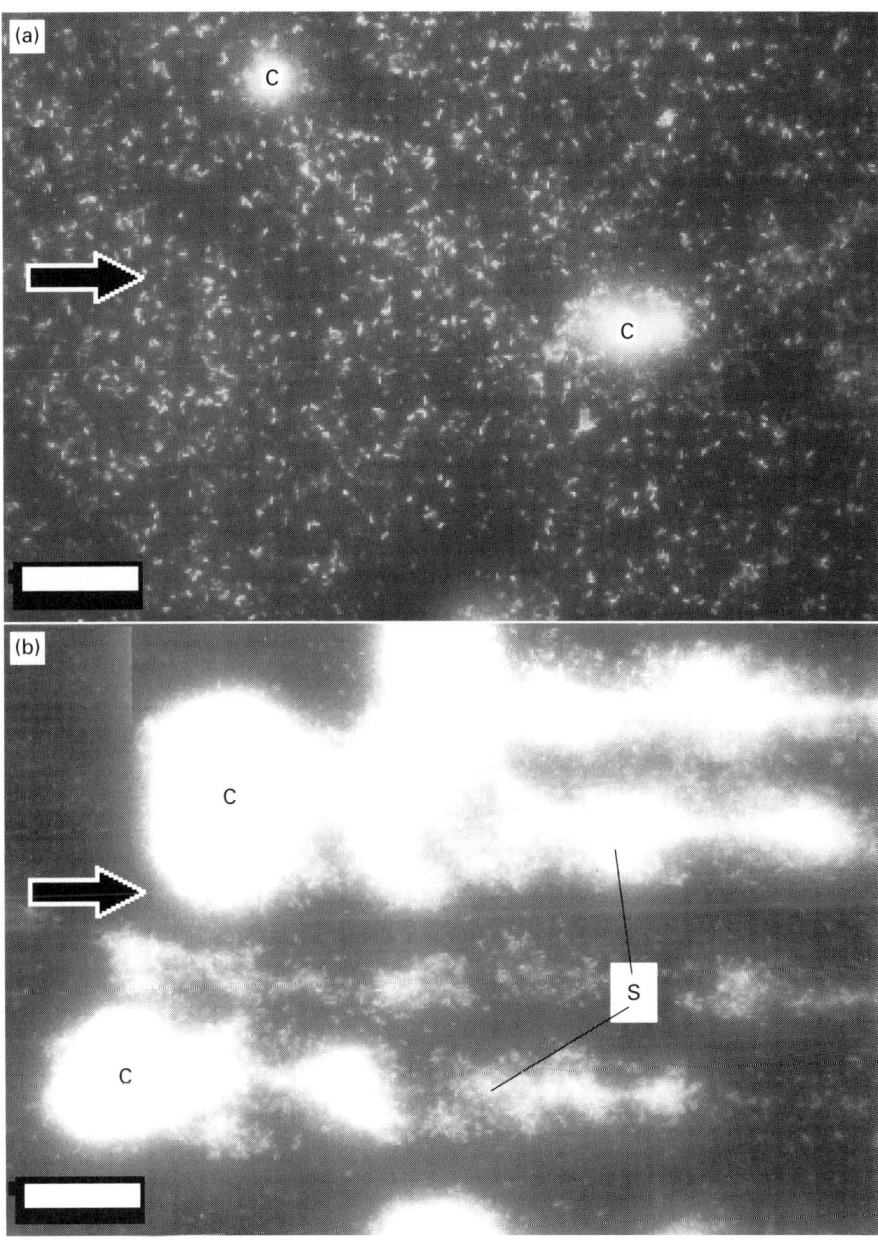

Fig. 1. Ten-day-old *P. aeruginosa* PAN067 biofilm grown under laminar (a) and turbulent (b) flow at *Re* 120 and 3600, respectively (Stoodley *et al.*, 1999b). The laminar grown biofilm was composed of single cells and small microcolonies (labelled 'C') while the turbulent grown biofilm microcolonies formed elongated streamers ('S') in the downstream direction. The biofilm was stained with the LIVE/DEAD Bac Light Bacterial Viability kit (Molecular Probes). Although not seen in this greyscale image, approximately 98 % of the cells were viable (green). Bar, 50 μm.

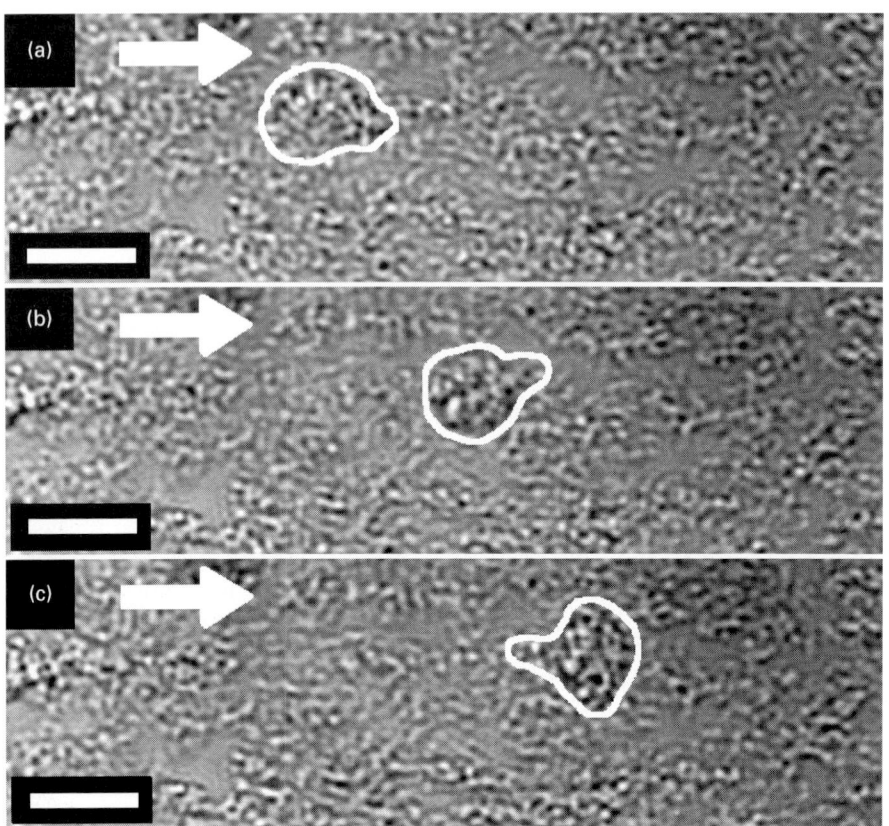

Fig. 2. Bacterial biofilm microcolony (outlined in white) moving downstream along the upper surface of a glass flow cell at a velocity of approximately 12 μm h^{-1} (Stoodley *et al.*, 1999d). The microcolony moved over the top of the surrounding monolayer of single cells. The bulk liquid velocity was 1 m s^{-1} in the direction shown by the arrow. The elapsed time between each panel was 50 min. Bar, 10 μm.

mechanism allows biofilm bacteria to colonize adjacent clean surfaces without depending on a planktonic phase, which is generally more susceptible to antimicrobial agents (Gilbert & Brown, 1995).

NUTRIENTS AND HYDRODYNAMICS

At higher nutrient concentrations and loading rates, biofilms tend to be thicker and denser than those grown in nutrient-poor environments (Characklis, 1990). However, less is known about the influence of nutrient type and concentration on the morphology of bacterial biofilms. Møller *et al.* (1997) reported that the morphology of an established undefined degradative community became more homogeneous when the nutrient source was changed from 2,4,6-trichlorobenzoic acid (2,4,6-TCB) to

Trypticase soy broth (TSB) while maintaining a constant carbon loading rate. They noted that the biofilm grown on TSB closely resembled the biofilms that they had previously grown exclusively on glucose and TSB. They hypothesized that mound-shaped microcolonies observed in the biofilms grown on 2,4,6-TCB may be characteristic of growth by their particular community on chlorinated substrates. It is likely that the morphological differences in their biofilms may have occurred due to population shifts in the community in response to changes in the enrichment conditions. Pure culture experiments on *Mycobacterium* spp. growing in laminar flow showed that, although biofilms took longer to accumulate on sterile tap water than on enrichment media, the morphology of the biofilms was similar (Hall-Stoodley *et al.*, 1999).

Stoodley *et al.* (1999c) have also shown that the morphology of an established biofilm can change significantly by varying the carbon concentration. In these experiments, the morphology of the microcolonies in a 21-d-old mixed-species biofilm grown under turbulent flow changed from that of ripples and streamers to large, closely packed, mound structures when the concentration of glucose (the sole carbon source) was increased by a factor of 10. The morphology was noticeably different within 10 h. In addition to the morphological change, there was also a change in the dynamic behaviour of the biofilm. At the low glucose concentration (40 p.p.m.), the biofilm appeared to flow downstream over the glass surface, but at 400 p.p.m. glucose, the downstream motion of biofilm microcolonies was much less evident. However, microcolonies could be observed to be continually growing and detaching using DTLM (Stoodley *et al.*, 1999f). When the glucose concentration was reduced back to 40 p.p.m., there was a net reduction of biomass and the ripples and streamers began to reform within approximately 48 h.

GENETIC REGULATION OF BIOFILM STRUCTURE

In the preceding sections, we have discussed some of the influences of the external environment on biofilm structure. Now we will turn to the influence that the biofilm micro-organisms themselves may have on the structure of the biofilms in which they live.

Cell signalling and quorum sensing

Quorum sensing (QS) is a mechanism used by both Gram-positive and Gram-negative bacteria to regulate their gene expression, and resulting phenotype, as a function of the density of the cell culture (Bassler, 1999). The cell culture density is 'sensed' through production of cell signalling molecules, which, once a threshold concentration is reached, initiate a signal transduction cascade, resulting in the expression of a number of target genes. In many Gram-negative bacteria, these cell signals commonly belong to a family of acylated-homoserine lactones (AHLs). However, cyclic dipeptides (Holden *et al.*, 1999) and quinolones (Pesci *et al.*, 1999) can also function as signalling molecules. QS in *P. aeruginosa* proceeds through the *lasI/lasR* system, which is homologous to the *luxI/luxR* system responsible for light production in some marine *Vibrio* species. However, in *P. aeruginosa*, instead of light production, high cell densities in stationary phase batch cultures can result in the production of virulence factors and secondary metabolites (Jones *et al.*, 1993; Latifi *et al.*, 1995, 1996; Winson *et al.*, 1995). The QS cascade in *P. aeruginosa* is activated by the cell signalling molecule OdDHL, whose synthesis is directed by *lasI*. At high concentrations, OdDHL binds with a transcriptional activator (the LasR protein), which further up-regulates *lasR* and *lasI* in addition to a number of other genes, including *lasB*, resulting in the production of elastase and other virulence factors (Pesci & Iglewski, 1997). The LasR–OdDHL complex also up-regulates *rhlI*, which produces another signalling molecule, N-butanoyl-L-homoserine lactone (BHL). BHL binds to RhlR and this complex up-regulates the *rhl* regulon, resulting in the production of rhamnolipid (Pearson *et al.*, 1997). Whiteley *et al.* (1999) have identified between 39 and 270 genes that are controlled by OdDHL- and BHL-activated QS mechanisms in *P. aeruginosa*.

It was suggested that QS may play a role in the development of biofilms which also exhibit high cell densities (Williams & Stewart, 1994). Davies *et al.* (1998) strengthened this hypothesis when they reported that the cell signal OdDHL was required for *P. aeruginosa* JP1, a *lasI* mutant (defective in the production of OdDHL), to develop the structurally complex biofilms which were formed by the parental wild-type (WT) PAO1 cells.

Cell signalling and hydrodynamics

Unlike suspended batch cultures, however, biofilms usually do not grow in completely mixed closed systems, and transport through biofilm microcolonies appears to be mainly through diffusion (Bryers & Drummond, 1998; deBeer *et al.*, 1997). In this case, it is not only the cell density that is important for the build-up of cell signalling molecules to concentrations at which QS mechanisms are activated, but also the production rate of signals, the rate of transport through the biofilm, the shape and dimensions of biofilm structures and the mass transport conditions outside the biofilm. The experiments by Davies *et al.* (1998) were conducted under very low laminar flows (*Re* 0.17). It is possible that under higher flows, cell signals may be diluted before they can reach QS concentrations within biofilm microcolonies. To investigate this further, we grew biofilms using *P. aeruginosa* PAOR, a *lasR* mutant (Latifi *et al.*, 1996), and the parental WT (PAO1) strain under laminar (*Re* 120) and turbulent (*Re* 3600) flow (Stoodley *et al.*, 1999b). Production of OdDHL was suppressed in the PAOR mutant, as demonstrated by biosensor assay (Winson *et al.*, 1998), which showed that the

OdDHL concentration in the spent medium was below detection limits (approx. 10^{-3} nM). We also used *P. aeruginosa* PAN067 (Jones *et al.*, 1993), a mutant deficient in the production of BHL, an N-acyl homoserine lactone which has been implicated in biofilm cell signalling (Davies *et al.*, 1998). BHL synthesis is directed by *rhlI*. In our experiments, we found that both the WT and the two mutant strains formed complex structures and it was the hydrodynamics that had the greatest influence on the observed microcolony structure (Fig. 3). In laminar flow, the microcolonies of both the mutant strains (PAOR and PAN067) and their parental strains were circular in shape but in turbulent flow they formed elongated streamers (Fig. 3e, f). The influence of the inability to produce AHLs on biofilm formation was more subtle than found by Davies *et al.* (1998) and appeared to be related more to the *rates* of growth and detachment than the ability to form complex structures (Stoodley *et al.*, 1999b). Clearly, further work is required to determine how the hydrodynamic conditions may influence QS mechanisms in biofilms, particularly those grown in well-mixed, open environments.

Biofilm structures formed through twitching motility

In addition to cell signalling mechanisms by which biofilm structures form through growth, time-lapse imaging has shown that microcolonies can also form from the co-ordinated movement of single attached cells to specific loci on a surface (Dalton *et al.*, 1996). In *P. aeruginosa*, such co-ordinated motion has been shown to be associated with twitching motility mediated by type IV pili (Semmler *et al.*, 1999). O'Toole & Kolter (1998) have shown that this type of motility is important for the formation of biofilm structures in the initial stages of biofilm development. However, since these studies are generally limited to the first few hours of biofilm development, it is not clear how twitching motility may influence the long-term structural arrangements of biofilms.

DISCUSSION

A more complete understanding of biofilm development and behaviour is essential if we are to predict, and ultimately control, biofilm processes. The use of confocal microscopy has documented some of the structural complexities of different types of biofilms, while time-lapse imaging is starting to reveal some of the dynamic behaviours occurring in biofilms.

Biofilm development and behaviour: nature or nurture?

Clearly, both environment and genotype have been shown to play a role in biofilm development and behaviour, but it is not so clear how the environmental conditions determine which factors dominate. Shear is one environmental condition we have studied that appears to be of fundamental significance. There are others yet to be elucidated, including nutrients and surface type to name a few. For example, the

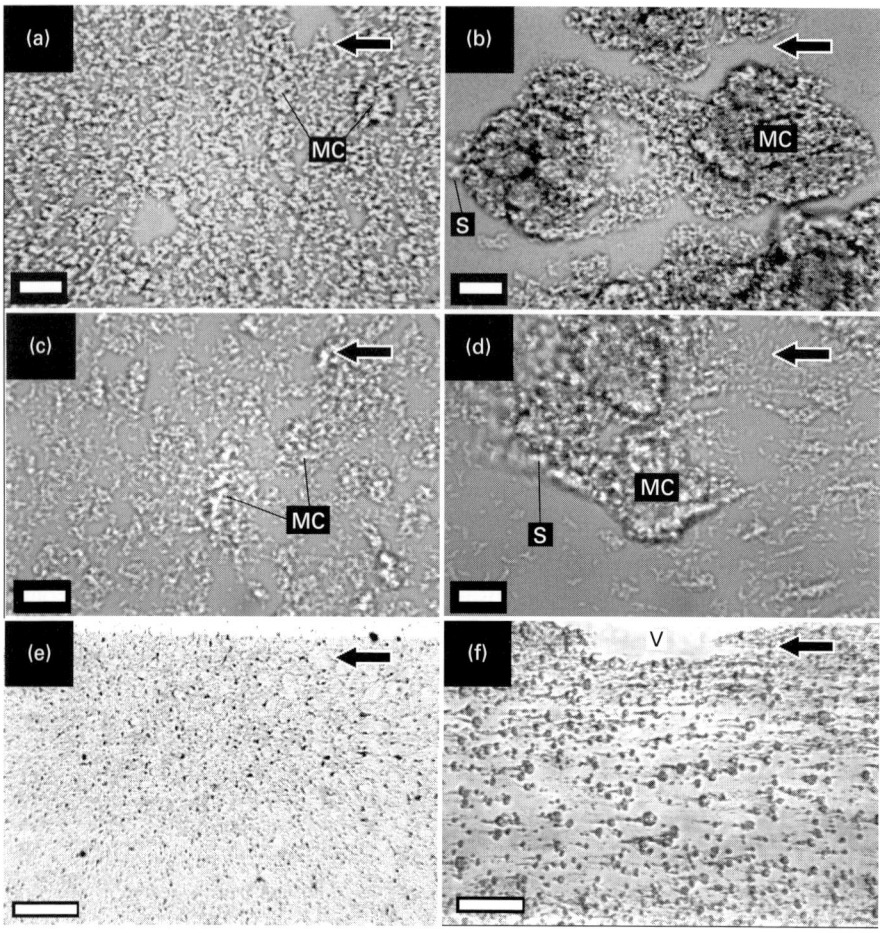

Fig. 3. Influence of cell signalling and hydrodynamics on biofilm structure after 6 days growth. (a) *P. aeruginosa* PAO1 grown under laminar flow (*Re* 120). The biofilm was composed of a monolayer of single cells interspersed with circular-shaped microcolonies (labelled 'MC'). Some void areas were devoid of cells. (b) *P. aeruginosa* PAO1 grown under turbulent flow (*Re* 3600). The microcolonies ('MC') were elongated in the downstream direction to form streamers ('S'). (c) *P. aeruginosa* PAOR, a *lasR* mutant (Latifi *et al.*, 1996) grown under laminar flow. The biofilm was similar in morphology to the parental PAO1 strain. (d) *P. aeruginosa* PAOR grown under turbulent flow. Again, the biofilm morphology was similar to the parental PAO1 strain grown under the corresponding flow velocity. (e) Low magnification image of the same PAOR biofilm as in (c) showing the overall pattern of the biofilm grown in laminar flow. (f) Low magnification image of the same PAOR biofilm as in (d) showing the influence of increased shear on biofilm morphology. The biofilm microcolonies formed elongated 'streamers'. A void area caused by localized sloughing detachment is indicated ('V'). All biofilms were grown on a minimal salts medium with glucose (400 p.p.m.) as the sole carbon source. The black arrow indicates the direction of bulk fluid flow in all panels. Bar, 10 μm (a, b, c, d) and 500 μm (e, f).

aggregation of single cells into microcolonies in the initial stages of biofilm formation in low shear environments appears to be controlled at the genetic level (O'Toole & Kolter, 1998), while the downstream motion of biofilm microcolonies in high shear flow appears to be a physical phenomenon related to the magnitude of the shear and the material properties of the biofilm exopolysaccharide matrix (Stoodley *et al.*, 1999d, f). Likewise, in low shear flows, cell signalling has been shown to play a significant role in the determination of biofilm structure (Davies *et al.*, 1998), while in high shear, the structures that develop appear to be shaped by the external shear and drag forces acting on the growing biofilm (Stoodley *et al.*, 1999b).

Increasingly, researchers are using genetic techniques to identify the role that individual genes may have on the phenotype of the individual cells and consequently the overall development of bacterial biofilms (O'Toole *et al.*, 1999). This approach has been advanced by the use of microtitre plates to assess biofilm accumulation. This technique allows rapid screening of large numbers of constructed mutants necessary for genetic analysis. However, these experiments are generally limited to studying biofilms in non-flowing, batch culture environments and in the very initial stages (hours) of biofilm development. In contrast, the microscopic monitoring of biofilms growing in flow cells allows long-term (days to months) experiments under flowing continuous culture conditions. However, this technique is limited by the number of replicates per experiment and in the total number of experiments that can be conducted.

An obvious approach is to use microtitre plates for rapid screening and then use flow cells to look at the longer term influence of a particular mutation on biofilm growth and behaviour in a flowing system. Presently, the construction of mutants deficient in specific phenotypes thought to be important for biofilm formation is proceeding at a much faster pace than can be studied in long-term flow cell experiments. To clear this backlog will require the development of biofilm flow cell systems capable of accommodating large numbers of replicates so that the influence of a particular mutation on biofilm development can be systematically assessed. It is only by the study of both the environmental *and* genetic influences on biofilm development that we will be able to begin to piece together how different biofilms behave in the real world, outside of the laboratory.

ACKNOWLEDGEMENTS

We thank A. Lazdunski from the Laboratoire d'Ingenierie et Dynamique des Systemes Membranaires, Marseille, France, for providing P. aeruginosa PAOR. Work in the laboratories of P. S. and H. M. L.-S. was supported by grants from the National Institutes of Health (1 RO1 GM60052–01) and by the co-operative agreement EEC-8907039 between the National Science Foundation and Montana State University – Bozeman.

REFERENCES

Bassler, B. L. (1999). How bacteria talk to each other: regulation of gene expression by quorum sensing. *Curr Opin Microbiol* **2**, 582–587.

Bryers, J. & Characklis, W. G. (1981). Early fouling biofilm formation in a turbulent flow system: overall kinetics. *Water Res* **15**, 483–491.

Bryers, J. D. & Drummond, F. (1998). Local macromolecule diffusion coefficients in structurally non-uniform bacterial biofilms using fluorescence recovery after photobleaching (FRAP). *Biotechnol Bioeng* **60**, 462–473.

Characklis, W. G. (1990). Microbial fouling. In *Biofilms*, pp. 523–584. Edited by W. G. Characklis & K. C. Marshall. New York: Wiley.

Christensen, B. E. & Characklis, W. G. (1990). Physical and chemical properties of biofilms. In *Biofilms*, pp. 93–130. Edited by W. G. Characklis & K. C. Marshall. New York: Wiley.

Costerton, J. W., Lewandowski, Z., Caldwell, D. E., Korber, D. R. & Lappin-Scott, H. M. (1995). Microbial biofilms. *Annu Rev Microbiol* **49**, 711–745.

Dalton, H. M., Goodman, A. E. & Marshall, K. C. (1996). Diversity in surface colonization behavior in marine bacteria. *J Ind Microbiol* **17**, 228–234.

Davies, D., Parsek, M. R., Pearson, J. P., Iglewski, B. H., Costerton, J. W. & Greenberg, E. P. (1998). The involvement of cell-to-cell signals in the development of a bacterial biofilm. *Science* **280**, 295–298.

deBeer, D. & Stoodley, P. (1995). Relation between the structure of an aerobic biofilm and mass transport phenomena. *Water Sci Technol* **32**, 11–18.

deBeer, D., Stoodley, P. & Lewandowski, Z. (1997). Measurement of local diffusion coefficients in biofilms by micro-injection and confocal microscopy. *Biotechnol Bioeng* **53**, 151–158.

Gilbert, P. & Brown, M. R. W. (1995). Mechanisms of the protection of bacterial biofilms from antimicrobial agents. In *Microbial Biofilms, Plant and Microbial Biotechnology Research Series 5*, pp. 118–130. Edited by H. M. Lappin-Scott & J. W. Costerton. Cambridge: Cambridge University Press.

Gjaltema, A., Arts, P. A. M., van Loosdrecht, M. C. M., Kuenen, J. G. & Heijnen, J. J. (1994). Heterogeneity of biofilms in rotating annular reactors: occurrence, structure, and consequences. *Biotechnol Bioeng* **44**, 194–204.

Hall-Stoodley, L., Keevil, C. W. & Lappin-Scott, H. M. (1999). *Mycobacterium fortuitum* and *Mycobacterium chelonae* biofilm formation under high and low nutrient conditions. *J Appl Microbiol* **85**, S60–S69.

Holden, M. T., Ram Chhabra, S., de Nys, R. & 14 other authors (1999). Quorum-sensing cross talk: isolation and chemical characterization of cyclic dipeptides from *Pseudomonas aeruginosa* and other gram-negative bacteria. *Mol Microbiol* **33**, 1254–1266.

Inglis, T. J. J. (1993). Evidence for dynamic phenomena in residual tracheal tube biofilm. *Br J Anaesth* **70**, 22–24.

Jones, S., Yu, B., Bainton, N. J. & 11 other authors (1993). The *lux* autoinducer regulates the production of exoenzyme virulence determinants in *Erwinia carotovora* and *Pseudomonas aeruginosa*. *EMBO J* **12**, 2477–2482.

Latifi, A., Winson, M. K., Bycroft, B. W., Stewart, G. S. A. B., Lazdunski, A. & Williams, P. (1995). Multiple homologues of LuxR and LuxI control expression of virulence determinants and secondary metabolites through quorum sensing in *Pseudomonas aeruginosa* PAO1. *Mol Microbiol* **17**, 333–344.

Latifi, A., Foglino, M., Tanaka, T., Williams, P. & Lazdunski, A. (1996). A hierarchical quorum sensing cascade in *Pseudomonas aeruginosa* links the transcriptional activators LasR and RhlR (VsmR) to the expression of the stationary phase sigma factor RpoS. *Mol Microbiol* **21**, 1137–1146.

McCoy, W. F., Bryers, J. D., Robbins, J. & Costerton, J. W. (1981). Observations of fouling biofilm formation. *Can J Microbiol* **27**, 910–917.

Møller, S., Korber, D. R., Wolfaardt, G. M., Molin, S. & Caldwell, D. E. (1997). Impact of nutrient composition on a degradative biofilm community. *Appl Environ Microbiol* **63**, 2432–2438.

Møller, S., Sternberg, C., Andersen, J. B., Christensen, B. B., Ramos, J. L., Givskov, M. & Molin, S. (1998). In situ gene expression in mixed-culture biofilms: evidence of metabolic interactions between community members. *Appl Environ Microbiol* **64**, 721–732.

Ohashi, A. & Harada, H. (1994). Adhesion strength of biofilm developed in an attached-growth reactor. *Water Sci Technol* **29**, 281–288.

O'Toole, G. A. & Kolter, R. (1998). Flagellar and twitching motility are necessary for *Pseudomonas aeruginosa* biofilm development. *Mol Microbiol* **30**, 295–304.

O'Toole, G. A., Pratt, L. A., Watnick, P. I., Newman, D. K., Weaver, V. B. & Kolter, R. (1999). Genetic approaches to study of biofilms. *Methods Enzymol* **310**, 91–109.

Pearson, J. P., Pesci, E. C. & Iglewski, B. H. (1997). Roles of *Pseudomonas aeruginosa las* and *rhl* quorum-sensing systems in control of elastase and rhamnolipid biosynthesis genes. *J Bacteriol* **179**, 5756–5767.

Pesci, E. C. & Iglewski, B. H. (1997). The chain of command in *Pseudomonas* quorum sensing. *Trends Microbiol* **5**, 132–134.

Pesci, E. C., Milbank, J. B., Pearson, J. P., McKnight, S., Kende, A. S., Greenberg, E. P. & Iglewski, B. H. (1999). Quinolone signaling in the cell-to-cell communication system of *Pseudomonas aeruginosa*. *Proc Natl Acad Sci USA* **96**, 11229–11234.

Semmler, A. B., Whitchurch, C. B. & Mattick, J. S. (1999). A re-examination of twitching motility in *Pseudomonas aeruginosa*. *Microbiology* **145**, 2863–2873.

Stoodley, P., deBeer, D., Boyle, J. D. & Lappin-Scott, H. M. (1999a). Evolving perspectives of biofilm structure. *Biofouling* **14**, 75–94.

Stoodley, P., Jørgensen, F., Williams, P. & Lappin-Scott, H. M. (1999b). The role of hydrodynamics and AHL signalling molecules as determinants of the structure of *Pseudomonas aeruginosa* biofilms. In *Biofilms: the Good, the Bad, and the Ugly*, pp. 323–330. Edited by J. W. T. Wimpenny, P. Gilbert, J. Walker, M. Brading & R. Bayston. Cardiff: BioLine.

Stoodley, P., Dodds, I., Boyle, J. D. & Lappin-Scott, H. M. (1999c). Influence of hydrodynamics and nutrients on biofilm structure. *J Appl Microbiol* **85**, 19S–28S.

Stoodley, P., Lewandowski, Z., Boyle, J. D. & Lappin-Scott, H. M. (1999d). The formation of migratory ripples in a mixed species bacterial biofilm growing in turbulent flow. *Environ Microbiol* **1**, 447–457.

Stoodley, P., Lewandowski, Z., Boyle, J. D. & Lappin-Scott, H. M. (1999e). Structural deformation of bacterial biofilms caused by short term fluctuations in flow velocity: an in-situ demonstration of biofilm viscoelasticity. *Biotechnol Bioeng* **65**, 83–92.

Stoodley, P., Boyle, J. D. & Lappin-Scott, H. M. (1999f). Biofilm structure and behaviour: influence of hydrodynamics and nutrients. In *Dental Plaque Revisited: Oral Biofilms in Health and Disease*, pp. 63–72. Edited by H. H. Newman & M. Wilson. Cardiff: BioLine.

Whiteley, M., Lee, K. M. & Greenberg, E. P. (1999). Identification of genes controlled by quorum sensing in *Pseudomonas aeruginosa*. *Proc Natl Acad Sci USA* **96**, 13904–13909.

Williams, P. & Stewart, G. S. A. B. (1994). Cell density dependent control of gene expression in bacteria – implications for biofilm development and control. In *Bacterial Biofilms and their Control in Medicine and Industry*, pp. 9–12. Edited by J. W. T. Wimpenny, W. Nichols, D. Stickler & H. M. Lappin-Scott. Cardiff: BioLine.

Winson, M. K., Camara, M., Latifi, A. & 10 other authors (1995). Multiple *N*-acyl-L-homoserine lactone signal molecules regulate production of virulence determinants and secondary metabolites in *Pseudomonas aeruginosa*. *Proc Natl Acad Sci USA* **92**, 9427–9431.

Winson, M. K., Swift, S., Fish, L., Throup, J. P., Jørgensen, F., Chhabra, S. R., Bycroft, B. W., Williams, P. & Stewart, G. S. A. B. (1998). Construction and analysis of luxCDABE-based plasmid sensors for investigating N-acyl homoserine lactone-mediated quorum sensing. *FEMS Microbiol Lett* **163**, 185–192.

Wolfaardt, G. M., Lawrence, J. R., Robarts, R. D., Caldwell, S. J. & Caldwell, D. E. (1994). Multicellular organization in a degradative biofilm community. *Appl Environ Microbiol* **60**, 434–446.

Coaggregation and coadhesion in oral biofilms

Paul E. Kolenbrander, Roxanna N. Andersen,
Karen M. Kazmerzak and Robert J. Palmer, Jr

Oral Infection and Immunity Branch, National Institute of Dental and Craniofacial Research, National Institutes of Health, Bethesda, MD 20892, USA

COAGGREGATION AND COADHESION

Certain molecules on the surfaces of human oral bacteria can be recognized by cognate surface components of genetically distinct cells, which bind to form networks of cell–cell interactions. When these interactions occur in suspension, they are called coaggregations (Kolenbrander, 1988). When the interaction occurs between suspended or planktonic cells and already adherent cells, it is called coadhesion (Bos *et al.*, 1994). Coadhesion may involve the accretion of an already formed coaggregate onto a biofilm, which is an assemblage of living cells on a substratum, or onto a virgin surface.

Coaggregation among human oral bacteria was first described 30 years ago (Gibbons & Nygaard, 1970). Coaggregation is measured by several methods, including visual inspection of clumps or coaggregates after mixing dense suspensions of two cell types (Gibbons & Nygaard, 1970), turbidometric measurement of supernatant after slow-speed centrifugation to pellet the coaggregates (McIntire *et al.*, 1978), filtration through specific pore size to separate single cells from coaggregates (Lancy *et al.*, 1980), distribution of radiolabelled cells of one cell type in coaggregates and supernatant after slow-speed centrifugation (Kolenbrander & Andersen, 1986) and binding of a radiolabelled cell type to partner cells immobilized on a nitrocellulose membrane (Lamont & Rosan, 1990). Coaggregations may be unimodal or bimodal (Kolenbrander, 1997). Unimodal coaggregations involve protease-sensitive molecules on the cell surface of one of the partners recognizing their cognate receptors (protease-insensitive) on the other partner's cell surface. Bimodal coaggregations involve more than one of the unimodal mechanisms. For example, one partner expresses both an

SGM symposium 59: Community structure and co-operation in biofilms. Editors D. Allison, P. Gilbert, H. Lappin-Scott, M. Wilson. Cambridge University Press. ISBN 0 521 79302 5 ©SGM 2000.

adhesin and a non-cognate receptor. Its partner expresses the respective cognates for the adhesin and receptor. Many of these coaggregations between oral bacteria are inhibited by simple sugars such as beta-galactosides and N-acetylneuraminic acid. Coaggregations have been proposed to contribute to the accretion of bacteria onto the tooth surface to form dental plaque (Kolenbrander & London, 1993). Early colonizers of the tooth surface coaggregate with other early colonizers but rarely with late colonizers. Likewise, late colonizers coaggregate with other late colonizers. Unusual among oral bacteria are the fusobacteria. These organisms coaggregate with both early and late colonizers, and they have been termed coaggregation bridge organisms that are involved in plaque maturation (Kolenbrander & London, 1993).

Nearly all, if not all, oral bacteria will exhibit coaggregation. Until recently, it was a phenomenon principally observed amongst human oral bacteria. It has now been reported to also occur between yeasts and bacteria (Grimaudo et al., 1996; Grimaudo & Nesbitt, 1997; Holmes et al., 1995, 1996; Jabra-Rizk et al., 1999; Millsap et al., 1999), between urogenital flora (Reid et al., 1988), flora from chicken crops (Vandevoorde et al., 1992) and freshwater bacteria (Buswell et al., 1997; Rickard et al., 2000). Whereas little if any difference was observed in the extent of the coaggregation between cells harvested in exponential versus stationary phase for human oral bacteria (Cisar et al., 1979), the coaggregations among aquatic bacteria were always maximal with cells harvested in their stationary phase of growth (Rickard et al., 2000). Other properties associated with the coaggregation of oral bacteria, such as inhibition by sugars, as well as heat- and protease-inactivated surface components, were also found amongst aquatic bacteria (Rickard et al., 2000). Intergeneric coaggregations have been observed with all oral bacteria so far examined, and intrageneric coaggregations between oral streptococci (Kolenbrander et al., 1990) and between oral fusobacteria (Andersen et al., 1998) have been reported, but intraspecies coaggregations were first reported among aquatic bacteria (Rickard et al., 2000). It appears, therefore, that as more ecosystems are examined for cellular interactions between genetically distinct bacteria, then additional examples of this dramatic phenomenon will be discovered.

COAGGREGATION AMONG HUMAN ORAL BACTERIA

The cell surface of oral bacteria is a complex compilation of macromolecules that present an astounding array of adhesins and receptors that are seemingly designed for specific participation in the central function of colonization of the oral substrata. Several reviews, published on various aspects of oral microbial adhesion, should be consulted in order to obtain a fuller appreciation of this critical property of oral bacteria (Bos et al., 1998; Bowden & Li, 1997; Ganeshkumar et al., 1998; Jenkinson & Demuth, 1997; Jenkinson & Lamont, 1997; Kolenbrander, 1997; Kolenbrander et al., 1999a; Scannapieco, 1994; Whittaker et al., 1996b).

Streptococcus

One of the many macromolecules presented on many streptococcal surfaces is lipoteichoic acid (LTA). This can be regarded as essential for cell viability since LTA-minus mutants cannot be isolated. D-Alanine, however, is one of the substitutions found on some LTA. Mutants unable to D-alanylate LTA have been isolated in *Streptococcus gordonii* DL1 (Clemans *et al.*, 1999). D-Alanyl LTA appears to be involved specifically in galactoside-inhibitable coaggregations among streptococci (Clemans *et al.*, 1999). Mutants unable to participate in these intrageneric coaggregations were obtained in the first instance by spontaneous mutation (Clemans & Kolenbrander, 1995a, b) and later by transposon Tn*916* mutagenesis of *S. gordonii* DL1 (Whittaker *et al.*, 1996a). Subsequently, regions identified as being involved in intrageneric coaggregation were further analysed by allelic replacement and insertion of an *ermAM* cassette into *dltA* (Clemans *et al.*, 1999), which encodes a D-alanine-D-alanyl carrier protein ligase (Heaton & Neuhaus, 1992, 1994). This enzyme catalyses the D-alanylation of the D-alanyl carrier protein (encoded by *dltC*; Debabov *et al.*, 1996), which in turn transfers the D-alanine to a membrane acceptor required for the D-alanylation of LTA. The *dltA* mutants are unable to exhibit intrageneric coaggregation and also show several other phenotypes (Clemans *et al.*, 1999). These include the failure to exhibit a parental pebbly surface as seen by scanning electron microscopy. Such cells exhibit aberrant cell septa in various positions and alignments along the cell surface rather than the normal parallel alignment of successive septa that is typical of streptococci. All intrageneric coaggregation mutants of *S. gordonii* DL1, including spontaneous, transposon-insertion and allelic replacement mutants, lacked a 100 kDa surface protein that is normally released with ease from parental cells by mild sonication. All of these mutants, however, retained robust intergeneric coaggregations with actinomyces and fusobacteria. Thus the D-alanylated LTA is hypothesized to provide both specific binding sites for the putative 100 kDa adhesin and scaffolding for the proper presentation of this adhesin, which is required to mediate intrageneric coaggregation (Clemans *et al.*, 1999).

The 100 kDa protein may bind to the D-alanyl LTA on *S. gordonii* DL1 (Clemans *et al.*, 1999) in a way that is analogous to the binding of choline-binding proteins of *Streptococcus pneumoniae* (Rosenow *et al.*, 1997; Yother & White, 1994). The choline-binding proteins are non-covalently bound to the phosphorylcholine of wall teichoic acid. By analogy, the 100 kDa protein may bind weakly to the D-alanylated LTA. Consistent with this idea is the fact that the 100 kDa protein is completely removed by a mild sonication treatment of *S. gordonii* DL1, and the sonicated cells are unable to coaggregate with the streptococcal partners (Clemans *et al.*, 1999). In sharp contrast, the treated cells retain their ability to coaggregate with actinomyces partner cells, indicating that the mediator of intergeneric coaggregation is still present on the cell surface. Retention of an ability to coaggregate after mild sonication is the usual observation when treating other oral bacteria such as *Veillonella atypica* and *Actinomyces* serovar WVA963 (Hughes *et al.*, 1992; Klier *et al.*, 1997). In these, only a portion of the adhesive activity is removed by mild sonication. This suggested that, unlike galactoside-inhibitable binding among streptococci, the binding of intergeneric coaggregation mediators is not by a weak non-covalent interaction.

Intrageneric coaggregation among streptococci is unimodal; a single kind of surface molecule on each partner cell, either adhesins (ADH) or their complementary receptor polysaccharides (RPS), mediates the coaggregation. Thus it seems that there are two distinct groups of oral streptococci (group ADH and group RPS) that can be distinguished functionally on the basis of their surface adhesion molecules. Streptococcal group ADH expresses LTA (the Lancefield group H antigen), and this group, coincidentally, is positive for galactoside-sensitive adhesins detected by intrageneric coaggregation (Hsu *et al.*, 1994; Kolenbrander *et al.*, 1990). The group H antigen occurs in most strains of *S. gordonii* and *Streptococcus sanguis* but not in strains of *Streptococcus oralis* and *Streptococcus mitis* (Kilian *et al.*, 1989). Streptococcal group RPS includes strains of *S. oralis* and *S. mitis* that synthesize N-acetylgalactosamine (GalNAc)-containing cell wall polysaccharides (Cisar *et al.*, 1995, 1997), which are receptors for the putative 100 kDa adhesin on *S. gordonii* DL1 and other members of group ADH (Clemans & Kolenbrander, 1995a). All group ADH strains that are positive for group H antigen are negative for the GalNAc-containing cell wall polysaccharides and vice versa (Cisar *et al.*, 1997; Hsu *et al.*, 1994). Further, an absorbed antiserum used to identify the 100 kDa protein in other streptococcal strains possessing group H antigen does not react with *S. oralis* strains (Clemans & Kolenbrander, 1995a). It appears, therefore, that streptococci of group ADH bearing the LTA group H antigen also express the 100 kDa putative adhesin. Likewise, streptococcal group RPS strains synthesize the complementary GalNAc-containing receptor polysaccharide. Coupling the unimodal nature of expressing one or the other of the cognate molecules with the fact that streptococci are the predominant bacteria on the surface of freshly cleaned tooth enamel suggests that interactions between the two distinct groups of early colonizers contribute to the streptococcal domination of initial dental plaque. Assessment of this hypothesis is possible with specific antisera recognizing either LTA or receptor polysaccharides and using the antisera as probes to identify early colonizers *in situ* on removable enamel chips placed *in vivo*.

Several other streptococcal surface molecules are known to be involved in coaggregation, haemagglutination and other adhesive functions. The Hs antigen of *S. gordonii*, a sialic-acid-binding lectin associated with specific fibrillar structures and haemagglutinating activity, migrates as a diffuse band above 200 kDa when examined under denaturing electrophoresis conditions (Takahashi *et al.*, 1997). The cell-wall-anchored CshA of *S. gordonii* is a large molecular size protein (259 kDa) and is the structural and functional polypeptide component of *S. gordonii* adhesive fibrils. It confers the hydrophobic as well as the adhesive properties of binding fibronectin and binding to *Actinomyces naeslundii* cells (McNab *et al.*, 1996, 1999). Upon transforming the gene that encodes CshA into *Enterococcus faecalis*, the enterococci expressed peritrichous 70-nm-long surface fibrils that reacted with anti-CshA antiserum. The enterococci also acquired the adhesive functions characteristic of *S. gordonii* bearing the CshA surface fibrils (McNab *et al.*, 1999). A second, antigenically related high-molecular-mass, surface protein called CshB is also expressed on *S. gordonii*, and both CshA and CshB are necessary for colonization of the murine oral cavity (McNab *et al.*, 1994). A relationship between the expression of CshA and the transport of peptides by HppA, an oligopeptide-binding protein of a hexa-heptapeptide permease, was shown by generating a chromosomal fusion between the *cshA* promoter and the chloramphenicol acetyltransferase gene, *cat*, and measuring Cat activity in parent and *hppA* mutant cells (McNab & Jenkinson, 1998). It appears that HppA is necessary to transport an extracellular modulator of *cshA* transcription and that this factor is present in increased amounts in conditioned culture media. It would be interesting if the modulator is identified as a quorum-sensing peptide such as those known to be essential for competence in streptococci (Håvarstein *et al.*, 1996, 1997). In this respect, a relationship between competence, a surface-relevant function for genetic exchange, and both ScaA, a Mn^{2+}-transporter (Kolenbrander *et al.*, 1998) located on the surface of *S. gordonii* DL1 (Kolenbrander, 2000), and the 100 kDa putative adhesin associated with D-alanylated LTA (Clemans *et al.*, 1999) has been reported. The absence of either the putative adhesin or ScaA reduces transformation frequency 10–20-fold (Clemans *et al.*, 1999; Kolenbrander *et al.*, 1998). Collectively, the diversity of functions for these surface molecules together with their functional inter-relationship suggest that molecules on the surface of a streptococcal cell can communicate to produce environmentally dynamic cells. Extending this idea to coaggregation and the potential interaction amongst bacteria in biofilms presents a broad picture of cell–cell communication and spatial organization of oral microbial biofilm communities.

Actinomyces

Actinomyces serovar WVA963 coaggregates with oral streptococci only by a lactose-inhibitable mechanism. The actinomyces 95 kDa putative lactose-sensitive adhesin requires the actinomyces type 2 fimbriae to mediate coaggregation (Klier *et al.*, 1997). A coaggregation-defective mutant that had lost the type 2 fimbriae but had retained the 95 kDa protein secretes the protein into the culture medium (Klier *et al.*, 1998). The putative adhesin, concentrated from the culture medium, binds to the coaggregation partner *S. oralis* 34 and is released when lactose is added. This indicates that the protein is biochemically active and capable of mediating the coaggregation. These results suggest that the putative adhesin is a minor subunit of a type 2 fimbrial complex and may be presented at the fimbrial tip as has been shown for enterobacterial adhesins (Hultgren *et al.*, 1989; Jones *et al.*, 1995) and the oral bacterium *Prevotella loescheii* (Weiss *et al.*, 1988).

The aggregation factor AnAF, purified from *A. naeslundii* KWS81, mediates coaggregation with *Porphyromonas gingivalis* KC409 (Yamaguchi *et al.*, 1998). AnAF is a glycoprotein with a molecular size greater than 200 kDa and binds directly to the partner cell, *P. gingivalis* KC409. Unlike the binding mediated by the *Actinomyces* serovar WVA963 95 kDa putative adhesin to partner cells (Klier *et al.*, 1998), the binding of AnAF to partner cells is not sensitive to lactose or galactose, although galactose is the major sugar constituent of AnAF (Yamaguchi *et al.*, 1998). The purified AnAF inhibits the coaggregations between *A. naeslundii* KWS81 and *P. gingivalis* KC409. It also inhibits coaggregations between the actinomyces and three other Gram-negative bacteria, *Fusobacterium nucleatum*, *Capnocytophaga ochracea* and *Prevotella intermedia*, but has no inhibitory effect on the coaggregation between *A. naeslundii* KWS81 and three Gram-positive partners, *Streptococcus intermedius*, *Streptococcus mutans* and *Streptococcus parasanguis*. These results suggest that AnAF recognizes its cognate coaggregation mediator on several Gram-negative bacteria and that the cognate mediator is functionally similar on these genetically distinct partner cell types. The idea of functional similarity of coaggregation mediators has been described in detail elsewhere (Kolenbrander *et al.*, 1999a). Functionally similar adhesins on five genera of partners recognize the same surface receptor polysaccharide on *S. oralis*, and functionally similar receptors on eight genera of Gram-negative partners of *F. nucleatum* are recognized by an adhesin on the fusobacterium (Kolenbrander *et al.*, 1999a).

Fusobacterium

The coaggregations of one of the fusobacteria, *F. nucleatum* PK1594, have been studied extensively (Kolenbrander & Andersen, 1989; Kolenbrander *et al.*, 1989). The galactose-binding adhesin of *F. nucleatum* PK1594 recognizes functionally similar cognate receptors on *P. gingivalis* PK1924, *Actinobacillus actinomycetemcomitans* JP2, *Capnocytophaga sputigena* ATCC 33612 and human erythrocytes, as evidenced by the ability of a single monoclonal antibody to block each of the interactions (Shaniztki *et al.*, 1997). The putative adhesin is a 30 kDa outer-membrane protein that binds to lactose-Sepharose affinity columns and is eluted by lactose. Besides the galactoside-inhibitable coaggregations, *F. nucleatum* PK1594 as well as other fusobacteria interact in numerous ways with oral surfaces (Bolstad *et al.*, 1996), and *F. nucleatum* PK1594 coaggregates with several bacteria by an N-acetylneuraminic-acid-inhibitable mechanism (Shaniztki *et al.*, 1998). These partners include *Actinomyces israelii* PK16, *C. ochracea* ATCC 33596, *S. mitis* J22 and *S. oralis* H1 (Shaniztki *et al.*, 1998). Monoclonal antibodies that were selected for their ability to inhibit coaggregation between *F. nucleatum* PK1594 and *A. israelii* PK16 also blocked the coaggregations between the fusobacterium and the other partners that exhibit N-acetylneuraminic-acid-inhibitable coaggregations. Monoclonal antibodies that block the N-acetylneuraminic-acid-inhibitable partnerships have no effect on galactoside-inhibitable coaggregations, and monoclonal antibodies that block the latter coaggregations have no effect on the N-acetylneuraminic-acid-inhibitable coaggregations, indicating the highly specific nature of these sugar-inhibitable interactions. Coaggregations between the fusobacterium and other partners, such as *A. israelii* PK14, *A. naeslundii* T14V, *A. naeslundii* PK984, *S. oralis* 34, *Streptococcus* SM PK509 and *V. atypica* PK1910, are not inhibited by either galactosides or N-acetylneuraminic acid, and none are inhibited by monoclonal antibodies that block these kinds (galactoside-inhibitable or N-acetylneuraminic-acid-inhibitable) of coaggregations (Shaniztki *et al.*, 1998). These data indicate that *F. nucleatum* PK1594 is capable of interacting with partners by at least three distinct mechanisms, two of which are highly specific, and they may explain why *F. nucleatum* is one of the most numerous bacteria found in subgingival samples of healthy and diseased sites (Beck *et al.*, 1992; Moore & Moore, 1994; Socransky *et al.*, 1998).

In a separate study of 19 strains of *F. nucleatum* and 12 strains of oral streptococci, some coaggregations were inhibited by L-arginine, L-lysine or N-acetyl-D-galactosamine; others were inhibited only if two of the inhibitors were present (Takemoto *et al.*, 1995). Inhibitor effects on coaggregations varied between strains, indicating significant coaggregation diversity in fusobacteria, as was reported earlier (Kolenbrander *et al.*, 1989). As a group, fusobacteria coaggregate with every species of oral bacteria tested so far (Andersen *et al.*, 1998; Kolenbrander *et al.*, 1989, 1995). The diverse adherence properties of fusobacteria as a group shown in these studies support the idea that fusobacteria are indeed a major influence in the community structure of dental plaque.

Reports of coaggregation between yeasts and oral bacteria showed that several factors influence these coaggregations. An extensive range of fusobacterial species was tested for coaggregation with *Candida albicans* strains; *F. nucleatum*, *Fusobacterium periodonticum* and *Fusobacterium sulci* coaggregated with the yeast, whereas *Fusobacterium alocis*, *Fusobacterium mortiferum* and *Fusobacterium simiae* strains did not (Grimaudo & Nesbitt, 1997). The fusobacteria were inactivated by heat, trypsin and proteinase K treatments. The yeasts, on the other hand, were inactivated by periodate oxidation, suggesting that the fusobacteria express a surface adhesin that recognizes a cognate carbohydrate surface receptor on the yeast (Grimaudo & Nesbitt, 1997). Coaggregation between *Candida dubliniensis* and *F. nucleatum* occurred when the yeast was grown at 25, 37 or 45 °C, but *C. albicans* only coaggregated when it was grown at either 25 or 45 °C (Jabra-Rizk *et al.*, 1999). A role for temperature was also seen in the interaction of *C. albicans* with *S. gordonii* in coadhesion experiments in a parallel plate flow chamber (Millsap *et al.*, 1999). Other physico-chemical mechanisms such as acid–base and hydrophobic interactions of yeast adhesion also have been investigated (Millsap, 1999). *S. gordonii* also coaggregates with *C. albicans*; a streptococcal cell wall polysaccharide receptor appears to mediate this coaggregation (Holmes *et al.*, 1995). Likewise, an actinomyces carbohydrate component may be involved in coaggregation between *Actinomyces* spp. and *C. albicans* (Grimaudo *et al.*, 1996). Thus the variety of bacterial coaggregation partners of the oral pathogen *C. albicans* suggests that there are multiple ways by which *C. albicans* adheres in the oral cavity. *C. albicans* may use these mechanisms to colonize a human host and to contribute to pathogenic sequelae in an immunocompromised host.

BACTERIAL ADHESION AND COADHESION

Planktonic cells bind to abiotic surfaces in bacterial adhesion, whereas coadhesion involves planktonic bacteria binding to cells that are already attached to the surface. In either adhesion or coadhesion, the ensuing growth of bacteria results in a biofilm community. Surfaces used to study oral bacterial adhesion and coadhesion are generally saliva-coated. Saliva contains a variety of components that act as receptors for bacterial adhesion, including α-amylase, fibronectin, lactoferrin, lysozyme, mucins, proline-rich proteins, agglutinins and secretory immunoglobulin A (Busscher & Evans, 1998; Scannapieco, 1994). Amylase-binding bacteria appear to colonize oral surfaces only of animals having salivary amylase activity (Scannapieco, 1994). A major amylase-binding protein of 20 kDa (AbpA) and the *abpA* gene encoding it have been identified in *S. gordonii* by conjugative transposon Tn916 insertional inactivation (Rogers *et al.*, 1998). Fap1 is about 200 kDa and is expressed on *S. parasanguis*, where it enables adhesion of cells to saliva-coated hydroxyapatite (Wu *et al.*, 1998). These two examples illustrate the recent progress made in identifying streptococcal surface molecules that mediate adhesion to saliva-coated surfaces.

The antigen I/II family of polypeptides is highly conserved among oral streptococci and is involved in binding salivary agglutinin molecules (Jenkinson & Demuth, 1997). *S. gordonii* is the only oral streptococcal species that expresses two antigen I/II polypeptides, SspA and SspB, which are expressed from tandemly arranged monocistronic chromosomal genes, *sspA* and *sspB* (Demuth *et al.*, 1996). This family is considered to be a group of colonization factors that mediate streptococcal interactions with a wide variety of surfaces, cell types and molecules, including salivary receptors, type I collagen (Love *et al.*, 1997; Sciotti *et al.*, 1997), *A. naeslundii* (Demuth *et al.*, 1996), *P. gingivalis* (Brooks *et al.*, 1997) and *C. albicans* (Holmes *et al.*, 1996; Jenkinson & Lamont, 1997). The SspB polypeptide has a higher affinity than SspA for binding type I collagen (Holmes *et al.*, 1998). Different regions of the antigen I/II molecule are thought to bind to particular receptors. The N-terminal Ala-rich region of antigen I/II from *S. mutans* serotype f binds collagen, laminin and fibronectin (Sciotti *et al.*, 1997), and the Ala-rich and extended N-terminal regions mediate the production of tumour necrosis factor alpha in a monocyte cell line (Chatenay-Rivauday *et al.*, 1998). The C-terminal 500 amino acids of *S. gordonii* I/II contain sites for binding salivary agglutinin glycoprotein, Ca^{2+} and *P. gingivalis* (Brooks *et al.*, 1997; Duan *et al.*, 1994). Although some binding functions of specific sequences of all members of the antigen I/II polypeptide family may be shared, others are different (Jenkinson & Lamont, 1997).

The type 1 fimbriae on the surface of oral actinomyces participate in a protein–protein interaction with the C-terminal amino acids of proline-rich proteins and statherin in the acquired pellicle coating the tooth surface. FimP is the structural subunit of type 1 fimbriae, and *fimP*, the gene encoding the subunit, of *A. naeslundii* has been cloned and sequenced (Yeung *et al.*, 1987). FimP is present on a wide variety of actinomyces, including many non-human strains (Yeung, 1992). Some strains, especially those of human origin, bind preferentially to the C-terminal Pro-Gln-containing decapeptide covalently linked to agarose beads, whereas other strains, especially those of rodent origin, bind preferentially to the C-terminal Thr-Phe-containing decapeptides (Li *et al.*, 1999). These results suggest that the adhesin function resides in the FimP structural subunit of type 1 fimbriae and may contribute to animal host tropism by the actinomyces. Anti-FimP monoclonal antibodies, however, do not block type-1-fimbriae-mediated adhesion to saliva-treated hydroxyapatite, whereas monoclonal antibodies reacting with the tip of the fimbriae do block adhesion. This suggests that additional minor subunits of the type 1 fimbrial structure are involved in this adhesion (Nesbitt *et al.*, 1996). Several genes located adjacent to *fimP* (Yeung & Ragsdale, 1997) are likely to be involved in the synthesis and assembly of the fimbriae and may include a distinct adhesin gene. There is considerable genomic diversity within *A. naeslundii*, as evidenced by ribotyping a large number of strains, which showed more than one ribotype colonized a single individual (Hallberg *et al.*, 1998a) or a particular root surface (Bowden *et al.*, 1999). Such diversity may also reveal itself in the structurally variant FimP (Hallberg *et al.*, 1998b) of type 1 fimbriae. This confers specific surface recognition properties on the actinomyces that contribute to its spatial arrangement in dental plaque. Additional studies will be necessary to sort out the two possibilities of adhesin-recognition sites in the FimP major structural subunit or in a tip-localized minor subunit.

RECEPTOR POLYSACCHARIDES

Lactose-sensitive coaggregations among oral viridans streptococci and other partner cell types are very common and are mediated by receptor polysaccharides on the streptococcal cell surfaces. Each streptococcal strain appears to express a single major kind of receptor polysaccharide in its cell wall polysaccharide layer (Cisar *et al.*, 1997). A receptor may mediate galactoside-inhibitable coaggregation with distinct but functionally similar adhesins borne on the surfaces of cells of many genera such as *Actinomyces*, *Haemophilus*, *Prevotella*, *Streptococcus* and *Veillonella* (Kolenbrander *et al.*, 1999a). The receptors are composed of distinct phosphodiester-linked hexa- or heptasaccharide repeating units (Cisar *et al.*, 1995; Whittaker *et al.*, 1996b) and are widely distributed on viridans streptococci, including *S. gordonii*, *S. mitis*, *S. oralis* and *S. sanguis* (Cisar *et al.*, 1997). The receptors contain either of two host-like motifs (GalNAcβ1→3Gal, called Gn; Galβ1→3GalNAc, called G) (Cisar *et al.*, 1997). These two motifs have been found in each of six structural types of receptor polysaccharides isolated from 22 strains of viridans streptococci (Cisar *et al.*, 1997), and all of the streptococci exhibit galactoside-inhibitable coaggregations with partners. Only the strains that express the Gn motif, however, participate in intrageneric coaggregations. The finding that there are only a limited number of structural types of receptor polysaccharide instead of a random assortment is surprising and suggests that the functional similarity of the cognate adhesins on diverse genera serves a critical role in the spatial organization of streptococci in dental plaque (Kolenbrander *et al.*, 1999a).

SPATIAL ARRANGEMENT OF CELLS ON SALIVA-COATED SURFACES

Oral microbial biofilms are formed on the tooth surface within minutes after the surface has been professionally cleaned. A diagrammatic representation of the colonization of the tooth surface is shown in Fig. 1. Streptococci are the primary colonizers of the pellicle-coated surface; actinomyces also colonize in high numbers in early plaque (Nyvad & Kilian, 1987). Both actinomyces and streptococci are capable of binding to proline-rich proteins that are found in the acquired pellicle (Gibbons *et al.*, 1988; Hsu *et al.*, 1994), and thus bind directly to the saliva-coated tooth (Fig. 1, stages 1–4). They also coaggregate with each other to form intergeneric coaggregations as well as intrageneric coaggregations among the streptococci. If left unattended, dental plaque acquires many additional species, especially fusobacteria, which coaggregate with the early colonizers (Fig. 1, stage 5) and late colonizers (Fig. 1, stages 6–8), forming coadhesion bridges between early and late colonizers. Complex community arrangements such as corncob formations occur between fusobacteria and streptococci (Fig. 1, stages 7 and 8) and also between fusobacteria and numerous other partners, including spirochaetes. Such a diagrammatic representation stimulates discussion of what happens in community development on the tooth surface. One approach to

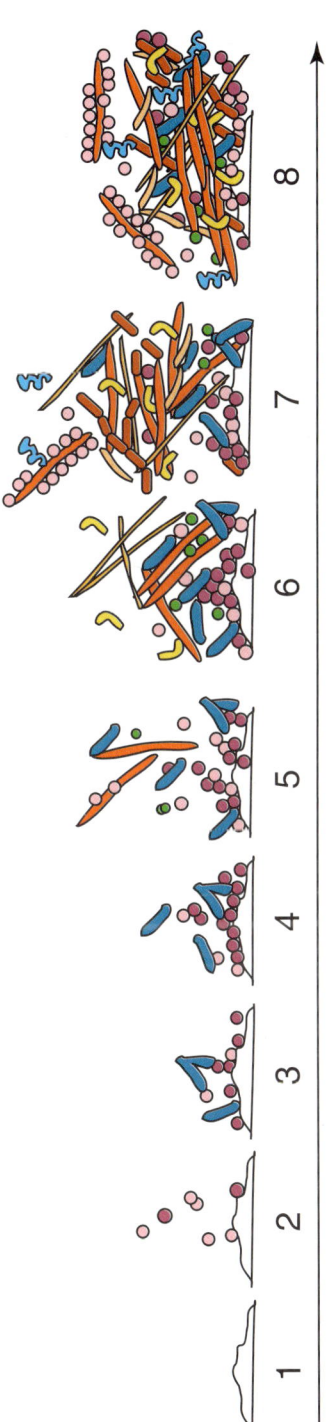

Fig. 1. Diagrammatic representation of biofilm formation on the tooth surface and the potential roles of coaggregation and coadhesion. Colonization is depicted in eight stages in the direction of the arrow. Each stage shows increasingly complex interbacterial associations. Different shapes and colours represent distinct bacterial cell types. The freshly cleaned tooth surface with its pellicle coating (stage 1) provides receptors for the initial colonizers, which are primarily streptococci (two colours of spherical shapes; stage 2) and actinomyces (blue rods; stage 3). The actinomyces may adhere directly to the acquired pellicle (single blue cell; stage 3) or coadhere with already attached streptococci (purple spheres; stage 3). Complementary to the accretion shown in stage 3, stage 4 illustrates another function that contributes to increasing the numbers of cells in biofilms; growth of attached cells is seen with the streptococci dividing and coating the surface. Also shown are the interactions between planktonic cells: (1) intrageneric coaggregation between streptococci; and (2) intergeneric coaggregation between streptococci and actinomyces. The dynamic nature of biofilms is depicted in stage 5 in that detachment of cells (purple and pink spheres) and exposure of the tooth surface is seen along with the appearance of new cell types (small green spheres representing veillonellae and long red rods representing fusobacteria; stage 5). The idea that fusobacteria form coaggregation/coadhesion bridges between early colonizers (purple, pink and green spheres and blue rods) and late colonizers (gold long rods representing capnocytophagae and gold curved rods representing selenomonads) is presented in stages 6 and 7. Stage 7 includes three additional late-colonizing cell types (chains of red short rods representing prevotellae, orange rods representing porphyromonads, and blue spiral-shaped cells representing spirochaetes), all of which coaggregate with fusobacteria. Whereas stage 7 depicts a loose arrangement of cells, the final stage depicts the close association of cells as has been reported for mature dental plaque. The corncob arrangements (long red cells surrounded by pink spherical cells) shown in stages 7 and 8 are at the periphery and are considered to be important in the accretion of mixed cell types onto oral biofilms.

characterize a plaque community is to examine the spatial arrangement of the bacteria *in situ* on removable enamel chips, using probes that identify bacteria *in situ*. Specific probes are being developed and tested *in vitro* in flow chambers with known species of oral bacteria. Cells expressing green fluorescent protein can be used in multi-species communities and can be located by using appropriate detector systems such as confocal scanning laser microscopy (CSLM) (M. Aspiras & P. Kolenbrander, unpublished results). Multi-species communities can be formed *in vitro* and the specificity and sensitivity of the probes can be tested.

Our laboratory has used a saliva-coated glass flow chamber (Palmer & Caldwell, 1995) to study initial adhesion events with two species of *Streptococcus* (Kolenbrander *et al.*, 1999b). The streptococci were suspended in saliva and added to the flowcell and allowed to adhere for 15 min before saliva flow was initiated. *S. gordonii* DL1 and *S. oralis* 34 bound approximately the same number of cells to the surface initially, but a majority of the *S. oralis* 34 cells appeared to detach after 1 h, suggesting weak attachment. *S. gordonii* DL1 bound tightly and formed biofilms of a mean depth of 25 μm. When introduced sequentially in either order and with a 15 min saliva wash between introductions, both streptococcal strains bound very well and formed approximately 20 μm biofilms during a 2 h incubation period (Kolenbrander *et al.*, 2000). Adhesion under these conditions was observed with continuous salivary flow at a rate of 0.2 ml min^{-1}. CSLM of adherent cells was accomplished by staining the cells with a nucleic-acid-binding fluorescent dye or by reacting cells with anti-streptococcal antibody and fluorescent-conjugated anti-antibody reagents. Different fluorescent dyes may be used to react with each streptococcal strain so that the location of each species in the biofilm is possible.

Oral bacteria are bathed in saliva. They are pioneer species when they bind to saliva-coated abiotic surfaces such as freshly cleaned tooth enamel, and they join biofilm communities when they coadhere to already attached bacterial cells. Saliva is a central component of oral bacterial communities. The importance of a saliva coating is shown in Fig. 2, where *A. naeslundii* T14V suspended in a dilute growth medium is placed in a flow chamber without a saliva coating (Fig. 2a) or with a saliva coating (Fig. 2b). *A. naeslundii* T14V expresses type 1 fimbriae that are required for binding to the proline-rich proteins present in saliva (Gibbons *et al.*, 1988). Clearly, the actinomyces bind significantly better to a saliva-coated surface than to an uncoated surface. Thus it seems critical, given the known binding of many oral bacteria to salivary receptors, that saliva be included in investigations of oral biofilm communities.

Although *S. gordonii* DL1 and *A. naeslundii* T14V coaggregate, they have not been tested for coadhesion in a saliva-coated flow chamber. Two antisera specific for each of

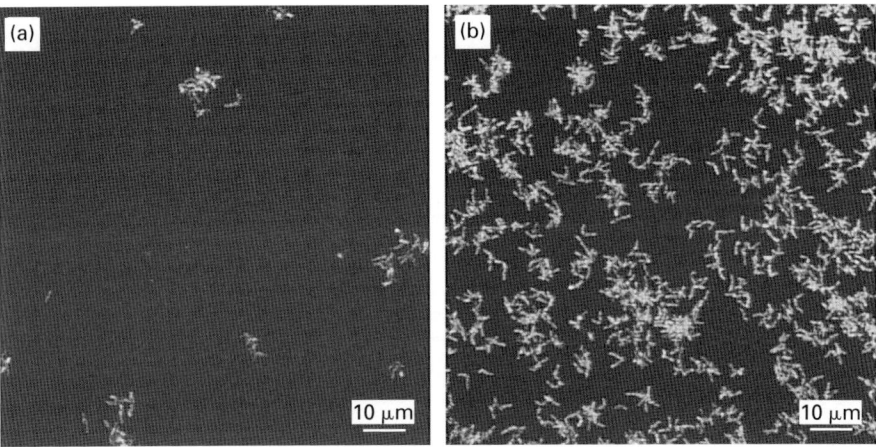

Fig. 2. Representative CSLM images of *Actinomyces naeslundii* T14V adherence to (a) uncoated glass and (b) saliva-coated glass surfaces statically incubated in a flowcell for 15 min before initiation of salivary flow at 0.2 ml min^{-1} for 15 min. Prior to flowcell incubation, *A. naeslundii* T14V was washed and resuspended in growth medium diluted to one-tenth normal concentration. *A. naeslundii* T14V was labelled with SYTO 9 green fluorescent nucleic acid stain (Molecular Probes). Microscopic observations and image acquisition were performed on a CSLM (Leica TCS 4D) with detectors and filters set for fluorescein detection. Extended focus images were generated by CSLM software, and images were processed for display using PhotoShop 5.5 and Illustrator 8.0 (Adobe Systems).

the species were used to show the spatial arrangement of the two cell types (Fig. 3a, b). An apparent close association of the two cell types is seen, suggesting that their ability to coaggregate is also observed in their ability to coadhere. As additional specific probes are developed, more species will be identified unambiguously in multi-species communities formed on the surfaces of saliva-coated flow chambers.

Another type of specific probe besides antisera useful for identifying bacteria in a multi-species community is based on the unique 16S rDNA sequences of bacterial species. A nucleotide probe can be fluorescence-labelled and reacted with cells *in situ*, a technique called fluorescence *in situ* hybridization (FISH). FISH is a powerful tool that provides taxonomic as well as positional data on biofilm organisms. Selectivity to the species level can be obtained using a set of fluorescently labelled rRNA-directed probes. Positional information on selected species can be obtained from natural samples (Moter *et al.*, 1998) as well as from *in vitro* biofilms (Christensen *et al.*, 1999). We wish to combine FISH procedures with the selective immunofluorescence (IF) protocols already in use in our laboratory. From a methodological standpoint, compatibility of IF with FISH could be problematic. To explore this issue, biofilms were constructed that consisted primarily of *S. gordonii* DL1 with small microcolonies of *A. naeslundii* T14V.

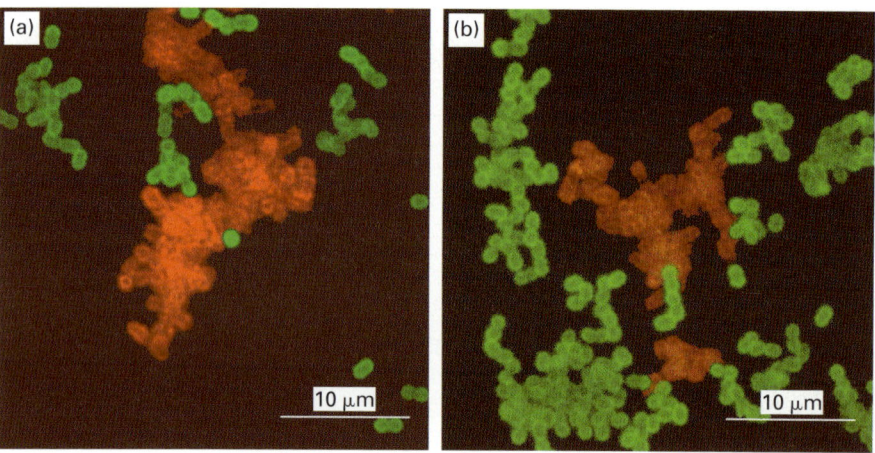

Fig. 3. Representative CSLM images of the mixed-species *Streptococcus gordonii* DL1 and *Actinomyces naeslundii* T14V biofilm at 1 h. *S. gordonii* DL1 was statically incubated in the saliva-coated flowcell for 15 min before initiation of salivary flow at 0.2 ml min^{-1} for 15 min. Subsequently, *A. naeslundii* T14V was added and statically incubated for 15 min followed by salivary flow for an additional 15 min. *S. gordonii* DL1 was labelled with rabbit anti-DL1 antibody, followed by FITC-conjugated (green) goat anti-rabbit antibody (Jackson ImmunoResearch Laboratories). *A. naeslundii* T14V was labelled with a mouse monoclonal antibody against type 1 fimbriae, followed by Cy3-conjugated (red) goat anti-mouse antibody (Jackson ImmunoResearch Laboratories). (a) Coadhesion of a single streptococcal cell (green) (centre) bound to a clump of actinomyces (red) and coadhesion of chains of streptococci to two different lobes of the clump of actinomyces (upper centre). (b) Coadhesion of streptococci and actinomyces (upper centre) and bridging of adherent streptococci by a small clump of actinomyces (lower centre).

These biofilms were stained with IF followed by FISH, or by FISH followed by IF. Indirect IF employed rabbit antiserum raised against whole cells of *S. gordonii* DL1 followed by the fluorescent Cy5-conjugated goat anti-rabbit IgG (for *S. gordonii*), and mouse anti-type-1-fimbriae IgG fraction followed by fluorescent Cy3-conjugated goat anti-mouse IgG (for *A. naeslundii*). FISH employed FluorX-conjugated EUB 338 (rRNA-directed probe hybridizing to all eubacteria; gift from U. B. Göbel, Humboldt University) after formaldehyde fixation and formamide permeabilization. When IF was performed prior to FISH, these procedures resulted in the expected staining pattern (Fig. 4a). However, when IF was performed subsequent to FISH, antibody specificity was greatly reduced (Fig. 4b): *S. gordonii*-directed as well as *A. naeslundii*-directed antibodies bound non-specifically to each cell type. We interpret these results as an indication of degradation of the cell surface epitopes – a consequence of the FISH fixation/hybridization protocol. Epitopes that have bound the antibody prior to FISH seem to be protected (Fig. 4a), and no loss in fluorescence was seen after fixation of IF-treated samples (R. Palmer & P. Kolenbrander, unpublished results). It should be noted

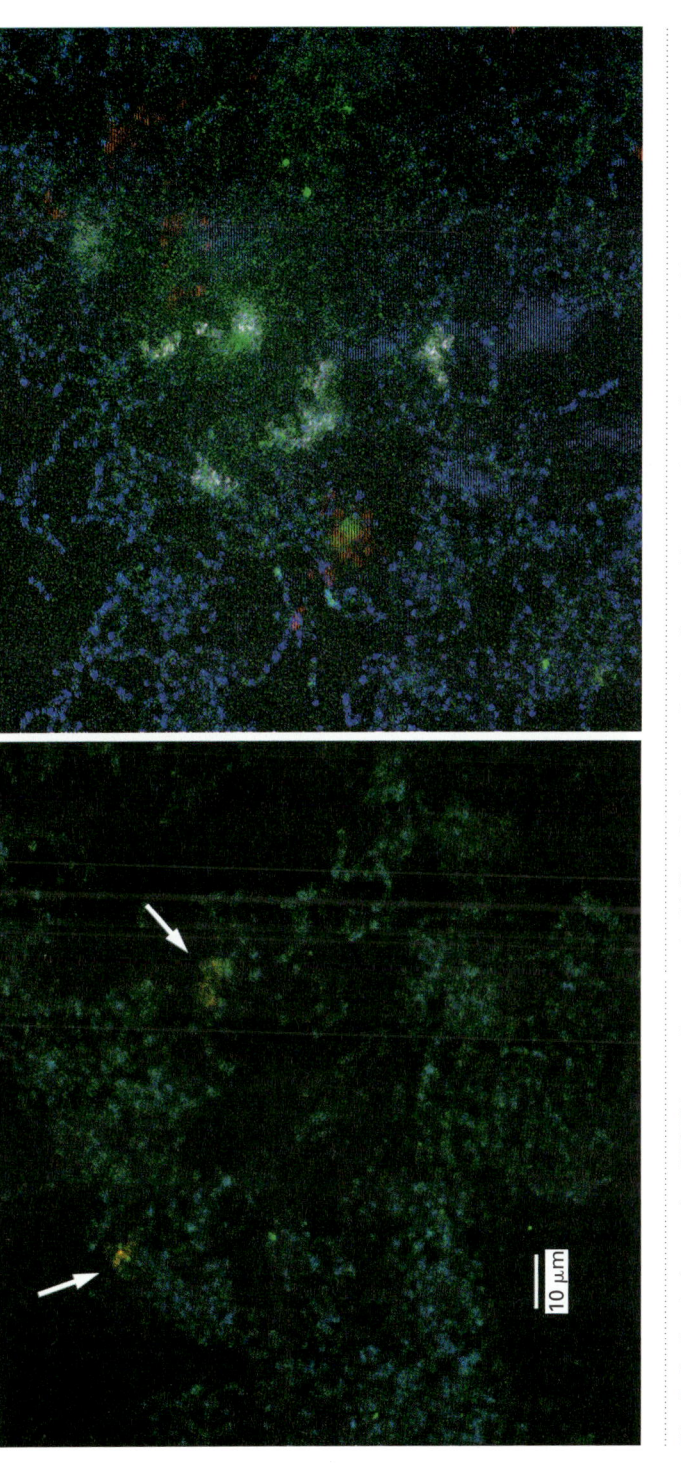

Fig. 4. Confocal micrographs of IF/FISH-treated two-species biofilms. Maximum projection of green (FluorX-conjugated EUB 338), red (Cy3-conjugated anti-*A. naeslundii* antibody) and blue (Cy5-conjugated anti-*S. gordonii* antibody) channels collected simultaneously on a Leica TCS 4D microscope using a ×40 (NA 1.0) oil-immersion lens at an Airy disc setting of 1.0. Biofilms were approximately 8 μm in thickness and were sectioned at 0.5 μm intervals. (a) IF followed by FISH. Yellow regions (arrows) show colocalization of FluorX (green; reacts with all eubacteria) and Cy3 (red; specific for *A. naeslundii*). Blue–green regions show colocalization of FluorX (green) and Cy5 (blue; specific for *S. gordonii*). (b) FISH followed by IF. White pixels (green + red + blue) represent colocalization of all three fluors (FluorX, Cy3 and Cy5). Prominent white areas in the middle of the image are *A. naeslundii* colonies that have stained with the two antibodies (*A. naeslundii*-directed and *S. gordonii*-directed) as well as the FISH probe. Some white pixels are also present in the streptococcal chains, and the green EUB probe is less prominent than in the previous panel, but quite obvious is the purple colour [red + blue; compare with blue–green colour in (a)] that also indicates labelling of the streptococci by the two antibodies. Thus specificity for the two organisms is lost when FISH precedes IF, but IF labelling does not hinder subsequent FISH labelling.

that no dehydration or stabilization of these samples was required because the biofilms were relatively thin (at most 8 μm thick) and contained little extracellular polymeric matrix. The combination of IF and FISH is a powerful tool to discover the spatial arrangement of cells in biofilms, and can be used to investigate the validity of diagrammatic representations like that shown in Fig. 1.

REFERENCES

Andersen, R. N., Ganeshkumar, N. & Kolenbrander, P. E. (1998). *Helicobacter pylori* adheres selectively to *Fusobacterium* spp. *Oral Microbiol Immunol* **13**, 51–54.

Beck, J. D., Koch, G. G., Zambon, J. J., Genco, R. J. & Tudor, G. E. (1992). Evaluation of oral bacteria as risk indicators for periodontitis in older adults. *J Periodontol* **63**, 93–99.

Bolstad, A. I., Jensen, H. B. & Bakken, V. (1996). Taxonomy, biology, and periodontal aspects of *Fusobacterium nucleatum*. *Clin Microbiol Rev* **9**, 55–71.

Bos, R., van der Mei, H. C., Meinders, J. M. & Busscher, H. J. (1994). A quantitative method to study co-adhesion of microorganisms in a parallel plate flow chamber: basic principles of the analysis. *J Microbiol Methods* **20**, 289–305.

Bos, R., van der Mei, H. C. & Busscher, H. J. (1998). On the role of co-aggregation and co-adhesion in dental plaque formation. In *Oral Biofilms and Plaque Control*, pp. 163–173. Edited by H. J. Busscher & L. V. Evans. Amsterdam: Harwood Academic Publishers.

Bowden, G. H. & Li, Y. H. (1997). Nutritional influences on biofilm development. *Adv Dent Res* **11**, 81–99.

Bowden, G. H. W., Nolette, N., Ryding, H. & Cleghorn, B. M. (1999). The diversity and distribution of the predominant ribotypes of *Actinomyces naeslundii* genospecies 1 and 2 in samples from enamel and from healthy and carious root surfaces of teeth. *J Dent Res* **78**, 1800–1809.

Brooks, W., Demuth, D. R., Gil, S. & Lamont, R. J. (1997). Identification of a *Streptococcus gordonii* SspB domain that mediates adhesion to *Porphyromonas gingivalis*. *Infect Immun* **65**, 3753–3758.

Busscher, H. J. & Evans, L. V. (1998). *Oral Biofilms and Plaque Control*. Amsterdam: Harwood Academic Publishers.

Buswell, C. M., Herlihy, Y. M., Marsh, P. D., Keevil, C. W. & Leach, S. A. (1997). Coaggregation amongst aquatic biofilm bacteria. *J Appl Microbiol* **83**, 477–484.

Chatenay-Rivauday, C., Yamodo, I., Sciotti, M. A., Ogier, J. A. & Klein, J. P. (1998). The A and the extended V N-terminal regions of streptococcal protein I/IIf mediate the production of tumour necrosis factor alpha in the monocyte cell line THP-1. *Mol Microbiol* **29**, 39–48.

Christensen, B. B., Sternberg, C., Andersen, J. B., Palmer, R. J., Jr, Nielsen, A. T., Givskov, M. & Molin, S. (1999). Molecular tools for study of biofilm physiology. *Methods Enzymol* **310**, 20–42.

Cisar, J. O., Kolenbrander, P. E. & McIntire, F. C. (1979). Specificity of coaggregation reactions between human oral streptococci and strains of *Actinomyces viscosus* or *Actinomyces naeslundii*. *Infect Immun* **24**, 742–752.

Cisar, J. O., Sandberg, A. L., Abeygunawardana, C., Reddy, G. P. & Bush, C. A. (1995). Lectin recognition of host-like saccharide motifs in streptococcal cell wall polysaccharides. *Glycobiology* **5**, 655–662.

Cisar, J. O., Sandberg, A. L., Reddy, G. P., Abeygunawardana, C. & Bush, C. A. (1997). Structural and antigenic types of cell wall polysaccharides from viridans group streptococci with receptors for actinomyces and streptococcal lectins. *Infect Immun* **65**, 5035–5041.

Clemans, D. L. & Kolenbrander, P. E. (1995a). Identification of a 100-kilodalton putative coaggregation-mediating adhesin of *Streptococcus gordonii* DL1 (Challis). *Infect Immun* **63**, 4890–4893.

Clemans, D. L. & Kolenbrander, P. E. (1995b). Isolation and characterization of coaggregation-defective (Cog-) mutants of *Streptococcus gordonii* DL1 (Challis). *J Ind Microbiol* **15**, 193–197.

Clemans, D. L., Kolenbrander, P. E., Debabov, D. V., Zhang, Q., Lunsford, R. D., Sakone, H., Whittaker, C. J., Heaton, M. P. & Neuhaus, F. C. (1999). Insertional inactivation of genes responsible for the D-alanylation of lipoteichoic acid in *Streptococcus gordonii* DL1 (Challis) affects intrageneric coaggregations. *Infect Immun* **67**, 2464–2474.

Debabov, D. V., Heaton, M. P., Zhang, Q., Stewart, K. D., Lambalot, R. H. & Neuhaus, F. C. (1996). The D-alanyl carrier protein in *Lactobacillus casei*: cloning, sequencing and expression of *dltC*. *J Bacteriol* **178**, 3869–3876.

Demuth, D. R., Duan, Y., Brooks, W., Holmes, A. R., McNab, R. & Jenkinson, H. F. (1996). Tandem genes encode cell-surface polypeptides SspA and SspB which mediate adhesion of the oral bacterium *Streptococcus gordonii* to human and bacterial receptors. *Mol Microbiol* **20**, 403–413.

Duan, Y., Fisher, E., Malamud, D., Golub, E. & Demuth, D. R. (1994). Calcium-binding properties of SSP-5, the *Streptococcus gordonii* M5 receptor for salivary agglutinin. *Infect Immun* **62**, 5220–5226.

Ganeshkumar, N., Hughes, C. V. & Weiss, E. I. (1998). Co-aggregation in dental plaque formation. In *Oral Biofilms and Plaque Control*, pp. 125–143. Edited by H. J. Busscher & L. V. Evans. Amsterdam: Harwood Academic Publishers.

Gibbons, R. J. & Nygaard, M. (1970). Interbacterial aggregation of plaque bacteria. *Arch Oral Biol* **15**, 1397–1400.

Gibbons, R. J., Hay, D. I., Cisar, J. O. & Clark, W. B. (1988). Adsorbed salivary proline-rich protein 1 and statherin: receptors for type 1 fimbriae of *Actinomyces viscosus* T14V-J1 on apatitic surfaces. *Infect Immun* **56**, 2990–2993.

Grimaudo, N. J. & Nesbitt, W. E. (1997). Coaggregation of *Candida albicans* with oral *Fusobacterium* species. *Oral Microbiol Immunol* **12**, 168–173.

Grimaudo, N. J., Nesbitt, W. & Clark, W. (1996). Coaggregation of *Candida albicans* with oral *Actinomyces* species. *Oral Microbiol Immunol* **11**, 59–61.

Hallberg, K., Holm, C., Hammarström, K.-J., Kalfas, S. & Strömberg, N. (1998a). Ribotype diversity of *Actinomyces* with similar intraoral tropism but different type of N-acetyl-β-D-galactosamine binding specificity. *Oral Microbiol Immunol* **13**, 188–192.

Hallberg, K., Holm, C., Ohman, U. & Strömberg, N. (1998b). *Actinomyces naeslundii* displays variant *fimP* and *fimA* fimbrial subunit genes corresponding to different types of acidic proline-rich protein and beta-linked galactosamine binding specificity. *Infect Immun* **66**, 4403–4410.

Håvarstein, L. S., Gaustad, P., Nes, I. F. & Morrison, D. A. (1996). Identification of the streptococcal competence-pheromone receptor. *Mol Microbiol* **21**, 863–869.

Håvarstein, L. S., Hakenbeck, R. & Gaustad, P. (1997). Natural competence in the genus *Streptococcus*: evidence that streptococci can change phenotype by interspecies recombinational exchanges. *J Bacteriol* **179**, 6589–6594.

Heaton, M. P. & Neuhaus, F. C. (1992). Biosynthesis of D-alanyl lipoteichoic acid: cloning, nucleotide sequence, and gene expression of the *Lactobacillus casei* gene for D-alanine activating enzyme. *J Bacteriol* **174**, 4707–4717.

Heaton, M. P. & Neuhaus, F. C. (1994). Role of D-alanyl carrier protein in the biosynthesis of D-alanyl-lipoteichoic acid. *J Bacteriol* **176**, 681–690.

Holmes, A. R., Gopal, P. K. & Jenkinson, H. F. (1995). Adherence of *Candida albicans* to a cell surface polysaccharide receptor on *Streptococcus gordonii*. *Infect Immun* **63**, 1827–1834.

Holmes, A. R., McNab, R. & Jenkinson, H. F. (1996). *Candida albicans* binding to the oral bacterium *Streptococcus gordonii* involves multiple adhesin-receptor interactions. *Infect Immun* **64**, 4680–4685.

Holmes, A. R., Gilbert, C., Wells, J. M. & Jenkinson, H. F. (1998). Binding properties of *Streptococcus gordonii* SspA and SspB (antigen I/II family) polypeptides expressed on the cell surface of *Lactococcus lactis* MG1363. *Infect Immun* **66**, 4633–4639.

Hsu, S. D., Cisar, J. O., Sandberg, A. L. & Kilian, M. (1994). Adhesive properties of viridans streptococcal species. *Microb Ecol Health Dis* **7**, 125–137.

Hughes, C. V., Andersen, R. N. & Kolenbrander, P. E. (1992). Characterization of *Veillonella atypica* PK1910 adhesin-mediated coaggregation with oral *Streptococcus* spp. *Infect Immun* **60**, 1178–1186.

Hultgren, S. J., Lindberg, F., Magnusson, G., Kihlberg, J., Tennent, J. M. & Normark, S. (1989). The PapG adhesin of uropathogenic *Escherichia coli* contains separate regions for receptor binding and for the incorporation into the pilus. *Proc Natl Acad Sci USA* **86**, 4357–4361.

Jabra-Rizk, M. A., Falkler, W. A., Jr, Merz, W. G., Kelley, J. I., Baqui, A. A. M. A. & Meiller, T. F. (1999). Coaggregation of *Candida dubliniensis* with *Fusobacterium nucleatum*. *J Clin Microbiol* **37**, 1464–1468.

Jenkinson, H. F. & Demuth, D. R. (1997). Structure, function and immunogenicity of streptococcal antigen I/II polypeptides. *Mol Microbiol* **23**, 183–190.

Jenkinson, H. F. & Lamont, R. J. (1997). Streptococcal adhesion and colonization. *Crit Rev Oral Biol Med* **8**, 175–200.

Jones, C. H., Pinkner, J. S., Roth, R., Heuser, J., Nicholes, A. V., Abraham, S. N. & Hultgren, S. J. (1995). FimH adhesin of type 1 pili is assembled into a fibrillar tip structure in the *Enterobacteriaceae*. *Proc Natl Acad Sci USA* **92**, 2081–2085.

Kilian, M., Mikkelsen, L. & Henrichsen, J. (1989). Taxonomic study of viridans streptococci: description of *Streptococcus gordonii* sp. nov. and emended descriptions of *Streptococcus sanguis* (White and Niven, 1946), *Streptococcus oralis* (Bridge and Sneath, 1982), and *Streptococcus mitis* (Andrewes and Horder, 1906). *Int J Syst Bacteriol* **39**, 471–484.

Klier, C. M., Kolenbrander, P. E., Roble, A. G., Marco, M. L., Cross, S. & Handley, P. S. (1997). Identification of a 95 kDa putative adhesin from *Actinomyces* serovar WVA963 strain PK1259 that is distinct from type 2 fimbrial subunits. *Microbiology* **143**, 835–846.

Klier, C. M., Roble, A. G. & Kolenbrander, P. E. (1998). *Actinomyces* serovar WVA963 coaggregation-defective mutant strain PK2407 secretes lactose-sensitive adhesin that binds to coaggregation partner *Streptococcus oralis* 34. *Oral Microbiol Immunol* **13**, 337–340.

Kolenbrander, P. E. (1988). Intergeneric coaggregation among human oral bacteria and ecology of dental plaque. *Annu Rev Microbiol* **42**, 627–656.

Kolenbrander, P. E. (1997). Oral microbiology and coaggregation. In *Bacteria as*

Multicellular Organisms, pp. 245–269. Edited by J. A. Shapiro & M. Dworkin. New York: Oxford University Press.

Kolenbrander, P. E. (2000). Oral microbial communities: biofilms, interactions, and genetic systems. *Annu Rev Microbiol* **54**, (in press).

Kolenbrander, P. E. & Andersen, R. N. (1986). Multigeneric aggregations among oral bacteria: a network of independent cell-to-cell interactions. *J Bacteriol* **168**, 851–859.

Kolenbrander, P. E. & Andersen, R. N. (1989). Inhibition of coaggregation between *Fusobacterium nucleatum* and *Porphyromonas* (*Bacteroides*) *gingivalis* by lactose and related sugars. *Infect Immun* **57**, 3204–3209.

Kolenbrander, P. E. & London, J. (1993). Adhere today, here tomorrow: oral bacterial adherence. *J Bacteriol* **175**, 3247–3252.

Kolenbrander, P. E., Andersen, R. N. & Moore, L. V. H. (1989). Coaggregation of *Fusobacterium nucleatum*, *Selenomonas flueggei*, *Selenomonas infelix*, *Selenomonas noxia*, and *Selenomonas sputigena* with strains from 11 genera of oral bacteria. *Infect Immun* **57**, 3194–3203.

Kolenbrander, P. E., Andersen, R. N. & Moore, L. V. H. (1990). Intrageneric coaggregation among strains of human oral bacteria: potential role in primary colonization of the tooth surface. *Appl Environ Microbiol* **56**, 3890–3894.

Kolenbrander, P. E., Parrish, K. D., Andersen, R. N. & Greenberg, E. P. (1995). Intergeneric coaggregation of oral *Treponema* spp. with *Fusobacterium* spp. and intrageneric coaggregation among *Fusobacterium* spp. *Infect Immun* **63**, 4584–4588.

Kolenbrander, P. E., Andersen, R. N., Baker, R. A. & Jenkinson, H. F. (1998). The adhesion-associated *sca* operon in *Streptococcus gordonii* encodes an inducible high-affinity ABC transporter for $Mn2+$ uptake. *J Bacteriol* **180**, 290–295.

Kolenbrander, P. E., Andersen, R. N., Clemans, D. L., Whittaker, C. J. & Klier, C. M. (1999a). Potential role of functionally similar coaggregation mediators in bacterial succession. In *Dental Plaque Revisited: Oral Biofilms in Health and Disease*, pp. 171–186. Edited by H. N. Newman & M. Wilson. Cardiff: BioLine.

Kolenbrander, P. E., Andersen, R. N., Kazmerzak, K. M., Wu, R. & Palmer, R. J., Jr (1999b). Spatial organization of oral bacteria in biofilms. *Methods Enzymol* **310**, 322–332.

Kolenbrander, P. E., Dü, L., Aspiras, M., Li, S., Kazmerzak, K., Wu, R. & Andersen, R. (2000). Spatial organization and contact-induced gene expression in biofilms of *Streptococcus gordonii* Challis. In *Streptococci and Streptococcal Diseases – Entering the New Millennium*. Edited by D. Martin & J. Tagg (in press).

Lamont, R. J. & Rosan, B. (1990). Adherence of mutans streptococci to other oral bacteria. *Infect Immun* **58**, 1738–1743.

Lancy, P., Jr, Appelbaum, B., Holt, S. C. & Rosan, B. (1980). Quantitative in vitro assay for "corncob" formation. *Infect Immun* **29**, 663–670.

Li, T., Johansson, I., Hay, D. I. & Strömberg, N. (1999). Strains of *Actinomyces naeslundii* and *Actinomyces viscosus* exhibit structurally variant fimbrial subunit proteins and bind to different peptide motifs in salivary proteins. *Infect Immun* **67**, 2053–2059.

Love, R. M., McMillan, M. D. & Jenkinson, H. F. (1997). Invasion of dentinal tubules by oral streptococci is associated with collagen recognition mediated by the antigen I/II family of polypeptides. *Infect Immun* **65**, 5157–5164.

McIntire, F. C., Vatter, A. E., Baros, J. & Arnold, J. (1978). Mechanism of coaggregation between *Actinomyces viscosus* T14V and *Streptococcus sanguis* 34. *Infect Immun* **21**, 978–988.

McNab, R. & Jenkinson, H. F. (1998). Altered adherence properties of a *Streptococcus gordonii hppA* (oligopeptide permease) mutant result from transcriptional effects on *cshA* adhesin gene expression. *Microbiology* **144**, 127–136.

McNab, R., Jenkinson, H. F., Loach, D. M. & Tannock, G. W. (1994). Cell-surface-associated polypeptides CshA and CshB of high molecular mass are colonization determinants in the oral bacterium *Streptococcus gordonii*. *Mol Microbiol* **14**, 743–754.

McNab, R., Holmes, A. R., Clarke, J. M., Tannock, G. W. & Jenkinson, H. F. (1996). Cell surface polypeptide CshA mediates binding of *Streptococcus gordonii* to other oral bacteria and to immobilized fibronectin. *Infect Immun* **64**, 4204–4210.

McNab, R., Forbes, H., Handley, P. S., Loach, D. M., Tannock, G. W. & Jenkinson, H. F. (1999). Cell wall-anchored CshA polypeptide (259 kilodaltons) in *Streptococcus gordonii* forms surface fibrils that confer hydrophobic and adhesive properties. *J Bacteriol* **181**, 3087–3095.

Millsap, K. W. (1999). *Adhesive interactions between yeasts and bacteria on biomaterials surfaces*. PhD thesis, University of Groningen.

Millsap, K. W., Bos, R., Busscher, H. J. & van der Mei, H. C. (1999). Surface aggregation of *Candida albicans* on glass in the absence and presence of adhering *Streptococcus gordonii* in a parallel-plate flow chamber: a surface thermodynamical analysis based on acid-base interactions. *J Colloid Interface Sci* **212**, 495–502.

Moore, W. E. C. & Moore, L. V. H. (1994). The bacteria of periodontal diseases. *Periodontol 2000* **5**, 66–77.

Moter, A., Hoenig, C., Choi, B.-K., Riep, B. & Göbel, U. B. (1998). Molecular epidemiology of oral treponemes associated with periodontal disease. *J Clin Microbiol* **36**, 1399–1403.

Nesbitt, W. E., Beem, J. E., Leung, K. P., Stroup, S., Swift, R., McArthur, W. P. & Clark, W. B. (1996). Inhibition of adherence of *Actinomyces naeslundii* (*Actinomyces viscosus*) T14V-J1 to saliva-treated hydroxyapatite by a monoclonal antibody to type 1 fimbriae. *Oral Microbiol Immunol* **11**, 51–58.

Nyvad, B. & Kilian, M. (1987). Microbiology of the early colonization of human enamel and root surfaces in vivo. *Scand J Dent Res* **95**, 369–380.

Palmer, R. J., Jr & Caldwell, D. E. (1995). A flowcell for the study of plaque removal and regrowth. *J Microbiol Methods* **24**, 171–182.

Reid, G., McGroarty, J. A., Angotti, R. & Cook, R. L. (1988). *Lactobacillus* inhibitor production against *Escherichia coli* and coaggregation ability with uropathogens. *Can J Microbiol* **34**, 344–351.

Rickard, A. H., Leach, S. A., Buswell, C. M., High, N. J. & Handley, P. S. (2000). Coaggregation between aquatic bacteria is mediated by specific-growth-phase-dependent lectin-saccharide interactions. *Appl Environ Microbiol* **66**, 431–434.

Rogers, J. D., Haase, E. M., Brown, A. E., Douglas, C. W., Gwynn, J. P. & Scannapieco, F. A. (1998). Identification and analysis of a gene (*abpA*) encoding a major amylase-binding protein in *Streptococcus gordonii*. *Microbiology* **144**, 1223–1233.

Rosenow, C., Ryan, P., Weiser, J. N., Johnson, S., Fontan, P., Ortquist, A. & Masure, H. R. (1997). Contribution of novel choline-binding proteins to adherence, colonization and immunogenicity of *Streptococcus pneumoniae*. *Mol Microbiol* **25**, 819–829.

Scannapieco, F. A. (1994). Saliva-bacterium interactions in oral microbial ecology. *Crit Rev Oral Biol Med* **5**, 203–248.

Sciotti, M. A., Yamodo, I., Klein, J. P. & Ogier, J. A. (1997). The N-terminal half part of the oral streptococcal antigen I/IIf contains two distinct binding domains. *FEMS Microbiol Lett* **153**, 439–445.

Shaniztki, B., Hurwitz, D., Smoradinsky, N., Ganeshkumar, N. & Weiss, E. I. (1997). Identification of a *Fusobacterium nucleatum* PK1594 galactose-binding adhesin which mediates coaggregation with periopathogenic bacteria and hemagglutination. *Infect Immun* **65**, 5231–5237.

Shaniztki, B., Ganeshkumar, N. & Weiss, E. I. (1998). Characterization of a novel *N*-acetylneuraminic acid-specific *Fusobacterium nucleatum* PK1594 adhesin. *Oral Microbiol Immunol* **13**, 47–50.

Socransky, S. S., Haffajee, A. D., Cugini, M. A., Smith, C. & Kent, R. L., Jr (1998). Microbial complexes in subgingival plaque. *J Clin Periodontol* **25**, 134–144.

Takahashi, Y., Sandberg, A. L., Ruhl, S., Muller, J. & Cisar, J. O. (1997). A specific cell surface antigen of *Streptococcus gordonii* is associated with bacterial hemagglutination and adhesion to alpha2-3-linked sialic acid-containing receptors. *Infect Immun* **65**, 5042–5051.

Takemoto, T., Hino, T., Yoshida, M., Nakanishi, K., Shirakawa, M. & Okamoto, H. (1995). Characteristics of multimodal co-aggregation between *Fusobacterium nucleatum* and streptococci. *J Periodontal Res* **30**, 252–257.

Vandevoorde, L., Christiaens, H. & Verstraete, W. (1992). Prevalence of coaggregation reactions among chicken lactobacilli. *J Appl Bacteriol* **72**, 214–219.

Weiss, E. I., London, J., Kolenbrander, P. E., Hand, A. R. & Siraganian, R. (1988). Localization and enumeration of fimbria-associated adhesins of *Bacteroides loescheii*. *J Bacteriol* **170**, 1123–1128.

Whittaker, C. J., Clemans, D. L. & Kolenbrander, P. E. (1996a). Insertional inactivation of an intrageneric coaggregation-relevant adhesin locus from *Streptococcus gordonii* DL1 (Challis). *Infect Immun* **64**, 4137–4142.

Whittaker, C. J., Klier, C. M. & Kolenbrander, P. E. (1996b). Mechanisms of adhesion by oral bacteria. *Annu Rev Microbiol* **50**, 513–552.

Wu, H., Mintz, K. P., Ladha, M. & Fives-Taylor, P. M. (1998). Isolation and characterization of Fap1, a fimbriae-associated adhesin of *Streptococcus parasanguis* FW213. *Mol Microbiol* **28**, 487–500.

Yamaguchi, T., Kasamo, K., Chuman, M., Machigashira, M., Inoue, M. & Sueda, T. (1998). Preparation and characterization of an *Actinomyces naeslundii* aggregation factor that mediates coaggregation with *Porphyromonas gingivalis*. *J Periodontal Res* **33**, 460–468.

Yeung, M. K. (1992). Conservation of an *Actinomyces viscosus* T14V type 1 fimbrial subunit homolog among divergent groups of *Actinomyces* spp. *Infect Immun* **60**, 1047–1054.

Yeung, M. K. & Ragsdale, P. A. (1997). Synthesis and function of *Actinomyces naeslundii* T14V type 1 fimbriae require the expression of additional fimbria-associated genes. *Infect Immun* **65**, 2629–2639.

Yeung, M. K., Chassy, B. M. & Cisar, J. O. (1987). Cloning and expression of a type 1 fimbrial subunit of *Actinomyces viscosus* T14V. *J Bacteriol* **169**, 1678–1683.

Yother, J. & White, J. M. (1994). Novel surface attachment mechanism of the *Streptococcus pneumoniae* protein PspA. *J Bacteriol* **176**, 2976–2985.

Cohesiveness in biofilm matrix polymers

Hans-Curt Flemming,[1] Jost Wingender,[1] Christian Mayer,[2] Volker Körstgens[2] and Werner Borchard[2]

[1] Department of Aquatic Microbiology, University of Duisburg, Germany
[2] Institute for Physical and Theoretical Chemistry, University of Duisburg, Germany

MICROBIAL AGGREGATES

The vast majority of micro-organisms live and grow in aggregated forms such as biofilms, flocs ('planktonic biofilms') and sludges. This form of growth is lumped in the somewhat inexact but generally accepted expression 'biofilm'. The feature which is common to all these phenomena is that the micro-organisms are embedded in a matrix of extracellular polymeric substances (EPS) which are responsible for morphology, structure, coherence, physico-chemical properties and activity of these aggregates (Wingender & Flemming, 1999). Biofilms are ubiquitously distributed in natural soil and aquatic environments, on tissues of plants, animals and man, as well as in technical systems such as filters and other porous materials, reservoirs, pipelines, ship hulls, heat exchangers, separation membranes, etc. (Costerton *et al.*, 1987; Flemming & Schaule, 1996); biofilms may also develop on medical devices, thus initiating persistent infections in humans (Costerton *et al.*, 1999). Biofilms develop adherent to a solid surface (substratum) at solid–water interfaces, but can also be found at water–air and at solid–air interfaces. They are accumulations of micro-organisms (prokaryotic and eukaryotic unicellular organisms), EPS, multivalent cations, inorganic particles, biogenic material (detritus) as well as colloidal and dissolved compounds. EPS are considered as the key components that determine the structural and functional integrity of microbial aggregates. EPS form a three-dimensional, gel-like, highly hydrated and locally charged biofilm matrix, in which the micro-organisms are more or less immobilized. EPS create a microenvironment for sessile cells which is conditioned by the nature of the EPS matrix. In general, the proportion of EPS in biofilms can vary between roughly 50 and 90% of the total organic matter (Christensen & Characklis,

SGM symposium 59: Community structure and co-operation in biofilms. Editors D. Allison, P. Gilbert, H. Lappin-Scott, M. Wilson. Cambridge University Press. ISBN 0 521 79302 5 ©SGM 2000.

1990; Nielsen *et al.*, 1997). In activated sludge and sewer biofilms, 85–90% and 70–98%, respectively, of total organic carbon was found to be extracellular, indicating that cell biomass may constitute only a minor fraction of the organic matter of microbial aggregates in wastewater environments (Frølund *et al.*, 1996; Jahn & Nielsen, 1998). EPS are attributed to play an essential role in wastewater treatment processes, since they are involved in floc formation of activated sludge and in the performance of fixed biofilms in reactors with attached microbial biomass, e.g. in trickling filters, rotating biological contactors, fluidized-bed reactors or submerged fixed-bed reactors (Bitton, 1994); in addition, EPS strongly influence the dewaterability of wastewater sludges (Poxon & Darby, 1997).

EXTRACELLULAR POLYMERIC SUBSTANCES AND THEIR ROLE AS MICROBIAL AGGREGATE CONSTRUCTION MATERIAL

Extracellular polymeric materials are of fundamental importance for microbial aggregates as they provide the cohesiveness in biofilms, flocs and sludge. The production of EPS is a general property of micro-organisms in most environments. The ability to form EPS is widespread among prokaryotic organisms (*Bacteria*, *Archaea*), but has also been shown to occur in eukaryotic (algae, fungi) micro-organisms (Cooksey, 1992; Sutherland, 1996). Microbial EPS are biosynthetic polymers (biopolymers). Geesey (1982) defined EPS as 'extracellular polymeric substances of biological origin that participate in the formation of microbial aggregates'. Another definition was given in a glossary to the report of the Dahlem Workshop on Structure and Function of Biofilms in Berlin, 1988 (Characklis & Wilderer, 1989); here, EPS were defined as 'organic polymers of microbial origin which in biofilm systems are frequently responsible for immobilizing cells and other particulate materials together (cohesion) and to the substratum (adhesion)'. In the context of a critical evaluation of EPS isolation techniques, Gehr & Henry (1983) described extracellular material as 'that material which can be removed from micro-organisms (and in particular, bacteria) without disrupting the cell, and without which the micro-organism is still viable'. This definition alludes to the observation that EPS are not essential structures of bacteria, since loss of EPS does not impair growth and viability of cells in laboratory cultures. Under natural conditions, however, EPS production seems to be an important feature of survival, as most environmental bacteria occur in microbial aggregates such as flocs and biofilms, whose structural and functional integrity is based essentially on the presence of an EPS matrix.

The acronym 'EPS' has been used specifically for 'extracellular polysaccharides' or 'exopolysaccharides' or as a collective term for 'exopolymeric substances', 'exopolymers' and 'extracellular polymeric substances'. Polysaccharides have often been assumed to be the most abundant components of EPS in early biofilm research (e.g.

Costerton *et al.*, 1981). That may be the reason why the term EPS has so frequently been used as an abbreviation for 'extracellular polysaccharides' or 'exopolysaccharides'. However, other macromolecules can also appear in significant amounts or even predominate in the EPS. In the following text, 'EPS' is used for 'extracellular polymeric substances' as a more general and comprehensive term for different classes of organic macromolecules, such as polysaccharides, proteins, nucleic acids, lipids/phospholipids or humic substances (Table 1), which have been described to occur in the intercellular spaces of microbial aggregates. The structure of EPS varies quite obviously. Fig. 1(a, b) shows scanning electron micrographs of biofouling layers on various reverse osmosis membranes. The supporting membrane materials and the preparation procedures were identical; both membranes were exposed to river water, but from different origins. In Fig. 1(a), the cells are embedded in a thick slime matrix while in Fig. 1(b), fibrillar structures are dominant. Although it is acknowledged that the dewatering procedure required for scanning electron microscopy imaging produces artefacts, morphological differences are still obvious and are attributed to the nature of the EPS.

Composition and properties of EPS

EPS consist of varying proportions of carbohydrates, proteins, nucleic acids, lipids/phospholipids and humic acids. Thus EPS contain a large variety of chemical structures. In many earlier studies, different types of species-specific and non-specific polysaccharides were considered to be the main constituents of EPS, so that isolation and purification procedures were often focused on the carbohydrate fraction of the EPS. However, when more extensive analyses of EPS were performed, proteins were frequently shown to be abundant in the EPS from pure cultures of Gram-negative and Gram-positive bacteria as well as from biofilms and flocs (see examples given by Wingender & Flemming, 1999). A number of publications have shown that proteins were even predominant over polysaccharides and represented the largest fraction in the EPS of biofilms and activated sludge from wastewater systems (Rudd *et al.*, 1983; Nielsen *et al.*, 1997; Bura *et al.*, 1998; Dignac *et al.*, 1998; Jorand *et al.*, 1998). Among the nucleic acids, DNA has regularly been found in the EPS from wastewater biofilms and flocs (e.g. Nielsen *et al.*, 1997) and also in the extracellular material from pure cultures (Platt *et al.*, 1985; Arvaniti *et al.*, 1994; Jahn & Nielsen, 1995; Watanabe *et al.*, 1998). The content of nucleic acids may even exceed that of proteins and polysaccharides, as has been shown for the EPS from the self-flocculating photosynthetic bacterium *Rhodovulum* sp. (Watanabe *et al.*, 1998, 1999): RNA was the major constituent, whereas DNA only appeared in minor quantities. Other macromolecular or oligomer components of EPS may be phospholipids (Gehrke *et al.*, 1998; Sand & Gehrke, 1999) and lipids (Goodwin & Forster, 1985). In addition, the accumulation of humic substances in the EPS matrix of wastewater biofilms and activated sludge seems to be common (Nielsen *et al.*, 1997; Jahn & Nielsen, 1998).

Table 1. General composition of some bacterial EPS (after Wingender et al., 1999)

EPS	Principal components (subunits, precursors)	Main type of linkage between subunits	Structure of polymer backbone	Substituents (examples)
Polysaccharides	Monosaccharides Uronic acids Amino sugars	Glycosidic bonds	Linear, branched, side chains	Organic: O-acetyl, N-acetyl, succinyl, pyruvyl Inorganic: sulfate, phosphate
Proteins (polypeptides)	Amino acids	Peptide bonds	Linear	Oligosaccharides (glycoproteins), fatty acids (lipoproteins)
Nucleic acids	Nucleotides	Phosphodiester bonds	Linear	—
(phospho)Lipids	Fatty acids Glycerol Phosphate Alcoholic compounds Isoprene	Ester bonds, ether bonds	Side chains	—
Humic substances	Phenolic compounds Simple sugars Amino acids	Ether bonds, C–C bonds, peptide bonds	Cross-linked	—

(a)

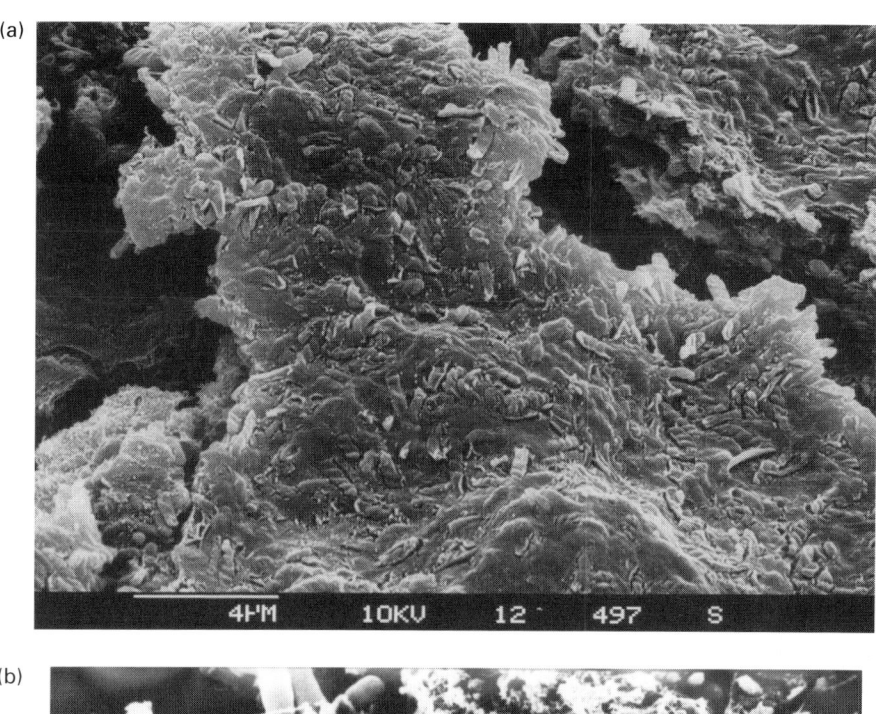

(b)

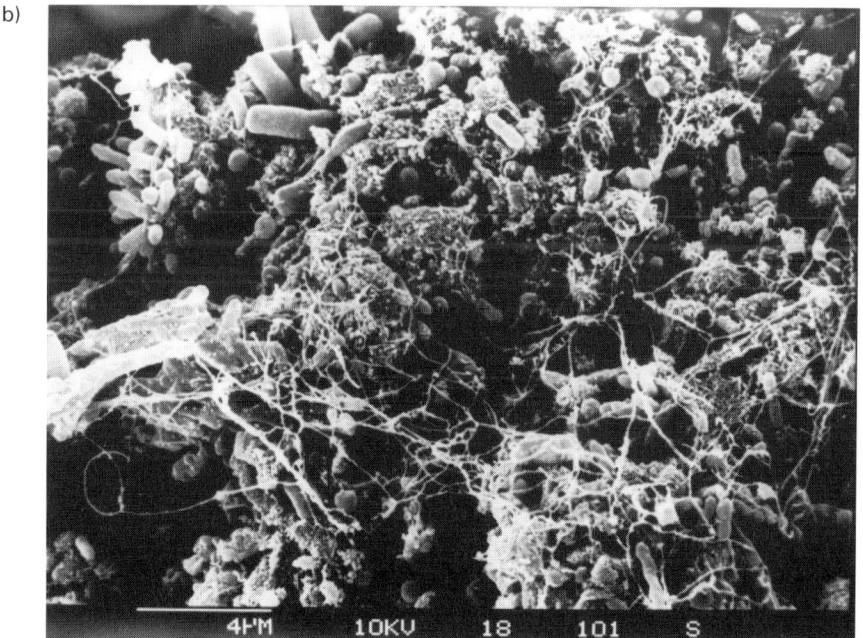

Fig. 1. Micro-organisms embedded in EPS, forming fouling layers on reverse osmosis membranes. (a) Confluent EPS matrix on membrane exposed to River Rhine water (reproduced from Flemming & Schaule, 1989, with permission); (b) filamentous EPS matrix on membrane exposed to River Seine water (courtesy of G. Schaule).

EPS only rarely consist of neutral molecules, such as, for example, the polysaccharides dextran and levan; more frequently, EPS represent polyelectrolytes with ionic groups which confer net negative or positive charges on the polymers, depending on the degree of dissociation. Anionic groups in EPS may be carboxyl, sulfate or phosphate groups. Carboxyl groups are mainly responsible for the negative charge of polysaccharides in the EPS. In addition to neutral sugars, extracellular polysaccharides often contain variable proportions of hexuronic acids such as glucuronic acid, galacturonic acid or mannuronic acid (Fazio *et al.*, 1982; Uhlinger & White, 1983). Exceptions are polysaccharides which completely consist of uronic acid residues, an example being the well-studied extracellular alginate-like polysaccharide of the Gram-negative bacterium *Pseudomonas aeruginosa*. Alginate is composed of mannuronate and guluronate residues. Uronic acids are mainly found as components of extracellular polymers; for example, in agar-grown biofilms of *P. aeruginosa*, 90–94% of the uronic acids (alginate) were found to be localized extracellularly (Mayer *et al.*, 1999). Sometimes, sulfate groups have been described to occur in sugar molecules such as in extracellular polysaccharides from the slime of *Staphylococcus epidermidis* strains (Arvaniti *et al.*, 1994). Cationic groups in polysaccharides may be due to the presence of amino sugars (Hejzlar & Chudoba, 1986; Veiga *et al.*, 1997).

Proteins can also contribute to the anionic properties of EPS. The negative charge of proteins is due to the presence of dibasic amino acids such as aspartic and glutamic acid, which have been found to be quantitatively important constituents of extracellular proteins extracted from activated sludge (Higgins & Novak, 1997; Dignac *et al.*, 1998). Nucleic acids are polyanionic due to the phosphate residues in the nucleotide moiety of the polymer molecule. Uronic acids, acidic amino acids and phosphate-containing nucleotides are components of EPS; they are crucial for the structure of flocs and biofilms, since they are expected to be involved in electrostatic interactions with multivalent cations (e.g. Ca^{2+}, Mg^{2+}, Fe^{3+}), thus mediating the formation and/or the stabilization of the network of the EPS matrix (see below).

EPS occur in chemically different modifications. Thus extracellular polysaccharides often carry organic substituents such as acetyl, succinyl or pyruvyl groups or inorganic substituents such as sulfate (Sutherland, 1994). These substituents greatly alter the structure and physico-chemical properties of the polysaccharides, such as water solubility and shape of the macromolecules. For example, acetyl groups in alginates decrease the capacity and selectivity of divalent cation binding (Geddie & Sutherland, 1994), increase solution viscosity of alginate (Lee *et al.*, 1996), enhance the water retention capacity of the polysaccharide (Skjåk-Bræk *et al.*, 1989), protect alginate from enzymic degradation by alginate lyases (Lange *et al.*, 1989) and affect biofilm formation (P. Tielen, J. Wingender & H.-C. Flemming, unpublished). Proteins can be substituted with fatty acids to form lipoproteins or can be glycosylated with oligosaccharides to form glycoproteins, which, for example, have been detected in the EPS from activated sludge (Horan & Eccles, 1986).

Extracellular proteins contribute to hydrophobic properties of EPS due to their high proportions of the hydrophobic amino acids alanine, leucine and glycine (Higgins & Novak, 1997; Dignac *et al.*, 1998). The combination of acidic and hydrophobic amino acids in extracellular proteins was believed to contribute to the electrostatic and hydrophobic interactions observed in flocs of activated sludge (Urbain *et al.*, 1993; Dignac *et al.*, 1998). Hydrophobic fractions of EPS extracted from activated sludge were made up of proteins but not carbohydrates, indicating that polysaccharides were not involved in hydrophobic interactions (Jorand *et al.*, 1998). In another study on mixed-culture biofilms, *in situ* characterization of biofilm EPS demonstrated the presence of positively and negatively charged microzones as well as hydrophobic regions that were spatially in close association and were distributed non-uniformly in the biofilm matrix (Jorand *et al.*, 1998). In a study of aerobic heterotrophic biofilms, the profiles of extracellular polysaccharides and proteins indicated a stratified biofilm structure with non-uniform EPS production, the greatest being formed in the upper layers of the biofilms (Zhang *et al.*, 1998). In conclusion, the hydrophilic/hydrophobic regions within EPS molecules and their distribution in the EPS matrix largely determine the ion-exchange potential and the sorption properties of biofilms and flocs. The differences in the composition and properties of EPS in microbial aggregates reflect a certain degree of spatial organization and differentiation within the community of floc and biofilm organisms.

Yield, composition and properties of EPS vary in response to the availability of nutrients and other environmental conditions. For example, in batch cultures of *Pseudomonas atlantica*, the carbohydrate composition of the EPS changed during the growth cycle, with a marked increase in the proportions and absolute amounts of uronic acids on prolonged incubation (Uhlinger & White, 1983). In sequencing batch reactors with sludge growing on synthetic wastewater, the C:N:P ratio was found to influence the hydrophobicity, surface charge and EPS composition of microbial flocs (Bura *et al.*, 1998). In batch cultures of a methanogenic bacterium, polymer production was enhanced under low-phosphate and low-nitrogen conditions (Veiga *et al.*, 1997); it was suggested that carbon utilization shifted towards EPS production when the C:N and/or C:P ratio was enhanced.

COHESIVENESS OF MATRIX POLYMERS

In primary biofilm formation, matrix polymers provide the 'glue' which attaches the cells to a given surface. In microbial aggregate development, the presence of EPS is considered as the basic prerequisite for the formation as well as for the maintenance of the integrity of biofilms or flocs (Wingender & Flemming, 1999). EPS are responsible for the mechanical stability of these microbial aggregates due to intermolecular interactions between many different macromolecules. Indeed, multivalent cations often promote and enforce the interactions between the EPS. The result of these interactions is the formation of a three-dimensional, gel-like matrix surrounding the EPS-producing cells. On first glimpse, the EPS matrix appears amorphous and without perceivable order. From a microbial ecology standpoint, however, the EPS have an important function: to create a gel-like matrix in which the organisms can be fixed, with a long retention time, next to each other. As indicated before, this is a prerequisite for the formation of stable microconsortia. This function should be achieved with the lowest expense of energy and nutrients. The fact that only 1–2% or an even lesser amount of organic matter is required to retain 98–99% of water and to form a stable gel (Christensen & Characklis, 1990) demonstrates how successful the strategy is.

What are the forces which keep this matrix together? Obviously, they are not covalent bonds. On the contrary, the cohesive forces are provided by weak interactions (Mayer *et al.*, 1999; Flemming, 1999). In principle, three types must be considered.

London (dispersion) forces

London (dispersion) forces occur mainly in hydrophobic areas and are not localized to functional groups; the binding energy is about 2.5 kJ mol^{-1}. These forces can be weakened by surface-active substances; this is the reason for the partial efficacy of surfactants in cleaners. However, biofilms cannot be fully dispersed by surfactants, and, in many practical cases, the effect is very weak. This is due to the other binding forces active in EPS molecule cohesion.

Electrostatic interactions

Electrostatic interactions are active between ions and between permanent and induced dipoles. The ionic interactions are relatively strong; it is mainly divalent cations that are involved, Ca^{2+} playing a particular role. These forces are responsible for a considerable proportion of the overall binding energy. The binding energy of non-ionic electrostatic bonds usually ranges between 12 and 29 kJ mol^{-1}. The binding force is strongly dependent upon the distance between the partners of the bond and is stronger over short distances. Electrostatic forces, of course, may also act repulsively. Thus they may help to maintain an 'inflated' structure of the EPS matrix and contribute to pores and channels in the sponge-like morphology of the matrix. Electrostatic interactions can be influenced by ionic strength, complexing agents and the pH value. These factors are the reason for the efficacy of, for example, citric acid in cleaners.

Hydrogen bonds

Hydrogen bonds are mainly active between hydroxyl groups, which are particularly frequent in polysaccharides, and water molecules. They also support the secondary and tertiary structure of proteins. The binding energy ranges between 10 and 30 kJ mol^{-1} and is stronger over short distances. These forces can be influenced by chaotropic agents which disturb the water structure; examples are urea, tetramethyl urea and guanidine hydrochloride. Interestingly, cleaning formulations usually contain no compounds which address this type of binding force.

The individual binding force of any type of these interactions is relatively small compared to a covalent C–C bond (about 250 kJ mol^{-1}). However, the total binding energies of weak interactions between EPS molecules add up to bond values equivalent to those of covalent C–C bonds. Hydrophilic and hydrophobic properties of EPS have been demonstrated, and on the basis of these results, all three types of binding forces are expected to contribute to the overall stability of floc and biofilm matrices, probably to various extents. The result is the formation of a three-dimensional, gel-like network of EPS, whose composition, structure and properties may vary dynamically as the micro-organisms respond to changes in environmental conditions. These types of interactions are symbolized in Fig. 2 (after Mayer *et al.*, 1999).

Entanglements

Entanglements may occur if the polymers in the matrix display molecular masses exceeding the critical molar mass required for entangling, which is the case above 2×10^5 g mol^{-1}. The viscosity of systems containing entangled macromolecules increases proportionally to the power of 3.4 of the molar mass and leads to time-dependent non-permanent knotting of the polymer chains at concentrations of 3–5% (w/w). In a short time mechanical stress response, the systems behave as being chemically cross-linked. On the long timescale, the systems are able to flow under stress. Quantitative investigations of the contribution of entanglements to the overall matrix cohesiveness are not known to the authors but are under investigation in their laboratories. In EPS matrices, however, in which the concentration of macromolecules is below 2% (w/w), entanglements must be considered of minor importance for the cohesiveness of the matrix.

Enzymes may cleave EPS molecules. This is the reason why they are sometimes part of cleaning compositions. However, the vast variability of EPS will prevent effective enzymic dispersal of biofilms. This is of considerable ecological relevance. An organism which can readily disperse biofilms would destroy ecologically important microbial communities in soil, sediments and wastewater treatment plants. Since biofilms have existed for apparently over 3.5 billion years (Schopf *et al.*, 1983), nature has had a long

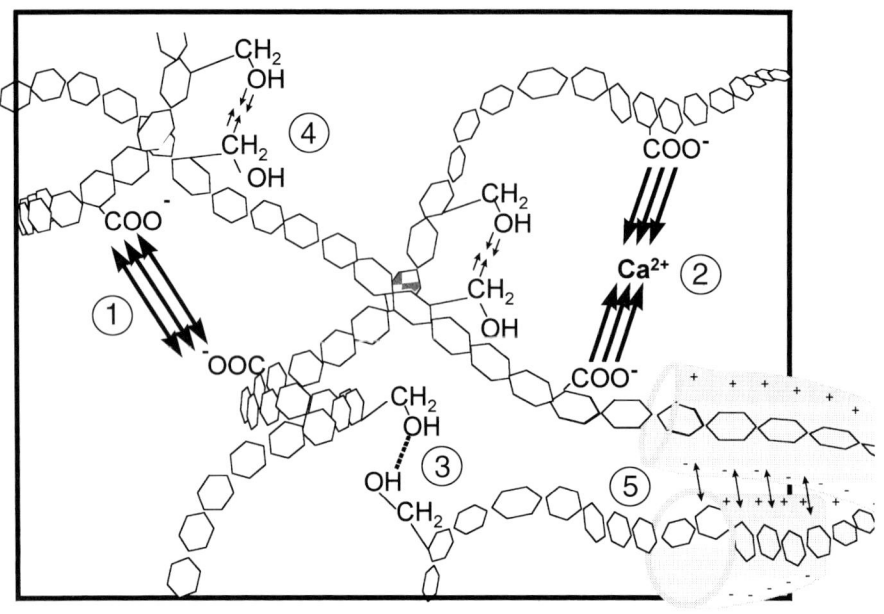

Fig. 2. Interactions between EPS molecules. 1, Repulsion of two carboxylic groups; 2, attraction of two carboxylic groups by a divalent cation; 3, hydrogen bond; 4, electrostatic attraction; 5, dispersion forces (after Mayer *et al.*, 1999). Figure modified from that published in Flemming (1999).

time already to prepare for such eventualities. Probably, the diversity of EPS has prevented the evolvement of an organism which produces such enzymes. This may explain the frequent failure of enzymes in attempts to disperse biofilms in technical systems.

The composition of EPS varies with the species present, the nutrient situation and other environmental conditions. This results in a variety of different biopolymers with different composition, structures and charges. This supports the assumption that under different conditions, different interaction forces may dominate – in other words, the matrix of one biofilm may be considerably different from that of another, grown at different sites and conditions.

An example: electrostatic interactions

The role of Ca^{2+} in the cohesiveness of the biofilm matrix has been addressed by means of a newly developed film rheometer (V. Körstgens & W. Borchard, unpublished), using a mucoid strain of *P. aeruginosa* SG 81 (Grobe *et al.*, 1995) as a model organism. Biofilms were grown in the presence and absence of 0.1 M $CaCl_2$ in *Pseudomonas*

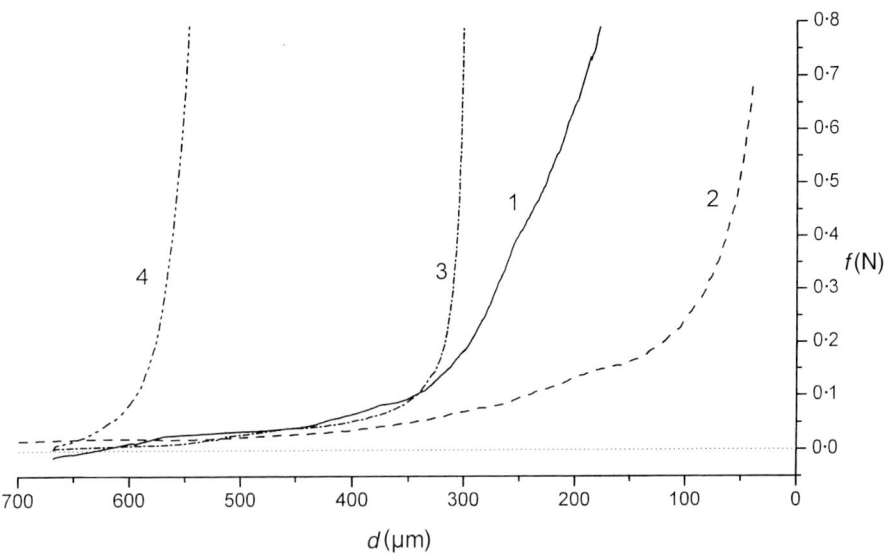

Fig. 3. Compression diagram of *P. aeruginosa* biofilms with force *f* (N) plotted over distance *d* (μm). 1, Film grown on *Pseudomonas* isolation agar (PIA); apparent elasticity modulus 5200 Pa. 2, Same as (1) after immersion for 10 min in water. 3, Film grown on PIA in the presence of 0.1 M $CaCl_2$; apparent elasticity modulus 54 300 Pa. 4, Same as (3) after immersion for 10 min in water; apparent elasticity modulus 41 300 Pa (after V. Körstgens and others, unpublished).

isolation agar (PIA) on membrane filters (0.45 μm pore diameter) on nutrient agar surface. Their apparent elasticity modulus was then determined (V. Körstgens, J. Wingender, W. Borchard & H.-C. Flemming, unpublished). In principle, the resistance of the film to axial compression was measured and the modulus was calculated. An example is shown in Fig. 3. Curve 1 shows the compression diagram for a biofilm grown without addition of Ca^{2+}. The apparent modulus is 5200 Pa. Curve 2 shows the diagram of the same biofilm after immersion in water for 10 min. The modulus cannot be calculated because the linear part of the curve is not long enough for reliable values. Curve 3 shows the diagram for a biofilm grown in the presence of Ca^{2+}. Under these conditions, the apparent modulus is as high as 54 300 Pa. In curve 4, the diagram for a calcium-grown biofilm after immersion in water for 10 min is shown (equivalent to curve 2). Obviously, the thickness of the biofilm has almost doubled (from 330 μm to 570 μm) and the apparent elasticity modulus is 41 300 Pa. The interpretation is simple, since the alginate of *P. aeruginosa* contains guluronate and mannuronate residues which can be bridged by divalent cations such as Ca^{2+}. However, the contribution of the considerable protein content in *P. aeruginosa* EPS to cohesiveness has not yet been elucidated and is under current investigation.

In another series of experiments, the effect of various cations and anions on the apparent viscosity of EPS of *P. aeruginosa* has been investigated (Mayer *et al.*, 1999). In this case, the EPS were isolated, diluted and the viscosity measurements were carried out with a rotational viscometer. It was assumed that, in principle, the same interactions are effective in the biofilm matrix and in the solution. Fig. 4(a) shows the diagrams for fluoride, chloride, bromide and sulphate whilst Fig. 4(b) shows the effect of sodium, potassium and ammonia.

It is evident that the monovalent ions all lead to a strong decrease of the viscosity, which suggests that electrostatic interactions are of clear importance for the cohesiveness of EPS produced by *P. aeruginosa*. By the addition of the low molecular mass salts, the charges of the polyelectrolytes are screened changing the conformation from a stretched macromolecule to a more coiled one. The increase of the curves at higher concentrations (Fig. 4a) can be interpreted by a stepwise folding process of the macromolecules. First, the molecules are dispersed and coiled because of intramolecular interactions. With increasing ionic strength, the coils aggregate and thus the apparent viscosity rises (Mayer *et al.*, 1999).

ROLE OF EPS IN MICROBIAL AGGREGATION

EPS may contribute to the attachment of cells to surfaces, but the extent of their involvement is still open to debate. However, there are indications that the production of certain EPS is triggered subsequently to initial adhesion and is necessarily involved in the development of biofilms (Allison & Sutherland, 1987; Vandevivere & Kirchman, 1993; Davies *et al.*, 1993). Thus network formation via the forces mentioned above seems to be essential for the development and maintenance of microbial aggregates. On the basis of these observations, EPS are included as key components in a number of models explaining the aggregation of micro-organisms as well as the physico-chemical properties of the extracellular matrix in flocs and biofilms (Wingender & Flemming, 1999; Wingender *et al.*, 1999).

In activated sludge flocs, EPS have been implicated in determining floc structure, floc charge, the flocculation process, floc settleability and dewatering properties. In the polymer-bridging model, floc formation is considered as the result of the interaction of high molar mass, long-chain EPS with microbial cells and other particles as well as with other EPS molecules, so that EPS bridge the cells into a three-dimensional matrix. Flocculation is associated with the formation of EPS. Cellular aggregation was found to depend on the physiological state of the micro-organisms; flocculation of cultures of mixed populations from domestic wastewater did not occur until they entered into a restricted state of growth (Pavoni *et al.*, 1972). There was a direct correlation between microbial aggregation and EPS accumulation; the ratio of EPS to micro-organism mass

(a)

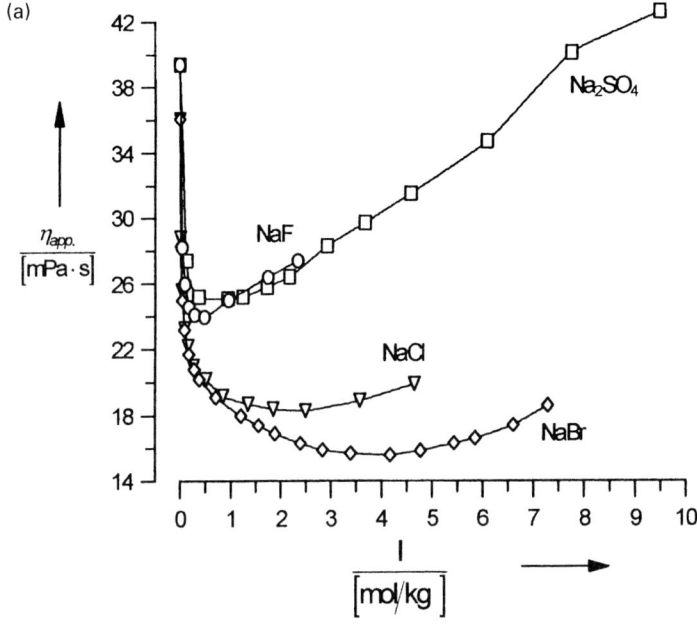

(b)

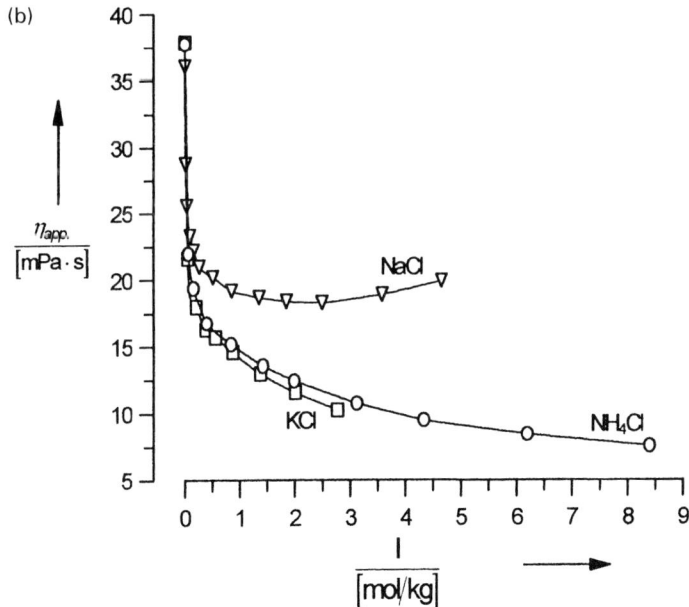

Fig. 4. Dependence of the apparent viscosity η_{app} of a cell-free solution of EPS from *P. aeruginosa* on the ionic strength of various anions (a) and cations (b): comparison of the influence of different electrolytes. Figures modified from those published in Mayer *et al.* (1999).

rapidly increased during cultivation. The major EPS were polysaccharides, proteins and nucleic acids (RNA, DNA). Surface charge was not considered a necessary prerequisite for flocculation, since it remained constant throughout all growth phases regardless of the flocculability of the culture. Bacteria washed free of EPS formed stable dispersions, but readdition of extracted EPS again resulted in flocculation.

In activated sludge cultivated on a carbon-limited medium, the release of EPS (carbohydrates, proteins, nucleic acids) by cell lysis resulted in an increase in flocculation (Vallom & McLoughlin, 1984). Addition of DNA to activated sludge or to pure cultures of bacteria isolated from activated sludge promoted flocculation, indicating that DNA can act as a polyelectrolyte flocculating agent (Vallom & McLoughlin, 1984); however, coacervation may also be an explanation. The importance of nucleic acids in microbial aggregation was also observed in the self-flocculating bacterium *Rhodovulum* sp., which produced EPS made up of carbohydrates, proteins and nucleic acids (Watanabe *et al.*, 1998). Treatment of flocculated cells with nucleolytic enzymes (RNase, DNase) resulted in deflocculation, while polysaccharide- and protein-degrading enzymes had no effect; it was argued that extracellular nucleic acids were active in flocculation (Watanabe *et al.*, 1998).

In batch cultures with *Zoogloea*, it was shown that production of an extracellular polysaccharide was accompanied by flocculation of the bacteria (Unz & Farrah, 1976). Polymer formation was initiated in mid-exponential growth phase and the quantity produced appeared to be influenced by the level of carbon and nitrogen in the medium. Methanogenic bacteria were shown to produce EPS, which were proposed to help in the formation and maintenance of anaerobic granules (Veiga *et al.*, 1997); high settleability of these granules is important for the performance of anaerobic sludge reactors. The detection of extracellular polymeric fibrils in natural and wastewater flocs by high-resolution transmission electron microscopy (TEM) confirmed the role of EPS as structural support to the microbial aggregates (Liss *et al.*, 1996).

The concept of EPS mediating microbial aggregation has been applied in field trials by using a slime-producing organism to enhance biomass settleability in aerobic wastewater treatment systems of the paper industry (Volpe *et al.*, 1998); addition of the EPS-producing bacteria (*Bacillus* sp.) to an aerated stabilization system and to an activated sludge unit resulted in an improvement of flocculation and effluent quality.

In addition to EPS, divalent cations are regarded as important constituents of microbial aggregates, since they bind to negatively charged groups present on bacterial surfaces, in EPS molecules and on inorganic particles entrapped in flocs and biofilms. It has been reported that extraction of Ca^{2+} from flocs and biofilms by displacement with monovalent cations or by chelation with the more general complexing agent EDTA or the more Ca^{2+}-specific chelant EGTA resulted in the destabilization of flocs (Bruus *et al.*, 1992; Higgins & Novak, 1997) and biofilms (Turakhia *et al.*, 1983). Practical implications are that weakening of activated sludge structure by removal of Ca^{2+} leads to an increase in the number of small particles with subsequent decrease of filterability and dewaterability. These observations suggest that divalent cations may be important for the maintenance of floc and biofilm structure by acting as bridging agents within the three-dimensional EPS matrix. Bruus *et al.* (1992) also integrated the role of divalent cations into their sludge floc model. The floc structure was proposed to be a three-dimensional EPS matrix held together by divalent cations with varying selectivity to the matrix ($Cu^{2+} > Ca^{2+} > Mg^{2+}$). It was argued that approximately half of the Ca^{2+} pool was associated with EPS, forming a matrix that resembled gels of carboxylate-containing alginates. Fe^{3+} may also be of importance in floc stabilization. Specific removal of Fe^{3+} from activated sludge flocs caused a weakening of floc strength, resulting in release of particles to bulk water, dissolution of EPS and partial floc disintegration (Nielsen & Keiding, 1998). Watanabe *et al.* (1999) reported that EPS-mediated flocculation of *Rhodovulum* sp. was promoted by Ca^{2+}, Mg^{2+}, Fe^{3+} and Al^{3+}, with trivalent cations having a stronger effect than divalent cations. The promotion of flocculation was explained by bridge formation between the EPS and the cations.

On the basis of investigations on laboratory-scale activated sludge reactors, Higgins & Novak (1997) emphasized the role of structural proteins in conjunction with divalent cations in flocculation. Increasing the concentrations of Ca^{2+} or Mg^{2+} resulted in an increase of bound protein, whereas there was little effect on bound polysaccharides. Addition of high concentrations of Na^+ led to a decrease of bound protein. It was proposed that the monovalent Na^+ displaced divalent cations from within the flocs. This displacement would reduce binding of protein within the floc and result in solubilization of protein. Further support for the involvement of extracellular protein in the aggregation of bacteria into flocs came from the observation that treatment of activated sludge flocs with a proteolytic enzyme (Pronase) resulted in deflocculation, in a shift to smaller particles in the 5–40 µm range, and in a release of polysaccharide. Gel electrophoretic analysis of extracted EPS from municipal, industrial and laboratory activated sludge revealed the presence of a single protein with a molecular mass of approximately 15 000 Da. Analysis of amino acid composition and sequence indicated that this protein displayed similarities to lectins; binding site inhibition studies demonstrated lectin-like activity of the 15 000 Da protein (Higgins & Novak, 1997). On the basis of these results, a model of flocculation was proposed. Lectin-like proteins specifically interact with polysaccharides that are cross-linked to adjacent proteins. Divalent cations bridge negatively charged functional groups on the EPS molecules.

The cross-linking of EPS and cation bridges leads to the stabilization of the biopolymer network mediating the immobilization of microbial cells.

Urbain *et al.* (1993) concluded from their studies on 16 activated sludge samples from different origins that internal hydrophobic bondings were involved in flocculation mechanisms and their balance with hydrophilic interactions determined the sludge settling properties. Hydrophobic areas in-between the cells were considered as essential adhesives within the floc structure and may be provided by the protein fraction of the EPS (Jorand *et al.*, 1998). Cell surface hydrophobicity was shown to be important for adhesion of bacteria to activated sludge flocs (Olofsson *et al.*, 1998). Cells with high cell surface hydrophobicity attached in greater numbers to the flocs than bacteria with a more hydrophilic surface. The hydrophobic cells not only attached on the surface of the flocs, but also penetrated the flocs through channels and pores, whereas hydrophilic cells did not. It was assumed that adhesion of hydrophobic bacteria within flocs would increase the potential of the flocs to clear free-living cells from the water phase (Olofsson *et al.*, 1998). Generally, microcolonies and single cells within flocs and biofilms are separated by microscopically transparent voids. However, specific staining (e.g. Thiéry's stain, ruthenium red) allowed microscopic detection of polysaccharidic material between microcolonies and single cells, indicating that the intercellular space was filled with EPS connecting the cells together (Jorand *et al.*, 1995). This was confirmed when flocs from natural riverine systems and from wastewater were viewed by high-resolution TEM (Liss *et al.*, 1996). Pores devoid of physical structures under optical microscopes were found to be filled with complex matrices of extracellular polymeric fibrils (4–6 nm diameter). These fibrils were found to represent the dominant bridging mechanism between organic and inorganic components of the flocs and contributed to the extensive surface area per unit volume of the flocs. It is speculated that this mechanism is also valid for biofilms.

REFERENCES

Allison, D. G. & Sutherland, I. W. (1987). The role of exopolysaccharides in adhesion of freshwater bacteria. *J Gen Microbiol* **133**, 1319–1327.

Arvaniti, A., Karamanos, N. K., Dimitracopoulos, G. & Anastassiou, E. D. (1994). Isolation and characterization of a novel 20-kDa sulfated polysaccharide from the extracellular slime layer of *Staphylococcus epidermidis*. *Arch Biochem Biophys* **308**, 432–438.

Bitton, G. (1994). *Wastewater Microbiology*. New York: Wiley-Liss.

Bruus, J. H., Nielsen, P. H. & Keiding, K. (1992). On the stability of activated sludge flocs with implications to dewatering. *Water Res* **26**, 1597–1604.

Bura, R., Cheung, M., Liao, B., Finlayson, J., Lee, B. C., Droppo, I. G., Leppard, G. G. & Liss, S. N. (1998). Composition of extracellular polymeric substances in the activated sludge floc matrix. *Water Sci Technol* **37**, 325–333.

Characklis, W. G. & Wilderer, P. A. (1989). *Structure and Function of Biofilms*. Chichester: Wiley.

Christensen, B. E. & Characklis, W. G. (1990). Physical and chemical properties of biofilms. In *Biofilms*, pp. 93–130. Edited by W. G. Characklis & K. C. Marshall. New York: Wiley.

Cooksey, K. E. (1992). Extracellular polymers in biofilms. In *Biofilms – Science and Technology*, pp. 137–147. Edited by L. F. Melo, T. R. Bott, M. Fletcher & B. Capdeville. Dordrecht: Kluwer.

Costerton, J. W., Irvin, R. T. & Cheng, K.-J. (1981). The bacterial glycocalyx in nature and disease. *Annu Rev Microbiol* 35, 299–324.

Costerton, J. W., Cheng, K.-J., Geesey, G. G., Ladd, T. I., Nickel, J. C., Dasgupta, M. & Marrie, T. J. (1987). Bacterial biofilms in nature and disease. *Annu Rev Microbiol* 41, 435–464.

Costerton, J. W., Stewart, P. S. & Greenberg, E. P. (1999). Bacterial biofilms: a common cause of persistent infections. *Science* 284, 1318–1322.

Davies, D. G., Chakrabarty, A. M. & Geesey, G. G. (1993). Exopolysaccharide production in biofilms: substratum activation of alginate gene expression by *Pseudomonas aeruginosa*. *Appl Environ Microbiol* 59, 1181–1186.

Dignac, M.-F., Urbain, V., Rybacki, D., Bruchet, A., Snidaro, D. & Scribe, P. (1998). Chemical description of extracellular polymers: implication on activated sludge floc structure. *Water Sci Technol* 38, 45–53.

Fazio, S. A., Uhlinger, D. J., Parker, J. H. & White, D. C. (1982). Estimations of uronic acids as quantitative measures of extracellular and cell wall polysaccharide polymers from environmental samples. *Appl Environ Microbiol* 43, 1151–1159.

Flemming, H.-C. (1999). The forces that keep biofilms together. In *Biofilms in Aquatic Systems*, pp. 1–12. Edited by W. Keevil, A. F. Godfree, D. M. Holt & C. S. Dow. Cambridge: Royal Society of Chemistry.

Flemming, H.-C. & Schaule, G. (1989). Biofouling auf Umkehrosmose- und Ultrafiltrationsmembranen. Teil II: Analyse und Entfernung des Belages. *Vom Wasser* 73, 287–301.

Flemming, H.-C. & Schaule, G. (1996). Biofouling. In *Microbially Influenced Corrosion of Materials – Scientific and Technological Aspects*, pp. 39–54. Edited by E. Heitz, W. Sand & H.-C. Flemming. Heidelberg: Springer.

Frølund, B., Palmgren, R., Keiding, K. & Nielsen, P. H. (1996). Extraction of extracellular polymers from activated sludge using a cation exchange resin. *Water Res* 30, 1749–1758.

Geddie, J. L. & Sutherland, I. W. (1994). The effect of acetylation on cation binding by algal and bacterial alginates. *Biotechnol Appl Biochem* 20, 117–129.

Geesey, G. G. (1982). Microbial exopolymers: ecological and economic considerations. *ASM News* 48, 9–14.

Gehr, R. & Henry, J. G. (1983). Removal of extracellular material. Techniques and pitfalls. *Water Res* 17, 1743–1748.

Gehrke, T., Telegdi, J., Thierry, D. & Sand, W. (1998). Importance of extracellular polymeric substances from *Thiobacillus ferrooxidans* for bioleaching. *Appl Environ Microbiol* 64, 2743–2747.

Goodwin, J. A. S. & Forster, C. F. (1985). A further examination into the composition of activated sludge surfaces in relation to their settlement characteristics. *Water Res* 19, 527–533.

Grobe, S., Wingender, J. & Trüper, H. G. (1995). Characterization of mucoid

Pseudomonas aeruginosa strains isolated from technical water systems. *J Appl Bacteriol* **79**, 94–102.

Hejzlar, J. & Chudoba, J. (1986). Microbial polymers in the aquatic environment – I. Production by activated sludge microorganisms under different conditions. *Water Res* **20**, 1209–1216.

Higgins, M. J. & Novak, J. T. (1997). Characterization of exocellular protein and its role in bioflocculation. *J Environ Eng* **123**, 479–485.

Horan, N. & Eccles, C. R. (1986). Purification and characterization of extracellular polysaccharides from activated sludge. *Water Res* **20**, 1427–1432.

Jahn, A. & Nielsen, P.-H. (1995). Extraction of extracellular polymeric substances (EPS) from biofilms using a cation exchange resin. *Water Sci Technol* **32**, 157–164.

Jahn, A. & Nielsen, P.-H. (1998). Cell biomass and exopolymer composition in sewer biofilms. *Water Sci Technol* **37**, 17–24.

Jorand, F., Zartarian, F., Thomas, F., Block, J. C., Bottero, J. Y., Villemin, G., Urbain, V. & Manem, J. (1995). Chemical and structural (2D) linkage between bacteria within activated sludge flocs. *Water Res* **29**, 1639–1647.

Jorand, F., Boué-Bigne, F., Block, J. C. & Urbain, V. (1998). Hydrophobic/hydrophilic properties of activated sludge exopolymeric substances. *Water Sci Technol* **37**, 307–315.

Lange, B., Wingender, J. & Winkler, U. K. (1989). Isolation and characterization of an alginate lyase from *Klebsiella aerogenes*. *Arch Microbiol* **152**, 302–308.

Lee, J. W., Ashby, R. D. & Day, D. F. (1996). Role of acetylation on metal induced precipitation of alginates. *Carbohydr Polym* **29**, 337–345.

Liss, S. N., Droppo, I. G., Flannigan, D. T. & Leppard, G. G. (1996). Floc architecture in wastewater and natural riverine systems. *Environ Sci Technol* **30**, 680–686.

Mayer, C., Moritz, R., Kirschner, C., Borchard, W., Maibaum, R., Wingender, J. & Flemming, H.-C. (1999). The role of intermolecular interactions: studies on model systems for bacterial biofilms. *Int J Biol Macromol* **26**, 3–16.

Nielsen, P.-H. & Keiding, K. (1998). Disintegration of activated sludge flocs in presence of sulfide. *Water Res* **32**, 313–320.

Nielsen, P.-H., Jahn, A. & Palmgren, R. (1997). Conceptual model for production and composition of exopolymers in biofilms. *Water Sci Technol* **36**, 11–19.

Olofsson, A.-C., Zita, A. & Hermansson, M. (1998). Floc stability and adhesion of green-fluorescent-protein-marked bacteria to flocs in activated sludge. *Microbiology* **144**, 519–528.

Pavoni, J. L., Tenney, M. W. & Echelberger, W. F. (1972). Bacterial exocellular polymers and biological flocculation. *J Water Pollut Control Fed* **44**, 414–431.

Platt, R. M., Geesey, G. G., Davis, J. D. & White, D. C. (1985). Isolation and partial chemical analysis of firmly bound exopolysaccharide from adherent cells of a freshwater bacterium. *Can J Microbiol* **31**, 657–680.

Poxon, T. L. & Darby, J. L. (1997). Extracellular polyanions in digested sludge: measurement and relationship to sludge dewaterability. *Water Res* **31**, 749–758.

Rudd, T., Sterritt, R. M. & Lester, J. N. (1983). Extraction of extracellular polymers from activated sludge. *Biotechnol Lett* **5**, 327–332.

Sand, W. & Gehrke, T. (1999). Analysis and function of the EPS from the strong acidophile *Thiobacillus ferrooxidans*. In *Bacterial Extracellular Polymeric Substances*, pp. 127–141. Edited by J. Wingender, T. Neu & H.-C. Flemming. Heidelberg & Berlin: Springer.

Schopf, J. W., Hayes, J. M. & Walter, M. R. (1983). Evolution on earth's earliest ecosystems: recent progress and unsolved problems. In *Earth's Earliest Biosphere*, pp. 361–384. Edited by J. W. Schopf. New Jersey: Princeton University Press.

Skjåk-Bræk, G., Zanetti, F. & Paoletti, S. (1989). Effect of acetylation on some solution and gelling properties of alginates. *Carbohydr Res* **185**, 131–138.

Sutherland, I. W. (1994). Structure-function relationships in microbial exopolysaccharides. *Biotechnol Adv* **12**, 393–448.

Sutherland, I. W. (1996). Extracellular polysaccharides. In *Biotechnology*, vol. 6, *Products of Primary Metabolism*, pp. 615–657. Edited by H.-J. Rehm & G. Reed. Weinheim: Verlag Chemie.

Turakhia, M. H., Cooksey, K. E. & Characklis, W. G. (1983). Influence of a calcium-specific chelant on biofilm removal. *Appl Environ Microbiol* **46**, 1236–1238.

Uhlinger, D. J. & White, D. C. (1983). Relationship between physiological status and formation of extracellular polysaccharide glycocalyx in *Pseudomonas atlantica*. *Appl Environ Microbiol* **45**, 64–70.

Unz, R. F. & Farrah, S. R. (1976). Exopolymer production and flocculation by *Zoogloea* MP6. *Appl Environ Microbiol* **31**, 623–626.

Urbain, V., Block, J. C. & Manem, J. (1993). Bioflocculation in activated sludge: an analytical approach. *Water Res* **27**, 829–838.

Vallom, J. K. & McLoughlin, A. J. (1984). Lysis as a factor in sludge flocculation. *Water Res* **18**, 1523–1528.

Vandevivere, P. & Kirchman, D. L. (1993). Attachment stimulates exopolysaccharide synthesis by a bacterium. *Appl Environ Microbiol* **59**, 3280–3286.

Veiga, M. C., Jain, M. K., Wu, W.-M., Hollingsworth, R. & Zeikus, J. G. (1997). Composition and role of extracellular polymers in methanogenic granules. *Appl Environ Microbiol* **63**, 403–407.

Volpe, G., Christiansen, J. A., Wescott, J., Leger, R. & Rumbaugh, E. (1998). Use of a slime producing organism to enhance biomass settleability in activated sludge and ASB systems. *Tappi J* **81**, 60–67.

Watanabe, M., Sasaki, K., Nakashimada, Y., Kakizono, T., Noparatnaraporn, N. & Nishio, N. (1998). Growth and flocculation of a marine photosynthetic bacterium *Rhodovulum* sp. *Appl Microbiol Biotechnol* **50**, 682–691.

Watanabe, M., Suzuki, Y., Sasaki, K., Nakashimada, Y. & Nishio, N. (1999). Flocculating property of extracellular polymeric substance derived from a marine photosynthetic bacterium, *Rhodovulum* sp. *J Biosci Bioeng* **87**, 625–629.

Wingender, J. & Flemming, H.-C. (1999). Autoaggregation in flocs and biofilms. In *Biotechnology*, vol. 8, pp. 63–86. Edited by J. Winter. Weinheim: VCH.

Wingender, J., Neu, T. & Flemming, H.-C. (1999). What are extracellular polymeric substances? In *Microbial Extracellular Polymeric Substances*, pp. 1–19. Edited by J. Wingender, T. Neu & H.-C. Flemming. Heidelberg, Berlin & New York: Springer.

Zhang, X., Bishop, P. L. & Kupferle, M. J. (1998). Measurement of polysaccharides and proteins in biofilm extracellular polymers. *Water Sci Technol* **37**, 345–348.

Microbial detachment from biofilms

Gillian F. Moore,[1] Braden C. Dunsmore,[1] Steven M. Jones,[1] Christopher W. Smejkal,[1] Jana Jass,[2] Paul Stoodley[3] and Hilary M. Lappin-Scott[1]

[1] Environmental Microbiology Research Group, Exeter University, Exeter, UK
[2] Department of Microbiology, Umeå University, Umeå, Sweden
[3] Center for Biofilm Engineering, Montana State University, Bozeman, MT, USA

INTRODUCTION

This chapter reviews the broad area of biofilm detachment, the mechanisms of detachment and the methods used to study this important process. Two case studies are included: the first of these focuses on the control of clinical biofilms; the second case study examines detachment in the water industry.

Biofilms are dynamic structures found in a wide variety of both natural and man-made environments. Their formation has been well studied; for example, Characklis (1990) described eight different stages of biofilm accumulation (Table 1 and Fig. 1). There has been much research into the initial attachment of micro-organisms to surfaces, including the effect of electrostatic interactions and electrochemical forces (Bos *et al.*, 1999). The physiological changes that attaching cells undergo have also been examined; for example, the production of surface appendages such as fimbriae (Austin *et al.*, 1998). In contrast to the work undertaken on attachment, detachment has received little attention although many researchers regard it as a crucial stage of biofilm development (Stewart, 1993; Allison *et al.*, 1999).

Bryers (1988) classified the detachment process into four separate groups: abrasion, grazing, erosion and sloughing. Detachment from the biofilm can be directly caused by the collision or rubbing together of surfaces on which the biofilm has developed, leading to abrasive detachment. Larger organisms feeding on the biofilm can indirectly cause detachment through grazing. Erosion and sloughing refer to physical or chemical processes, which indirectly affect the biofilm structure, leading to detachment. Erosion

SGM symposium 59: Community structure and co-operation in biofilms. Editors D. Allison, P. Gilbert, H. Lappin-Scott, M. Wilson. Cambridge University Press. ISBN 0 521 79302 5 ©SGM 2000.

Table 1. Biofilm formation processes (Characklis, 1990)

Stage	Process
1	Organic preconditioning of the substratum
2	Transport of microbial cells to the substratum
3	Reversible adsorption of the cells to the substratum
4	Desorption of cells from substratum to bulk liquid
5	Irreversible adsorption of cells on the substratum
6	Growth of cells adsorbed to the substratum and production of extracellular polymeric substances
7	Attachment of other cells and particles from the bulk liquid to the biofilm
8	Detachment of portions of the biofilm

refers to the continual removal of cells or small groups of cells from the biofilm, whereas sloughing is the loss of discrete amounts of biofilm.

The detachment of pathogenic micro-organisms from a biofilm can have serious implications, such as the contamination of a water supply. Alternatively, increasing the rate of detachment by various types of intervention to control biofilm growth is one way of preventing harmful biofilm accumulation. To achieve this control, more must be understood about the detachment process within biofilms.

MECHANISMS OF BIOFILM DETACHMENT
The environment around and within the biofilm plays a key role in any detachment process. The physical and chemical properties of the environment can influence the whole biofilm or change the biological properties of the cells within the biofilm and this may lead to detachment. Abrasion, erosion and sloughing are all types of detachment processes. They may be regarded as distinct from grazing in that, once understood, they may be manipulated to alter detachment rates and control biofilms.

Detachment processes
Unlike erosion and sloughing, abrasion is caused by direct physical contact with the biofilm structure, for example, scraping a biofilm from a surface with cleaning utensils. The effect of abrasive detachment is particularly important in the water industry as it can seriously affect the performance of the fluidized-bed reactors that utilize biofilms to treat wastewater. Within these reactors, biofilms are developed on particles of an inert material and, in the treatment process, these can collide and rub together causing detachment of the biofilm (Chang *et al.*, 1991).

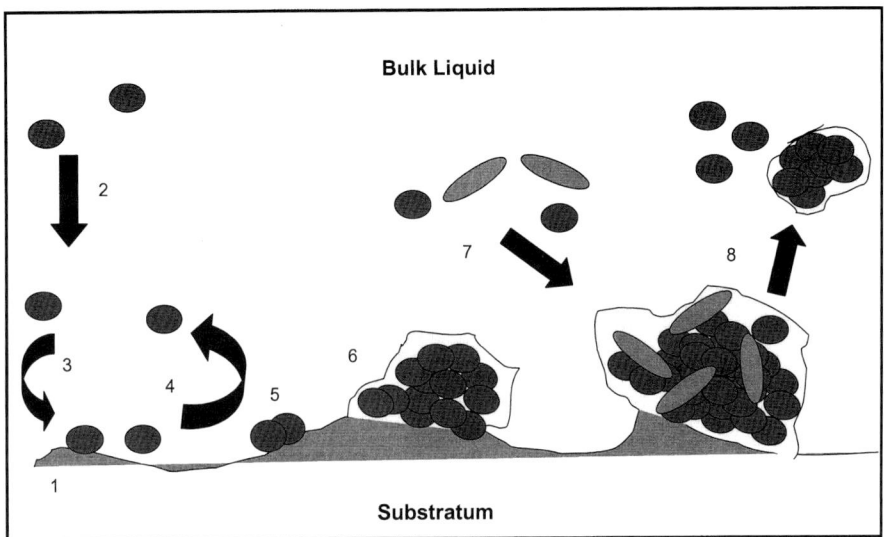

Fig. 1. Biofilm formation process (Characklis, 1990). Characklis (1990) described biofilm formation in eight stages: 1, preconditioning; 2, transport of cells to substratum; 3, reversible adsorption; 4, desorption; 5, irreversible adsorption; 6, growth and extracellular polymeric substances production; 7, attachment by other micro-organisms; 8, detachment.

Erosive detachment can be prompted by the fluid shear stress at the biofilm–fluid interface. Here there is a laminar sub-layer where shear forces are low compared to those further above the interface in the bulk liquid (Lappin-Scott *et al.*, 1994). When the biofilm grows and increases in thickness, its irregular surface grows above the laminar sub-layer and into more turbulent flows where cells are continually removed by the higher shear stresses (Characklis, 1990; Brading *et al.*, 1995).

It has been noted that chemical changes within the biofilm structure or in the surrounding bulk liquid can often cause sloughing of the biofilm. For example, Ohashi & Harada (1994) demonstrated that gas vacuoles of nitrogen formed in a mature biofilm of denitrifying bacteria. These were produced within the biofilm and weakened its structure.

Factors influencing detachment

Detachment processes can be linked to the environment within the biofilm or within the bulk liquid. These influences can be physical, chemical or biological and may lead to changes in the physiology of the cells in the biofilm and to eventual detachment.

Physical properties. Physical influences are mainly in the bulk liquid surrounding the biofilm, for example the liquid velocity. Chang *et al.* (1991) examined detachment rates in a liquid fluidized-bed reactor and found that increases in liquid velocity and particle concentration lead to faster detachment rates.

The shear stress at the biofilm–liquid interface influences the erosion of a biofilm (Characklis, 1990). However, Stoodley *et al.* (1999a) suggested that the frequency of changes in shear, as well as the magnitude of shear, can lead to detachment. Between infrequent changes in stress, the biofilm may have an opportunity to repair itself, whereas in more frequent fluctuations, the biofilm reduces in thickness and detachment occurs.

There are a number of reports on the effect of physical forces on biofilm structure (Chang *et al.*, 1991; Nicolella *et al.*, 1997; Kwok *et al.*, 1998; Stoodley *et al.*, 1999b). Tijhuis *et al.* (1996) reported that with high detachment forces, the biofilm became smooth and strong. This was in contrast to the rough and weaker biofilms formed under low-force conditions.

Chemical properties. Changes in the chemical properties within the biofilm or the bulk liquid can cause detachment. Biofilm detachment can be influenced by changes in the availability of nutrients (Stoodley *et al.*, 1999b). Sawyer & Hermanowicz (1998) found that nutrient limitation at the biofilm–liquid interface of *Aeromonas hydrophila* increased the rate of detachment. However, other studies have shown that an increase in nutrient availability can lead to an increase in the detachment rate (Characklis *et al.*, 1990).

The electrochemical properties within the biofilm have been suggested to be a significant factor in detachment. Changes in these, associated with the requirement for protons needed for lactose transport into the cell, could cause the extracellular polysaccharide matrix to expand or contract with the change in potential and so lose its cohesive properties (Characklis *et al.*, 1990).

The effect of substrate limitation on detachment rates has also been investigated. Peyton & Characklis (1993) reported that substrate limitation decreased the rate of detachment in a *Pseudomonas aeruginosa* biofilm by limiting the growth rate of the cells within the biofilm. Other work has shown that the surface properties of bacterial cells change during different stages of growth and this may affect detachment rates (Gilbert *et al.*, 1991). Therefore, detachment may be mediated by the cell cycle (Allison *et al.*, 1990).

There are many different regulatory systems that control the expression of genes within bacteria. Some of these have been studied in the context of biofilm growth and offer insights into detachment processes. The stationary phase sigma factor RpoS has been found to be involved in the regulation of the transcription of a variety of genes when the cell is subjected to stresses, such as nutrient limitation, that lead to the stationary phase of growth (Loewen *et al.*, 1998). The involvement of RpoS in the biofilm development of both *Escherichia coli* and *Salmonella enteritidis* PT4 has been reported (Adams & McLean, 1999; Moore *et al.*, 1999). The role of RpoS in biofilm formation may include the control of gene expression for fimbriae (Hammer *et al.*, 1995) and extracellular polymeric substances. However, the growth rate of the cells within the biofilm is increased after substrate loading, leading to the inactivation of RpoS and therefore the loss of activation of the proteins it regulates. This could result in a weakening of biofilm cohesion and detachment may occur.

Other investigators have also highlighted the relationship between growth phase and detachment. For example, *Clostridium thermocellum* has been shown to detach from the substratum during the stationary phase of growth (Lamed & Bayer, 1986). This bacterium adsorbs strongly to cellulose, which it then utilizes as a substrate. A cellulosome is produced and this enhances attachment to the substratum. When growth reaches the stationary phase, the cellulosome is released and the cell is free to attach elsewhere. It has been shown that, under certain conditions, bacteria within the biofilm can produce chemicals that destabilize the biofilm structure, leading to sloughing. Applegate & Bryers (1991) showed that in an oxygen-limited *Pseudomonas putida* biofilm, there was an increase in the calcium ion concentration and the production of extracellular polymeric substance. Calcium ions act as cross-linking agents within the biofilm and, along with the extracellular polymeric substance production, increase cohesiveness. Therefore, when sloughing occurred in oxygen-limited biofilms, it was on a large scale as the cells were firmly aggregated.

Biological properties. Some of the biological mechanisms involved in detachment from biofilms have been studied. *P. aeruginosa* biofilms produce alginate, a polysaccharide that enhances attachment of the cells to surfaces (Mai *et al.*, 1993). The bacterium also produces alginate lyase, which induces sloughing by degrading the polysaccharide that holds the biofilm together (Boyd & Chakrabarty, 1994). *Pseudomonas fluorescens* has also been reported to produce lyase, which degrades exopolysaccharides under starvation conditions and therefore increases detachment under these conditions (Allison *et al.*, 1998).

Enzymes produced by a range of organisms similarly bring about detachment. *Streptococcus mutans* produces surface protein releasing enzyme (SPRE) which releases proteins from the surface of the cell. The activity of SPRE increases the detachment of the bacterium from the biofilm (Lee *et al.*, 1996).

The importance of regulatory systems that control gene expression by bacteria has already been noted. The influence of other regulatory systems on biofilm formation and detachment has also been investigated. Davies *et al.* (1998) reported that mutants lacking the ability to synthesize specific signalling molecules formed undifferentiated biofilm, and these were sensitive to SDS. Other investigators have also demonstrated that homoserine lactones play a role in the formation, structure and detachment events by *P. aeruginosa* (Stoodley *et al.*, 1999c; Allison *et al.*, 1999). Homoserine lactones have also been reported to play a role in the biofilm detachment of *A. hydrophila* (Lynch *et al.*, 1999). Puckas *et al.* (1997) reported that *Rhodobacter sphaeroides* produces the homoserine lactone 7,8-*cis*-tetra-decenoylhomoserine lactone. When cell aggregation by the bacterium was examined, it was noted that the wild-type grew as individual cells in suspension whereas a mutant deficient in the production of the homoserine lactone produced large aggregates.

The ability of cell–cell communication to influence biofilm structure suggests that it may regulate a number of physiological functions. Reports have implicated quorum sensing in the direct or indirect regulation of RpoS activity (Sperandio *et al.*, 1999). More research is needed in this new and complex area of regulatory systems in biofilms to clarify the specific roles they play. However, such systems could be the key to the development of novel agents to control biofilm growth.

METHODS FOR THE STUDY OF BIOFILM DETACHMENT

Many experimental systems are available to study biofilm growth (Ladd & Costerton, 1990). In this section, methods used for the study of the relatively new area of detachment from biofilms will be described. These methods can be broken down using a framework to assist the process of investigation (Fig. 2).

'Three choice' experimenting

When designing an experiment to study biofilm detachment, three choices must be made, that is, a culture method, a sample method and a method (or methods) of analysis (Fig. 2). The culture method refers to the situation in which the biofilm is established. The biofilm may already be established within its natural environment (a 'real' biofilm) or within a laboratory model system. Biofilms formed in the laboratory can be cultured in either a batch or flowing system to mimic their natural environment. If a flowing system is chosen, the rate of fluid flow is important to the experimental design as hydrodynamics have a profound effect on biofilms (Stoodley *et al.*, 1999d, e). Flow rate can affect the structure of the biofilm and lead to markedly different detachment events. Turbulent flow biofilms have been shown to be denser than, or of a different shape from, those grown under laminar flow (Chang *et al.*, 1991; Stoodley *et*

Culture Method	Sample Method	Method of Analysis
'Real' Environmental samples	Destructive	Macro (e.g. Plate counts, turbidity)
Batch system $\rightarrow$	$\rightarrow$	
	Non-Destructive	Micro (e.g. Light, epifluorescent and scanning electron microscopy)
Flowing system (Laminar/turbulent)		

Fig. 2. 'Three choice' experimental design: the necessary choices for design of experiments to study biofilm detachment.

al., 1999b). Finally, the flow rate should be chosen with reference to the environment that is being modelled.

Once a biofilm is established, the detachment events must be monitored. These methods can be either destructive or non-destructive. Non-destructive techniques allow *in situ*, real-time analysis, whereas destructive sampling involves the removal of the biofilm from its environment. Analysis of the sample is performed using either a macro-method, such as plate counting or turbidity measurement, or a micro-method involving the use of microscopy.

Critical evaluation of methods used for detachment studies

Macroscopic. A method which 'reproducibly quantifies the ease of removal of micro-organisms from surfaces' (Eginton *et al.*, 1995) gives a useful indication of cell detachment from a biofilm by measuring the strength of attachment. Here, tiles of different materials were suspended in media and biofilms developed. Tiles were then sequentially placed onto several agar plates and the resulting c.f.u. were counted. This methodology uses a batch system of growth followed by destructive sampling to allow detached cells to be measured macroscopically. The merits of this method are its simplicity and reproducibility.

Lee *et al.* (1996) investigated the role of surface proteins in detachment. In this study, hydroxyapatite rods were suspended in a chemostat containing *S. mutans*. The addition of SPRE increased the detachment from the biofilm. Detachment was measured by removal of the rods into buffer; these were then treated by gentle rocking or sonication. Viable cell counts were made from each treatment with detachment being measured as the difference between cells detached 'naturally' and the total of natural and sonicated counts. In the study, a batch culture was coupled to a destructive sampling technique and analysed macroscopically: it is one of the simplest, yet effective methods of studying detachment.

Nicolella *et al.* (1996, 1997) identified several of the parameters significant to detachment of bacteria from a fluidized bed. A laboratory-scale reactor, consisting of a glass column packed with sand, was used to establish a biofilm. The fluidized bed was designed to mimic the real system with fluid flow through the reactor. Suspended solids were drawn off and filtered to assess detachment. This method was non-destructive and the analysis was macroscopic.

Microscopic. Stoodley *et al.* (1999a) devised a real time, *in situ* flow cell system for non-destructive studies of biofilm processes, including detachment. Time-lapse image analysis was used to investigate mixed-species biofilms *in situ* within a glass flow cell (Fig. 3). Surface area measurements allowed both attachment and detachment to be monitored (Stoodley *et al.*, 1999c). The effect of specific conditions, such as changes in nutrient concentration or fluid flow, on detachment was captured and image subtraction was used to calculate detachment rates from the biofilm.

Sawyer & Hermanowicz (1998) used digital image analysis with time-lapse photography to demonstrate increased detachment rates of *A. hydrophila* under limiting nutrient conditions in a glass flow cell. Continuous video capture was also employed (Ohashi & Harada, 1994) to perform *in situ* analysis of a denitrifying bacterial biofilm formed on polyvinylchloride plates within a rectangular open channel reactor. Different types of detachment were described during the progression of biofilm formation. By combining a flowing culture system with non-destructive image analysis by microscopy, an effective tool was created for detachment studies. Images taken *in situ* may be used to describe properties of the biofilm such as morphology, surface area covered, as well as the mechanisms of detachment.

Combined macroscopic and microscopic analysis. Biofilms formed on basalt particles in airlift reactors were studied by Kwok *et al.* (1998). Biofilm was established under reproducible conditions, then the effect of changing the substrate loading and fluid force on detachment was investigated. Detachment was assessed by sonicating the

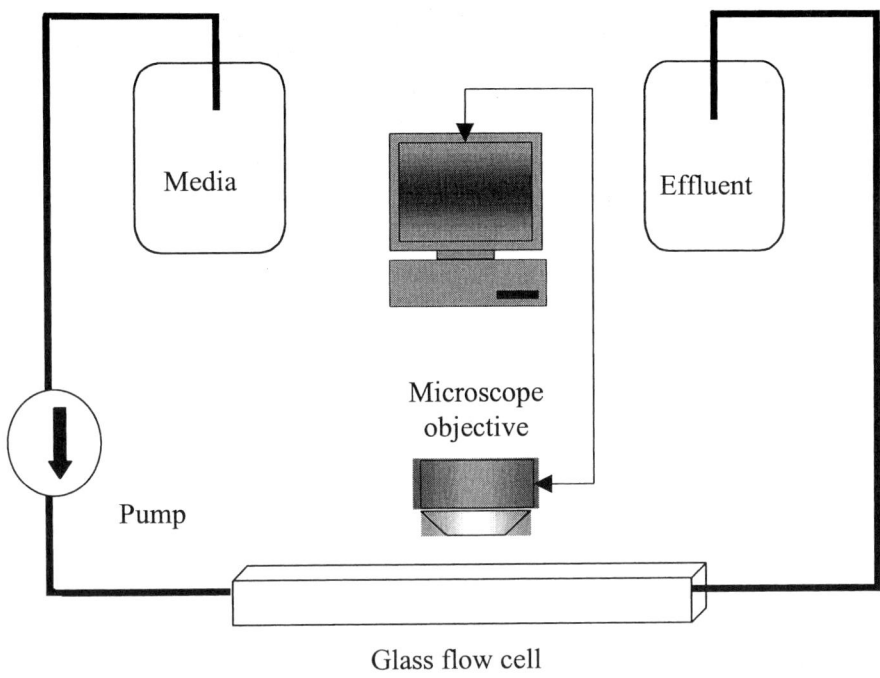

Fig. 3. Schematic of a simple image analysis system to study biofilm detachment. Systems can be more sophisticated to include: recirculating flow (useful for high flow rates), non-pulsed flow (useful for slow flow rates) and multiple flow cells.

particles to remove the biofilm, then the organic carbon was determined by dry weight. Biofilm morphology was also monitored using an image analysis system.

In summary, few methods are available for the quantification of detachment in biofilms. However, there are examples of experimental design that enable the researcher to measure detachment. The culture method chosen to develop the biofilm will largely depend on the system to be modelled.

Destructive sampling can provide insight into detachment and may be currently the most fruitful way of studying real biofilms. The danger of using such sample methods for measuring detachment is that the destructive nature of the sampling may alter the process of detachment and, for this reason, non-destructive sampling is preferable. The strengths of macroscopic sampling lie with the ability to take large samples. This can be especially useful when sampling industrial systems. Macro-analysis is highly reproducible when compared to microscopic analysis. The latter does, however, have the advantage of providing greater insight into the mechanisms behind detachment.

CASE STUDY: THE CONTROL OF CLINICAL BIOFILMS

In clinical situations, biofilms can colonize human tissues and man-made surfaces introduced into the human body, such as medical prostheses. These biofilms can prove potentially fatal to the patient. It is important that control measures are formulated specifically to detach and eradicate these biofilms. The following section discusses commonly applied antimicrobial agents used to control biofilms by inducing detachment.

Man-made surfaces

The introduction of foreign materials into the human body, such as contact lenses or catheters, can provide new surfaces for biofilm formation. *P. aeruginosa* (Stapleton & Dart, 1995), *Staphylococcus epidermidis* (Gabriel *et al.*, 1996), *Bacillus* sp. (Gopinathan *et al.*, 1997) and *Acanthamoeba* (Marciano-Cabral *et al.*, 2000) are commonly associated with biofilm formation on contact lenses. If these biofilms are not removed, and the user continues to wear the lenses, conditions such as keratitis and acute conjunctivitis can occur (Stamler, 1998). Biofilms on contact lenses are removed by subjecting the lens to a rigorous cleaning regime. Research by Landa *et al.* (1998) examined the efficacy of ophthalmic solutions, such as all-in-one solutions and a detergent mixture. The latter consisted of 0.25% (w/v) sodium lauryl sulphate (SLS) and 0.2% (w/v) sodium methyl cocoyl taurate (Tauranol). These solutions were tested against attached *P. aeruginosa* biofilms on gas permeable and soft contact lenses. The results showed that the all-in-one solutions were successful at inducing minor bacterial detachment of *P. aeruginosa* biofilms and the SLS/Tauranol mixture was shown to detach 95% of the *P. aeruginosa* biofilm. A study by Gavin *et al.* (2000) demonstrated that even after 30 min exposure to 3% (w/v) hydrogen peroxide, 11–13% of the *P. aeruginosa* biofilm cells were actively respiring and recoverable by standard culture methods.

In addition to the bacterial presence, protozoa can also exist on contact lens surfaces. The most significant of these protozoa is *Acanthamoeba polyphaga*, which grazes upon the biofilm. Gorlin *et al.* (1996) reported that, in the absence of a biofilm, the detaching fluids or saline rinses essentially removed all amoebae from the lens surface. However, the cleaning solutions were less effective at controlling the *Acanthamoeba* in the presence of a biofilm of *P. aeruginosa*.

Methicillin-resistant *Staphylococcus aureus* (MRSA) can colonize catheters (Steinberg *et al.*, 1996). The effects of antimicrobial agents on MRSA biofilms grown on silastic are being studied in our laboratory (S. M. Jones, T. J. Humphrey & H. M. Lappin-Scott, unpublished observations). Detachment of a 48 h MSRA biofilm established on silastic rubber was induced after exposing the biofilm to 100 μg vancomycin ml^{-1}. The biofilm detached, the surface area coverage being reduced by 4.76% min^{-1} after administering the antibiotic (Figs 4 and 5), thus demonstrating that high concentrations of vancomycin are effective at increasing detachment of MRSA biofilms.

Human tissues

Biofilms can form on natural surfaces within the human body and lead to serious health problems. For example, endocarditis is a life-threatening infection of the heart valves. Once established, the mortality rate may be as high as 70%. The bacteria commonly associated with endocarditis are viridans streptococci (Douglas *et al.*, 1993), *S. aureus* (Kelly & Barnass, 1999), *Enterococcus faecium* (Houry & Crisman, 1999) and *S. epidermidis* (Costa *et al.*, 1999). Research by Entenza *et al.* (1999) examined the efficacy of trovafloxacin, vancomycin and ciprofloxacin in the treatment of experimental staphylococcal and streptococcal endocarditis. Results showed that trovafloxacin and ciprofloxacin were successful in inducing detachment of staphylococcal biofilms from heart valves and were comparable to vancomycin in terms of effectiveness. However, the use of ciprofloxacin resulted in the selection of resistant strains of staphylococci (Entenza *et al.*, 1999). These results showed that the antibiotics tested have varying degrees of success for treating endocarditis.

The recent discovery of MRSA strains that have a reduced susceptibility to vancomycin (VISA) have also reduced its effectiveness in treating heart valve colonization. Several studies have demonstrated that the administration of vancomycin against a VISA biofilm situated on a rabbit heart valve was ineffective, with no evidence of detachment (Patron *et al.*, 1999; Backo *et al.*, 1999). Probiotics, such as *Lactobacillus* spp., have been suggested as a novel treatment strategy. Such biofilms are encouraged to develop on natural surfaces and reduce their availability for pathogenic organisms. In addition, these harmless species may produce substances that prevent adhesion of pathogens (Hawthorn & Reid, 1990; Reid *et al.*, 1990).

Antimicrobial agents have been shown to have some effect on biofilm detachment. However, it is well documented that biofilms have a greater resistance to antimicrobial agents than their planktonic counterparts (Costerton *et al.*, 1995). An improved understanding of biofilm detachment will inform the search for new agents that are more effective against biofilms.

CASE STUDY: BIOFILM DETACHMENT – THE POTENTIAL HAZARDS IN THE WATER INDUSTRY

Biofilms in water systems

Many processes in the water industry rely on biofilms for water treatment. In such cases, biofilms are beneficial (Costerton, 1999) as the micro-organisms degrade soluble

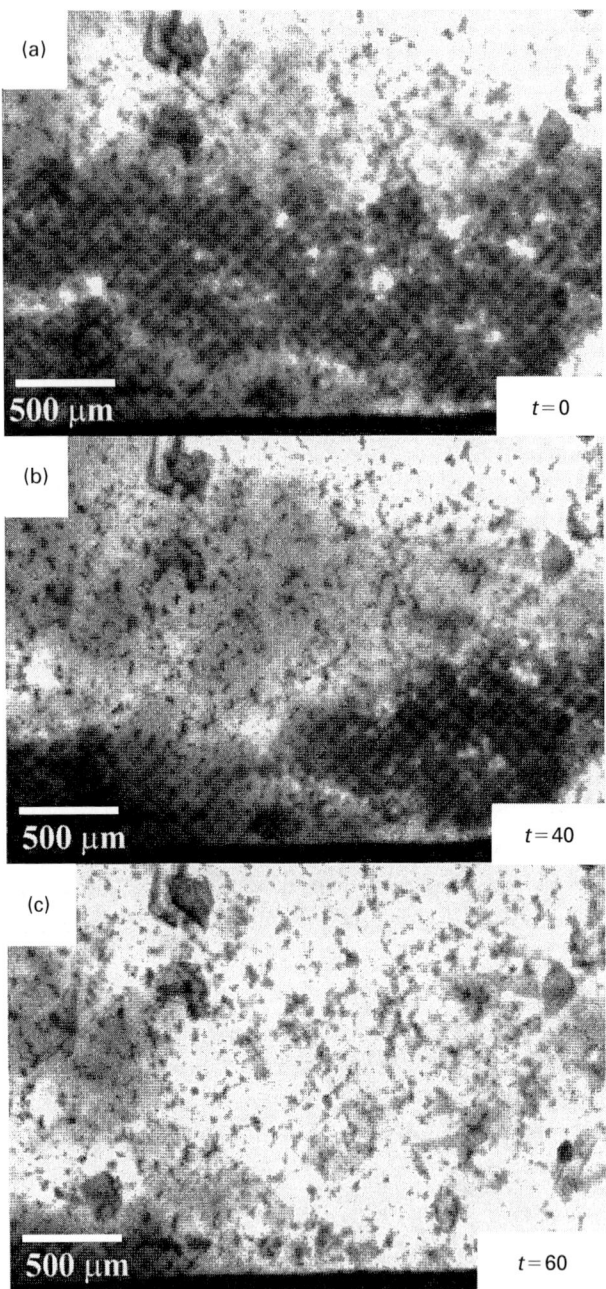

Fig. 4. Detachment of MRSA biofilm after administration of vancomycin. A 48 h MRSA biofilm formed on silastic rubber was exposed to 100 μg vancomycin ml^{-1}. (a) MRSA biofilm at $t=0$, before exposure to vancomycin; (b) biofilm after 40 min exposure; (c) biofilm after 60 min exposure (S. M. Jones, T. J. Humphrey & H. M. Lappin-Scott, unpublished results).

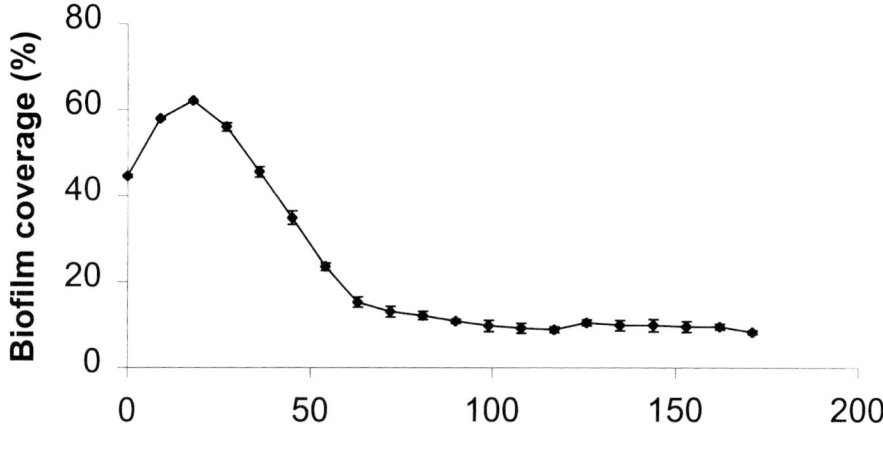

Fig. 5. Per cent surface biofilm coverage of an MRSA biofilm after exposure to vancomycin. A 48 h MRSA biofilm was established and exposed to 100 μg vancomycin ml^{-1}. Detachment was monitored by calculating biofilm coverage (%) (S. M. Jones, T. J. Humphrey & H. M. Lappin-Scott, unpublished results).

organic and nitrogenous waste materials and utilize their by-products as carbon sources. However, there are instances where biofilms have adverse effects such as in the biodeterioration, biofouling and contamination of water supplies. The harbouring of potential pathogens within biofilms has been described (Lappin-Scott & Costerton, 1989) and this can lead to detrimental effects. These pathogens can be lethal to humans and other animals because many release endotoxins which can be introduced into the water system during the sloughing of the biofilm (Rioufol *et al.*, 1999). Therefore, detachment of biofilms may involve the release of a whole array of micro-organisms, including potential pathogens, adsorbing onto microbial flocculents (flocs) and surfaces. The pathogens that are released may include *Entamoeba coli*, *Giardia lamblia*, *Shigella* spp., *Salmonella* spp. (Daly *et al.*, 1998) and *E. coli* (Mackerness *et al.*, 1993).

Biofilms in water treatment systems consist mainly of bacteria; however, it is important to consider constituent eukaryotic organisms that may be harmful or influence detachment. Microalgae form a major component in water exposed to light (Christensen & Characklis, 1990), and in hard-water reservoirs and river systems, the light and water chemistry enable the biofilm to become calcitic, leading to substantial corrosion of water pipes (Callow *et al.*, 1995). Filamentous fungi are also important in water treatment plants and there is now an improved understanding of their ability to form biofilms (Elvers *et al.*, 1998; Roberts *et al.*, 1999a, b). Phycomycetes are found in water and are primarily responsible for corrosion and deterioration in that environment. In addition, the unicellular protozoa are frequently found grazing on bacteria, algae and other particulate matter in trickling filters.

Methods of water treatment

The principal methods of water treatment include: the reverse osmosis system, the activated sludge method and the fluidized-bed system. The reverse osmosis system is designed to decontaminate water by membrane separation of organic solutes (Altman *et al.*, 1999; Koyuncu *et al.*, 2000), inorganic solutes (Buhrmann *et al.*, 1999; Jaouen *et al.*, 1999; Padilla & Tavani, 1999) and bacteriophages (Governal & Gerba, 1999). The effectiveness of this process is reduced by biofilm formation as the semi-permeable membrane surfaces are particularly susceptible to colonization (Bremere *et al.*, 1999).

The activated sludge method of water treatment uses a mixed microbial population in the form of flocs. Suspended and colloidal material in waste water adhere and adsorb to the flocs. The flocs are held together by extracellular polymeric substances produced by many micro-organisms such as *Zoogloea ramigera*. The waste materials are then digested and the sludge is separated from the water by sedimentation. Floc stability is determined by the ionic strength of the medium and can be explained by the Derjaugin, Landau, Verwey and Overbeek theory (Zita & Hermansson, 1994). When the ionic strength increases above 0.1, floc stability decreases and flocs become detached (Zita & Hermansson, 1994). This theory has been demonstrated with *Achromobacter* sp. and *Pseudomonas* sp., where reversible adsorption was shown when cells were displaced by gentle rinsing or fluid shear (Marshall *et al.*, 1971). In severe cases, increases in the ionic strength lead to heavy loss of micro-organisms (deflocculation) into the treated effluent. This in turn causes the release of toxic xenobiotic and nitrogenous compounds into the water supply.

Deflocculation or 'wash-out' can also be caused by the presence of low levels of dissolved oxygen, shock toxic loads, low pH, floc age and turbulence in waste water reactors. Many laboratory studies support the role of abiotic factors in determining the extent of biofilm detachment. For example, when an *A. hydrophila* biofilm was exposed to nutrient-limiting conditions there was an increase in detachment rates (Sawyer & Hermanowicz, 1998). In addition, Melo & Vieira (1999) demonstrated that high liquid velocity and turbulent flow in waste water reactors induced detachment due to the hydrodynamic forces.

The fluidized-bed system (Fig. 6) comprises a combination of both fixed and suspended bacterial cells where, through a constantly formed layer of extracellular polymeric

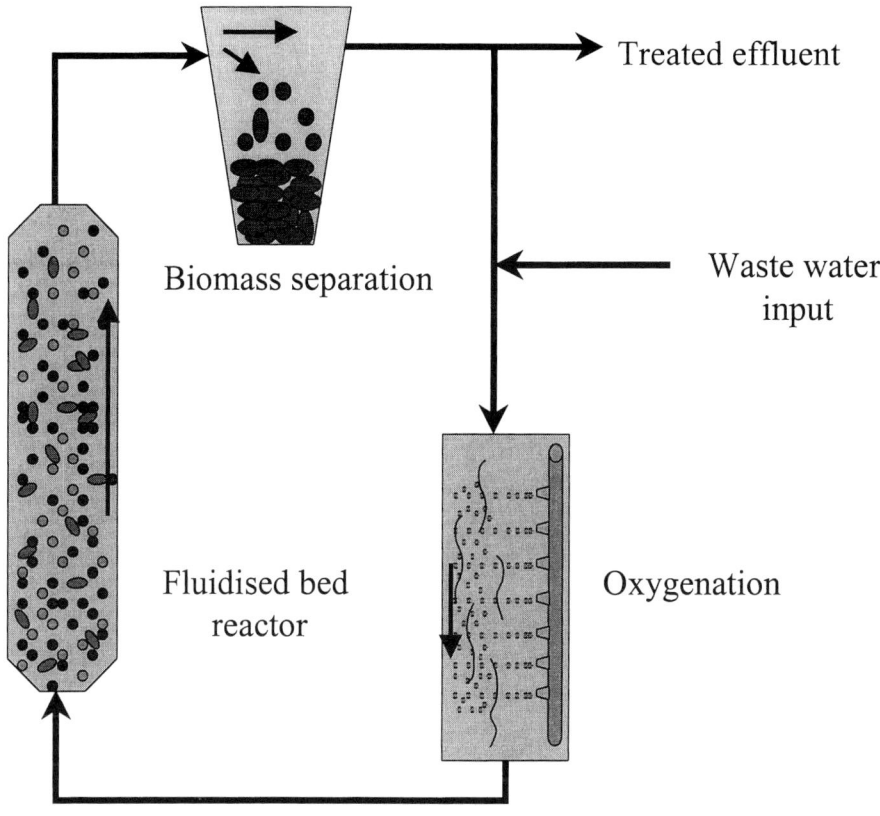

Fig. 6. Schematic representation of a fluidized-bed reactor.

substances, an upward flow of waste water can be effectively treated. This system has advantages over the floc growth system as a very high density of biomass can be formed and, due to the free movement of particles, the system does not have the blockages associated with the fixed-bed process. In addition, the use of solid support materials such as sand, carbon, glass or anthracite particles, ranging in size from 0.2 to 3 mm, prevents 'wash-out' as encountered in the floc growth system (Cooper & Wheeldon, 1980). Finally, the rate of biomass removal through the effluent stream never exceeds the rate at which biomass accumulates, thus the flow rate is low, leading to minimal detachment through erosion. However, there is a small degree of biofilm detachment as a consequence of abrasive forces.

Biofilm control in water systems
The control of biofilms in industrial water distribution systems is very challenging (Holt, 1995) as it is essential to optimize chemical or physical treatments to gain

complete depletion of the biofilm, without compromising water quality. There are many strategies employed in treating contamination by biofilms (biofouling) in industrial water processes; most of these involve the detachment of the biofilm from its surface. Biocide regimes using chlorine dioxide and ultraviolet irradiation are two such methods that have proved successful in biofilm detachment and control in water systems (Walker *et al.*, 1995). Shear force is another technique and involves the physical removal of the biofilm from its surface under many conditions, including alterations in fluid flow and changes in hydrodynamic conditions (Stoodley *et al.*, 1999b). Such erosive methods to combat and thus deform biofilms can also be achieved by many other processes, including flushing, backwashing, sand scouring and the use of non-abrasive or abrasive sponge balls (Rittmann, 1989).

In summary, it is evident that there are many modes of biofilm detachment in the water industry. The influence of factors such as the system used, nutrient limitations and physical stresses all have a role in influencing detachment rates. This is of significance in terms of both biofilm control and the potential hazardous implications that may arise. Of the limited research conducted on biofilm detachment, most studies focus on laboratory observations but there are a few studies of the natural detachment processes in water systems.

CONCLUSION

This chapter has highlighted the importance of understanding the detachment process in biofilms and indicated that this is, as yet, a relatively new but crucial area of research. However, through new experimental designs and methodology, progress is being made with establishing the mechanisms of biofilm detachment. Understanding detachment mechanisms is crucial for controlling biofilms. In the water industry, this knowledge is fundamental for the optimum performance of biofilm reactors to stabilize beneficial biofilms. In clinical situations, it is imperative to minimize biofilm formation and understand factors influencing the detachment of detrimental biofilms. Studies have shown that regulatory systems such as quorum sensing are involved in detachment and further research in this area could lead to the production of novel agents for biofilm control.

REFERENCES

Adams, J. L. & McLean, R. J. C. (1999). Impact of *rpoS* deletion on *Escherichia coli* biofilms. *Appl Environ Microbiol* **65**, 4285–4287.

Allison, D. G., Evans, D. G., Brown, M. R. W. & Gilbert, P. (1990). Possible involvement of the division cycle in dispersal of *Escherichia coli* from biofilms. *J Bacteriol* **172**, 1667–1669.

Allison, D. G., Ruiz, B., Sanjose, C., Jaspe, A. & Gilbert, P. (1998). Extracellular products as mediators of the formation and detachment of *Pseudomonas fluorescens* biofilms. *FEMS Microbiol Lett* **167**, 179–184.

Allison, D. G., Heys, S. J. D., Willcock, L., Holah, J. & Gilbert, P. (1999). Cellular detachment and dispersal from bacterial biofilms: a role for quorum sensing? In *Biofilms – The Good, the Bad and the Ugly*, pp. 279–286. Edited by J. Wimpenny, P. Gilbert, J. Walker, M. Brading & R. Bayston. Cardiff: BioLine.

Altman, M., Semiat, R. & Hasson, D. (1999). Removal of organic foulants from feed waters by dynamic membranes. *Desalination* **125**, 65–75.

Applegate, D. H. & Bryers, J. D. (1991). Effects of carbon and oxygen limitations and calcium concentrations on biofilm removal processes. *Biotechnol Bioeng* **37**, 17–25.

Austin, J. W., Sanders, G., Kay, W. W. & Collinson, S. K. (1998). Thin aggregative fimbriae enhance *Salmonella enteritidis* biofilm formation. *FEMS Microbiol Lett* **162**, 295–301.

Backo, M., Gaenger, E., Burkart, A., Chai, Y. L. & Bayer, A. S. (1999). Treatment of experimental staphylococcal endocarditis due to a strain with reduced susceptibility in vitro to vancomycin: efficacy of ampicillin-sulbactam. *Antimicrob Agents Chemother* **43**, 2565–2568.

Bos, R., van der Mei, H. C. & Busscher, H. J. (1999). Physico-chemistry of initial microbial adhesive interactions – its mechanisms and methods for study. *FEMS Microbiol Rev* **23**, 179–230.

Boyd, A. & Chakrabarty, A. M. (1994). Role of alginate lyase in cell detachment of *Pseudomonas aeruginosa*. *Appl Environ Microbiol* **60**, 2355–2359.

Brading, M. G., Boyle, J. D. & Lappin-Scott, H. M. (1995). Biofilm formation in laminar flow using *Pseudomonas fluorescens* EX101. *J Ind Microbiol* **15**, 297–304.

Bremere, I., Kennedy, M., Michel, P., vanEmmerik, R., Witkamp, G. J. & Schippers, J. (1999). Controlling scaling in membrane filtration systems using a desupersaturation unit. *Desalination* **124**, 51–62.

Bryers, J. D. (1988). Modeling biofilm accumulation. In *Physiology Models in Microbiology*, vol. 2, pp. 109–144. Edited by M. J. Bazin & J. I. Prosser. Boca Raton, FL: CRC.

Buhrmann, F., VanderWaldt, M., Hanekom, D. & Finlayson, F. (1999). Treatment of industrial wastewater for reuse. *Desalination* **124**, 263–269.

Callow, M. E., Heath, C. R., Hartley Barry, A., Leadbeater, S. C. & House, W. A. (1995). Calcification of algal biofilms in hard waters. In *The Life and Death of Biofilm*, pp. 113–117. Edited by J. Wimpenny, P. Handley, P. Gilbert & H. Lappin-Scott. Cardiff: BioLine.

Chang, H. T., Rittman, B. E., Amar, D., Heim, R., Ehlinger, O. & Lesty, Y. (1991). Biofilm detachment mechanisms in a liquid-fluidized bed. *Biotechnol Bioeng* **38**, 499–506.

Characklis, W. G. (1990). Biofilm process. In *Biofilms*, pp. 195–231. Edited by W. G. Characklis & K. C. Marshall. New York: Wiley.

Characklis, W. G., McFeters, G. S. & Marshall, K. C. (1990). Physiological ecology in biofilm systems. In *Biofilms*, pp. 341–394. Edited by W. G. Characklis & K. C. Marshall. New York: Wiley.

Christensen, B. E. & Characklis, W. G. (1990). Physical and chemical properties of biofilms. In *Biofilms*, pp. 93–130. Edited by W. G. Characklis & K. C. Marshall. New York: Wiley.

Cooper, P. F. & Wheeldon, D. H. V. (1980). Fluidised and expanded bed reactors for waste-water treatment. *Water Pollut Control* **79**, 286–306.

Costa, G. M., Pizzi, C., Leone, C., Borghi, A., Cordioli, E. & Bugiardini, R. (1999). Thrombosis of a mitral valve prosthesis resulting from *Staphylococcus epidermidis* endocarditis. *Cardiologia* **44**, 675–678.

Costerton, J. W. (1999). The role of bacterial exopolysaccharides in nature and disease. *J Ind Microbiol Biotechnol* **22**, 551–563.

Costerton, J. W., Lewandowski, Z., Caldwell, D. E., Korber, D. R. & Lappin-Scott, H. M. (1995). Microbial biofilms. *Annu Rev Microbiol* **49**, 711–745.

Daly, B., Betts, W. B., Brown, A. P. & O'Neil, J. F. (1998). Bacterial loss from biofilms exposed to free chlorine. *Microbios* **96**, 7–21.

Davies, D. G., Parsek, M. R., Pearson, J. P., Iglewski, B. H., Costerton, J. W. & Greenberg, E. P. (1998). The involvement of cell-to-cell signals in the development of a bacterial biofilm. *Science* **280**, 295–298.

Douglas, C. W. I., Heath, J., Hampton, K. K. & Preston, F. E. (1993). Identity of viridans streptococci isolated from cases of infective endocarditis. *J Med Microbiol* **39**, 179–182.

Eginton, P. J., Gibson, H., Holah, J., Handley, P. S. & Gilbert, P. (1995). Quantification of the ease of removal of bacteria from surfaces. *J Ind Microbiol* **15**, 305–310.

Elvers, K. T., Leeming, K., Moore, C. P. & Lappin-Scott, H. M. (1998). Bacterial-fungal biofilms in flowing water photo-processing tanks. *J Appl Microbiol* **84**, 607–618.

Entenza, J. M., Vouillamoz, J., Glauser, M. P. & Moreillon, P. (1999). Efficacy of trovafloxacin in treatment of experimental staphylococcal or streptococcal endocarditis. *Antimicrob Agents Chemother* **43**, 77–84.

Gabriel, M. M., Schultz, C. L., Wilson, L. A. & Ahearn, D. G. (1996). Effect of *Staphylococcus epidermidis* on hydrogel contact lens retention on the rabbit eye. *Curr Microbiol* **32**, 176–178.

Gavin, J., Button, N. F., Watson-Craik, I. A. & Logan, N. A. (2000). Observation of soft contact lens disinfection with fluorescent metabolic stains. *Appl Environ Microbiol* **66**, 874–875.

Gilbert, P., Evans, D. J., Evans, E., Duguid, I. G. & Brown, M. R. W. (1991). Surface characteristics and adhesion of *Escherichia coli* and *Staphylococcus epidermidis*. *J Appl Bacteriol* **71**, 71–77.

Gopinathan, U., Stapleton, F., Sharma, S., Willcox, M. D., Sweeney, D. F., Rao, G. N. & Holden, B. A. (1997). Microbial contamination of hydrogel contact lenses. *J Appl Microbiol* **82**, 653–658.

Gorlin, A. I., Gabriel, M. M., Wilson, L. A. & Ahearn, D. G. (1996). Binding of Acanthamoeba to hydrogel contact lenses. *Curr Eye Res* **15**, 151–155.

Governal, R. A. & Gerba, C. P. (1999). Removal of MS-2 and PRD-1 bacteriophages from an ultrapure water system. *J Ind Microbiol Biotechnol* **23**, 166–172.

Hammer, M., Arnqvist, A., Bian, Z., Olsen, A. & Normark, S. (1995). Expression of two *csg* operons is required for production of fibronectin- and congo red-binding curli polymers in *Escherichia coli*. *Mol Microbiol* **18**, 661–670.

Hawthorn, L. E. & Reid, G. (1990). Exclusion of uropathogen adhesion to polymer surfaces by *Lactobacillus acidophilus*. *J Biomed Mater Res* **24**, 39–46.

Holt, D. (1995). Challenge of controlling biofilms in water distribution systems. In *The Life and Death of Biofilm*, pp. 161–166. Edited by J. Wimpenny, P. Handley, P. Gilbert & H. Lappin-Scott. Cardiff: BioLine.

Houry, D. & Crisman, T. (1999). Bivalve polymicrobial infective endocarditis. *South Med J* **92**, 1098–1099.

Jaouen, P., Lepine, B., Rossignol, N., Royer, R. & Quemeneur, F. (1999). Clarification and concentration with membrane technology of a phycocyanin solution extracted from *Spirulina platensis*. *Biotechnol Tech* **13**, 877–881.

Kelly, J. & Barnass, S. (1999). *Staphylococcus aureus* endocarditis presenting as meningitis and mimicking meningococcal sepsis. *Int J Clin Pract* **53**, 306–307.

Koyuncu, I., Turan, M., Topacik, D. & Ates, A. (2000). Application of low pressure nanofiltration membranes for the recovery and reuse of dairy industry effluents. *Water Sci Technol* **41**, 213–221.

Kwok, W. K., Picioreanu, C., Ong, S. L., Van Loosdrecht, M. C. M., Ng, W. J. & Heijnen, J. J. (1998). Influence of biomass production and detachment forces on biofilm structures in a biofilm airlift suspension reactor. *Biotechnol Bioeng* **58**, 400–407.

Ladd, T. I. & Costerton, J. W. (1990). Methods for studying biofilm bacteria. *Methods Microbiol* **22**, 285–310.

Lamed, R. & Bayer, E. A. (1986). Contact and cellulolysis in *Clostridium thermocellum* via extensile surface organelles. *Experientia* **42**, 72–73.

Landa, A. S., van der Mei, H. C., van Rij, G. & Busscher, H. J. (1998). Efficacy of ophthalmic solutions to detach adhering *Pseudomonas aeruginosa* from contact lenses. *Cornea* **17**, 293–300.

Lappin-Scott, H. M. & Costerton, J. W. (1989). Bacterial biofilms and surface fouling. *Biofouling* **1**, 323–342.

Lappin-Scott, H. M., Brading, M. G. & Jass, J. (1994). How do bacteria reach surfaces? In *Bacterial Biofilms*, pp. 19–24. Edited by J. Wimpenny, W. Nichols, D. Stickler & H. M. Lappin-Scott. Cardiff: BioLine.

Lee, S. F., Li, Y. H. & Bowden, G. H. (1996). Detachment of *Streptococcus mutans* biofilm cells by an endogenous enzymatic activity. *Infect Immun* **64**, 1035–1038.

Loewen, P. C., Hu, B., Strutinsky, J. & Sparling, R. (1998). Regulation in the *rpoS* regulon of *Escherichia coli*. *Can J Microbiol* **44**, 707–717.

Lynch, M. J., Swift, S., Kirke, D., Dodd, C. E. R., Keevil, C. W., Stewart, G. S. A. B. & Williams, P. (1999). Investigation of quorum sensing in *Aeromonas hydrophila* biofilms formed on stainless steel. In *Biofilms – The Good, the Bad and the Ugly*, pp. 209–223. Edited by J. Wimpenny, P. Gilbert, J. Walker, M. Brading & R. Bayston. Cardiff: BioLine.

Mackerness, C. W., Colbourne, J. C., Dennis, P. J., Rachwal, A. J. & Keevil, C. W. (1993). Formation and control of coliform biofilms in drinking water distribution systems. *Society for Applied Bacteriology Technical Series: Microbial Biofilms*. pp. 217–226.

Mai, G. T., McCormack, J. G., Seow, W. K., Pier, G. B., Jackson, L. A. & Thong, Y. H. (1993). Inhibition of adherence of mucoid *Pseudomonas aeruginosa* by alginase specific monoclonal antibodies and antibiotics. *Infect Immun* **61**, 4338–4343.

Marciano-Cabral, F., Puffenbarger, R. & Cabral, G. A. (2000). The increasing importance of Acanthamoeba infections. *J Eukaryot Microbiol* **47**, 29–36.

Marshall, K. C., Stout, R. & Mitchell, R. (1971). Mechanism of the initial events in the sorption of marine bacteria to surfaces. *J Gen Microbiol* **68**, 337–348.

Melo, L. & Vieira, M. J. (1999). Physical stability and biological activity of biofilms under turbulent flow and low substrate concentration. *Bioprocess Eng* **20**, 363–368.

Moore, G. F., Humphrey, T. J., Holah, J., Woodward, M. J. & Lappin-Scott, H. M. (1999). The effect of adverse conditions on the biofilm formation of *Salmonella enteritidis*. In *Biofilms – The Good, the Bad and the Ugly*, pp. 15–22. Edited by J. Wimpenny, P. Gilbert, J. Walker, M. Brading & R. Bayston. Cardiff: BioLine.

Nicolella, C., DiFelice, R. & Rovatti, M. (1996). An experimental model of biofilm detachment in liquid fluidized bed biological reactors. *Biotechnol Bioeng* **51**, 713–719.

Nicolella, C., Chiarle, S., DiFelice, R. & Rovatti, M. (1997). Mechanisms of biofilm detachment in fluidized bed reactors. *Water Sci Technol* **36**, 229–235.

Ohashi, A. & Harada, H. (1994). Characterization of detachment mode of biofilm developed in an attached-growth reactor. *Water Sci Technol* **30**, 35–45.

Padilla, A. P. & Tavani, E. L. (1999). Treatment of an industrial effluent by reverse osmosis. *Desalination* **126**, 219–226.

Patron, R. L., Climo, M. W., Goldstein, B. P. & Archer, G. L. (1999). Lyostaphin treatment of experimental aortic valve endocarditis caused by a *Staphylococcus aureus* isolate with reduced susceptibility to vancomycin. *Antimicrob Agents Chemother* **43**, 1754–1755.

Peyton, B. M. & Characklis, W. G. (1993). A statistical analysis of the effect of substrate utilization and shear-stress on the kinetics of biofilm detachment. *Biotechnol Bioeng* **41**, 728–735.

Puckas, M. R., Greenberg, E. P., Kaplan, S. & Schaefer, A. L. (1997). A quorum sensing system in the free-living photosynthetic bacterium *Rhodobacter sphaeroides*. *J Bacteriol* **179**, 7530–7537.

Reid, G., McGroarty, J. A., Domingue, P. A. G., Chow, A. W., Bruce, A. W., Eisen, A. & Costerton, J. W. (1990). Coaggregation of urogenital bacteria *in vitro* and *in vivo*. *Curr Microbiol* **20**, 47–52.

Rioufol, C., Devys, C., Meunier, G., Perraud, M. & Goullet, D. (1999). Quantitative determination of endotoxins released by bacterial biofilms. *J Hosp Infect* **43**, 203–209.

Rittmann, B. E. (1989). Detachment from biofilms. In *Structure and Function of Biofilms*, pp. 49–58. Edited by W. G. Characklis & P. A. Wilderer. New York: Wiley.

Roberts, S. K., Lappin-Scott, H. M. & Leeming, K. (1999a). The control of bacterial-fungal biofilms. In *Biofilms – The Good, the Bad and the Ugly*, pp. 93–104. Edited by J. Wimpenny, P. Gilbert, J. Walker, M. Brading & R. Bayston. Cardiff: BioLine.

Roberts, S. K., Bass, C., Brading, M., Lappin-Scott, H. M. & Stoodley, P. (1999b). Biofilm formation and structure: what's new? In *Dental Plaque Revisited – Oral Biofilms in Health and Disease*, pp. 15–35. Edited by H. N. Newman & M. Wilson. Cardiff: BioLine.

Sawyer, L. K. & Hermanowicz, S. W. (1998). Detachment of biofilm bacteria due to variations in nutrient supply. *Water Sci Technol* **37**, 211–214.

Sperandio, V., Mellies, J. L., Nguyen, W., Shin, S. & Kaper, J. B. (1999). Quorum sensing controls expression of type III secretion gene transcription and protein secretion in enterohemorrhagic and enteropathogenic *Escherichia coli*. *Proc Natl Acad Sci USA* **96**, 15196–15201.

Stamler, J. F. (1998). The complications of contact lens wear. *Curr Opin Ophthalmol* **9**, 66–71.

Stapleton, F. & Dart, J. (1995). Pseudomonas keratitis associated with biofilm formation on a disposable soft contact lens. *Br J Ophthalmol* **79**, 864–865.

Steinberg, J. P., Clark, C. C. & Hackman, B. O. (1996). Nosocomial and community-acquired *Staphylococcus aureus* bacteremias from 1980 to 1993: impact of intravascular devices and methicillin resistance. *Clin Infect Dis* **23**, 255–259.

Stewart, P. S. (1993). A model of biofilm detachment. *Biotechnol Bioeng* **41**, 111–117.

Stoodley, P., Lewandowski, Z., Boyle, J. D. & Lappin-Scott, H. M. (1999a). Structural deformation of bacterial biofilms caused by short-term fluctuations in fluid shear: an *in situ* investigation of biofilm rheology. *Biotechnol Bioeng* **65**, 83–92.

Stoodley, P., Dodds, I., Boyle, J. D. & Lappin-Scott, H. M. (1999b). Influence of hydrodynamics and nutrients on biofilm structure. *J Appl Microbiol* **85**, 19S–28S.

Stoodley, P., Jørgensen, F., Williams, P. & Lappin-Scott, H. M. (1999c). The role of hydrodynamics and AHL signalling molecules as determinants of the structure of *Pseudomonas aeruginosa* biofilms. In *Biofilms – The Good, the Bad and the Ugly*, pp. 223–230. Edited by J. Wimpenny, P. Gilbert, J. Walker, M. Brading & R. Bayston. Cardiff: BioLine.

Stoodley, P., Boyle, J. D., DeBeer, D. & Lappin-Scott, H. M. (1999d). Evolving perspectives in biofilm structure. *Biofouling* **14**, 75–90.

Stoodley, P., Lewandowski, Z., Boyle, J. D. & Lappin-Scott, H. M. (1999e). The formation of migratory ripples in a mixed species bacterial biofilm growing in turbulent flow. *Environ Microbiol* 447–455.

Tijhuis, L., Hijman, B., van Loosdrecht, M. C. M. & Heijnen, J. J. (1996). Influence of detachment, substrate loading and reactor scale on the formation of biofilms in airlift reactors. *Appl Environ Microbiol* **45**, 7–17.

Walker, J. T., Mackerness, C. W., Mallon, D., Makin, T., Williets, T. & Keevil, C. W. (1995). Physico-chemical eradication in the field. In *The Life and Death of Biofilm*, pp. 167–171. Edited by J. Wimpenny, P. Handley, P. Gilbert & H. Lappin-Scott. Cardiff: BioLine.

Zita, A. & Hermansson, M. (1994). Effects of ionic-strength on bacterial adhesion and stability of flocs in a waste-water activated-sludge system. *Appl Environ Microbiol* **60**, 3041–3048.

Modelling and predicting biofilm structure

Cristian Picioreanu, Mark C. M. van Loosdrecht and
Joseph J. Heijnen

Kluyver Institute for Biotechnology, Delft University of Technology, Julianalaan 67, 2628 BC Delft,
The Netherlands

INTRODUCTION TO BIOFILM STRUCTURE

The general view on biofilm structure has dramatically changed during the last decade. It had been previously assumed that most biofilms are more or less homogeneous layers of micro-organisms in a slime matrix. The use of confocal scanning laser microscopy (CSLM) and computerized image analysis tools has revealed a more complex picture of biofilm morphology (Lawrence *et al.*, 1991; Caldwell *et al.*, 1993) and structural heterogeneities (Costerton *et al.*, 1994; Gjaltema *et al.*, 1994). Cell clusters may be separated by interstitial voids and channels, which create a characteristic porous structure. In some cases, biofilms grow as clusters taking a 'mushroom' shape, whilst in others, more compact and homogeneous layers can be observed.

Types of biofilm heterogeneity

In many biofilms, the reported non-uniformities are unidirectional and perpendicular to the substratum. Three-dimensional variation of microbial species, biofilm porosity, substrate concentration and diffusivity have been repeatedly reported in biofilm. It is becoming clear that there are many forms of heterogeneity in biofilms, and a definition of *biofilm heterogeneity* is needed. According to Bishop & Rittmann (1995), heterogeneity may be defined as 'spatial differences in any parameter we think is important'. An adapted list from Bishop & Rittmann (1995) summarizes a few examples of possible biofilm heterogeneity:

(1) *Geometrical heterogeneity*: biofilm thickness, biofilm surface roughness, biofilm porosity and substratum surface coverage.

SGM symposium 59: Community structure and co-operation in biofilms. Editors D. Allison, P. Gilbert, H. Lappin-Scott, M. Wilson.
Cambridge University Press. ISBN 0 521 79302 5 ©SGM 2000.

(2) *Chemical heterogeneity*: diversity of chemical solutes (nutrients, metabolic products, inhibitors), pH variations, diversity of reactions (aerobic/anaerobic, etc.).

(3) *Biological heterogeneity*: microbial diversity of species and their spatial distribution, differences in activity [growing cells, extracellular polymeric substances (EPS) producers, dead cells, etc.].

(4) *Physical heterogeneity*: density, permeability, visco-elasticity, viscosity, properties of EPS, strength, solute concentration, solute diffusivity, presence of abiotic solids.

Geometrical heterogeneity is a primary cause of many other kinds of heterogeneity and was therefore our primary focus. A proper description of biofilm geometrical morphology is probably the basis for a good description of the other types of heterogeneity.

Quantifying biofilm geometrical heterogeneity

Geometrical properties of biofilm surface and volume have to be defined. As Lewandowski *et al.* (1999) clearly recognized, 'quantifying the structure of biofilms is an important step towards understanding and describing biofilm systems'. Different measures of biofilm geometrical heterogeneity have been proposed. Zhang & Bishop (1994a) measured biofilm density, porosity, specific surface area and mean pore radius, adding two further properties, tortuosity and pore length (Zhang & Bishop, 1994b), to account for the non-uniform effective diffusivities in the biofilm depth. Hermanowicz *et al.* (1995, 1996) used fractal dimensions to describe the morphology of a biofilm whilst Gibbs & Bishop (1995) characterized biofilm surface by surface roughness and fractal dimension. Murga *et al.* (1995) introduced a coefficient of surface roughness. Irregularity of biofilm particles in an airlift reactor was differentiated by Kwok *et al.* (1998) using a shape factor. Recently, Lewandowski *et al.* (1999) defined new biofilm parameters such as textural entropy, arial porosity, fractal dimension and maximum diffusion distance. All of these researchers studied structural elements of real biofilms. It is, at present, not clear which of these parameters are most appropriate. For purposes of describing the structure of computer-simulated biofilms, Picioreanu *et al.* (1998b) defined measures such as biofilm surface roughness, surface area enlargement, fractal dimension of biofilm surface, biofilm porosity and compactness.

Importance of biofilm structure upon properties

An improved knowledge of biofilm structure will lead to better understanding, interpretation and prediction of the influence that biofilm may have on systems (Stoodley *et al.*, 1997) and to an increased ability to manipulate structures, in order to create desirable biofilm properties.

Overall transformation rates in a bioreactor. The morphological characteristics of biofilms (thickness, density and surface shape) are very important for the overall performance of biofilm reactors. These factors affect strongly the biomass hold-up and mass transfer in a biofilm reactor (Garrido *et al.*, 1997; Tijhuis *et al.*, 1995). The biofilm density has a direct effect on the achievable biomass concentration in the reactor, and consequently on the overall substrate conversion rate. For aerobic processes, thin biofilms (<150 μm) are favourable (Tijhuis *et al.*, 1994). Limited substrate penetration depth leads to a high fraction of non-active biomass in thick biofilms. The shape of the biofilm surface influences the mass transfer properties of nutrients and antimicrobial agents to biofilm cells, by affecting the external mass transfer resistance. The effect of biofilm surface heterogeneity on mass transfer boundary layers was recently studied with microelectrode techniques by several research groups (De Beer *et al.*, 1994, 1996; De Beer & Stoodley, 1995; Yang & Lewandowski, 1995; Zhang *et al.*, 1994; Bishop *et al.*, 1997). Similarly, biofilm porosity and gel matrix composition affect the internal resistance to solute transport, by influencing the diffusivities of soluble compounds. A review of experimental determinations of effective diffusion coefficients within biofilms is provided by Stewart (1998).

Stability of bioreactor operation. Sloughing phenomena occur more frequently with thicker biofilms (Beeftink, 1987). The control of biofilm thickness is therefore an important aspect for the stable operation of biofilm reactors. Biofilm surface shape is also an important parameter for the stability of the reactor. This is especially the case in particle biofilm processes (e.g. fluid-bed reactors), where fluffy biofilms and outgrowths lead to instabilities with respect to the separation ('settleability') of biofilm particles from the treated water (Tijhuis *et al.*, 1995). Moreover, due to sloughing, filamentous biofilms and excessively thick biofilms often lead to lower effluent quality. On the other hand, suspended particles in the wastewater are more easily filtered from the wastewater by filamentous than by smooth biofilms.

Fluid frictional resistance. Stoodley *et al.* (1999b) suggest that 'fluid shear stresses acting on individual cell clusters in a heterogeneous biofilm can be much more complex than for a simple planar biofilm' and propose that the three types of flow past a rigid rough surface, identified by Nowell & Church (1979), can be applicable to biofilms grown in flow cells, as a function of surface coverage. First, if the cell clusters are very isolated, then the velocity profile behind a cluster will completely recover to the profile in the empty channel. Second, a 'wake interaction flow' may occur when cell clusters are close enough so that each cluster is in a flow wake (or vortex) created by an up-stream cluster. Finally, when the biofilm completely covers the surface, vortices are less likely to occur, and the flow may 'skim' over the biofilm surface. Furthermore, in turbulent conditions, the biofilm can develop filamentous structures ('streamers') that are free to oscillate in flowing conditions (Stoodley *et al.*, 1998). As Stoodley *et al.* (1998) clearly demonstrated, streamer oscillations are directly related to vortex generation and drag. Their data (Stoodley *et al.*, 1999a) support the hypothesis that both viscous and elastic behaviour of biofilms may explain the large pressure drops observed in biofilm-fouled pipes.

Biofilm detachment. Biofilms become more susceptible to sloughing, not only with increased thickness, but also with increased geometrical heterogeneity. As Stoodley *et al.* (1997) suggested, in a very heterogeneous biofilm, large chunks are more likely to be broken away. Moreover, Stoodley *et al.* (1999b) hypothesized that the detachment of individual clusters from the biofilm may be dependent on their shape and location in the biofilm. On the other hand, uniformly thick biofilms may loose biomass mainly by a continuous process of superficial erosion. All these features can be found in the detachment model presented by Picioreanu (1999). A direct implication of biofilm detachment is microbial contamination and the spread of biofilm in drinking-water systems or in diverse medical applications. It is not only shape and other geometrical heterogeneities that influence biofilm detachment, but also differences in matrix strength. Ohashi & Harada (1994, 1996) and Ohashi *et al.* (1999) found that the biofilm strength increased with an increasing biofilm dry density. If cells are decaying in the biofilm depth, due to ageing or starvation, this may lead to lessening of the adhesive strength, and a faster biofilm removal rate.

Biofilm ecology. Natural biofilms are consortia of many different microbial species. In many cases, multi-degradation processes occur in a food chain, for instance, in nitrification or in methanogenesis. Based on confocal laser microscope observations after *in situ* hybridization, Schramm *et al.* (1996) hypothesized that ammonia- and nitrite-oxidizing colonies in a biofilm grew in close proximity as a direct result of the sequential metabolism of ammonia to nitrate. It is believed that, by growing in this spatial arrangement, the very short diffusion path from *Nitrosomonas* clusters to *Nitrobacter* species facilitates an efficient transfer of the intermediate nitrite.

Factors influencing biofilm structure formation

There is a multitude of causes that, together, can lead to emergence of a certain type of biofilm. However, the reasons for biofilm structural heterogeneity are largely unknown. Because biofilm accumulates as the result of an interaction between micro-organisms and the growth medium, then these causes must relate to both biotic and abiotic factors. While microbiologists, in general, tend to look more for biotic reasons of biofilm heterogeneity, engineers are inclined to study the abiotic. As also Stoodley *et al.* (1997) recognized, there are at least four major influences on biofilm structure: (1) geometrical characteristics of substratum surface; (2) characteristics of micro-organisms constituting the biofilm; (3) hydrodynamic conditions; (4) nutrient availability.

(1) *Substratum characteristics* play an important role, especially in early stages of biofilm formation. Geometrical characteristics of the carrier surface are important because its roughness promotes bacterial colonization (Verran *et al.*, 1991; Gjaltema *et al.*, 1997). Fox *et al.* (1990) observed that carrier surface roughness was critical to biofilm development during the start-up period of an expanded-bed reactor. They hypothesized that biofilm growth begins in the crevices where adherent cells are protected from shear forces. As the biofilm fills the crevices, cell clusters from neighbouring crevices join, and a mature biofilm completely covers the rough surface.

(2) *Biological characteristics* of micro-organisms forming a biofilm affect the biofilm accumulation in many ways. First, differences in kinetics of microbial growth (growth rates and yields of biomass on substrate) lead to a difference in biofilm structure. Research on nitrifying and heterotrophic bacteria shows that under similar airlift reactor conditions (temperature, shear), the slow-growing nitrifiers form a much denser biofilm than fast-growing heterotrophs (Tijhuis *et al.*, 1994, 1995). Similar observations are reported for methanogenic and acidifying bacteria (Alphenaar, 1995). This fact could also be explained by a second important aspect: the rate of EPS production. Heterotrophs (e.g. *Pseudomonas aeruginosa*) can produce copious amounts of polymeric matrix, which decrease the mean cell density in the biofilm. Finally, other researchers argue that quorum sensing and cell signalling might play major roles in biofilm detachment (Allison *et al.*, 1999), by inducing production of enzymes that cause a rapid cleavage of the EPS chains and thus releasing cells from the polymeric matrix. Conversely, surface attachment can induce alginate expression in *P. aeruginosa*, reinforcing the bonds between cell and substratum (Davies *et al.*, 1994). Davies *et al.* (1998) reported recently that homoserine lactones were required for the development of heterogeneous biofilms, whereas mutants lacking these lactones formed compact biofilms. Other researchers (Stoodley *et al.*, 1999c), however, found no marked difference in the structural complexity of mutant biofilm as compared to the wild-type, but rather hydrodynamics had the more profound influence on the emerged structure.

(3) *Hydrodynamics* influences biofilm development in several ways. Firstly, the flow pattern determines the magnitude of external mass transfer of nutrients to, and of products from, the biofilm. The higher the liquid velocity, the higher the mass transfer rate can be, and, consequently, the faster the potential biofilm accumulation rate. Secondly, faster fluid flow implies larger shear forces at the biofilm surface, leading to higher biofilm detachment rate. Finally, a new influence of flow on the structure of bacterial biofilms was signalled by Stoodley *et al.* (1999a). Their creep experiments demonstrated that the polymeric matrix of *Pseudomonas* biofilms was exhibiting permanent strain when subjected to variable fluid shear. This material thinning could lead to streamer formation.

(4) *Nutrient availability* is reflected in the flux of substrate received by biofilm bacteria from the environment. Because biofilm formation is determined by the balance between biofilm growth rate and detachment rate, effects of nutrient availability and hydrodynamics on biofilm structure must be analysed together. Results from biofilm experiments in rotating cylinder (RotoTorque) and in tubular reactors clearly indicate the effect of fluid shear rate and of substrate loading rate on biofilm density and biofilm thickness achieved in 'steady state'. The biofilm density increased (Characklis *et al.*, 1982; Christensen & Characklis, 1990) and biofilm thickness decreased (Characklis, 1981) with increasing fluid shear stress, at constant substrate loading rate. These observations can be attributed to increased coefficients of substrate transfer from bulk liquid to the biofilm surface. With an increasing substrate loading rate, but keeping the shear constant, both biofilm density (Characklis *et al.*, 1982) and maximum achieved thickness (Characklis, 1981) increased. A faster growth of biofilm thickness at higher oxygen concentration and at higher flow velocity, during biofilm development in membrane-aerated biofilm reactors, was reported by Casey (1999).

In particle-type biofilm reactors [e.g. biofilm airlift suspension (BAS) reactor, fluidized-bed reactor], similar dependencies of biofilm structural parameters on substrate loading rate and carrier concentration have been observed. Kwok *et al.* (1998) reported that biofilm thickness on basalt carrier particles decreased with a decreasing substrate loading rate, at the same carrier concentration in the reactor. An increasing carrier concentration, keeping the loading rate constant, led to a decreased biofilm thickness. This was obviously caused by a higher detachment force generated by particle–particle collisions. Also the biofilm density clearly increased with increasing carrier concentration (heterotrophic micro-organisms) in Kwok's experiments. Similar observations have been made in fluidized-bed reactors (Chang *et al.*, 1991) and in BAS reactors with nitrifying biofilms (Van Benthum *et al.*, 1996).

Evolution of biofilm models

Mathematical models have been used for the last three decades as tools to simulate the behaviour of microbial biofilms. The initial models described biofilms as uniform steady-state films containing a single type of organism (Fig. 1a). These were governed exclusively by one-dimensional mass transport and biochemical transformations (Atkinson & Davies, 1974; Rittmann & McCarty, 1980). Later, stratified dynamic models (Fig. 1b) able to represent multisubstrate–multispecies biofilms (Wanner & Gujer, 1986; Wanner & Reichert, 1996) were developed. Although these one-dimensional models were advanced descriptions of multispecies interactions within the biofilm, they were not able to provide the information concerning biofilm morphology. Biofilm morphology is an input in these models, not an output. Structural heterogeneity in biofilms was already known at that time, but has been recently underlined through

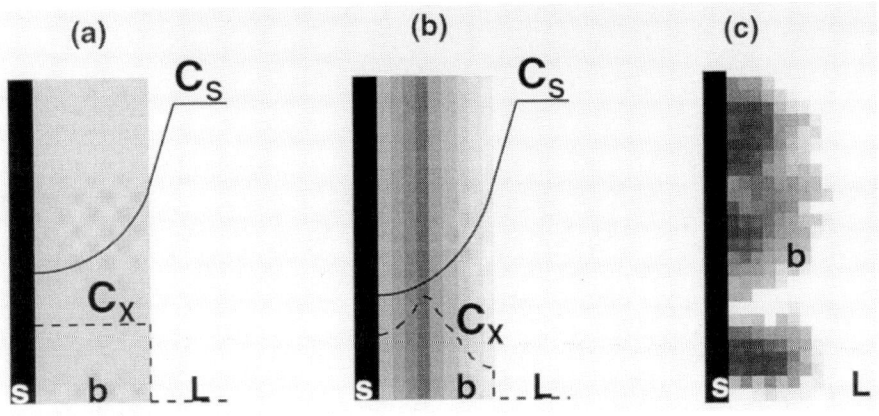

Fig. 1. Evolution of biofilm models from (a) uniform biomass distribution and one-dimensional substrate gradient in the 1970s to (b) one-dimensional stratified biomass and multisubstrate–multispecies biofilms in the 1980s, to (c) multidimensional distribution of biomass and substrates at the end of the 1990s. L, Bulk liquid; b, biofilm; s, substratum. C_s represents the concentration of substrate, and C_x is the concentration of cells in the biofilm, both being one-dimensional representations of laterally averaged values from slices parallel to the substratum surface.

numerous experimental observations. New biofilm models are needed now, to provide more complex two- and three-dimensional descriptions of microbial biofilm (Fig. 1c) that incorporate solute mass transport and transformation, population dynamics and hydrodynamics. This evolution in model complexity has paralleled the advances in computational tools. While hand calculators were the tools used in the 1970s, the biofilm models of today reflect the availability of fast personal computers and advanced parallel processing. Some of the important contributions to biofilm structure modelling are summarized in Table 1.

The amount of experimental evidence describing some biofilms as heterogeneous entities in structure and composition contradicts the simplifying assumptions of the original one-dimensional models. This has challenged engineers to create a more accurate mathematical description of biofilms. The challenge has resulted in an increasing model complexity, derived from the inclusion of an ever increasing number of parameters to explain the biofilm structure. The new generation of structural biofilm models should describe/predict the formation of microcolonies, the development of heterogeneous colonization patterns and the sloughing of large biofilm sections. They could be further expanded to simulate experimentally observed phenomena such as formation of streamers and advective flux through microchannels. Nevertheless, the real challenge to the modeller is not to create models that include as many parameters as possible, but rather to determine the level of significance of these parameters in the

Table 1. Multidimensional mathematical models that generate heterogeneous microbial aggregates

Authors	Processes* Continuous and Discrete								Quantitative	Spatial scale	Application
	F	C	D	R	G	B	A	M			
Ben-Jacob et al. (1994)	—	—	C	D	D	—	—	D	Yes	Cell clusters	Colonies on agar plate
Barker & Grimson (1993)	—	—	D	D	D	—	—	—	Semi	Cell clusters	Colonies in gels
Takács & Fleit (1995)	—	—	C	C	D	—	—	—	Yes	Small cell clusters	Activated sludge flocs
Dillon et al. (1996)	C	C	C	C	—	D	D	D	Yes	Individual cells	Microbial colonization
Kreft et al. (1998, 1999)	—	—	C	C	D	—	—	—	Yes	Individual cells	Colonies on agar plates; **biofilms**
Wimpenny & Colasanti (1997)	—	—	D	D	D	—	—	—	No	Individual cells	**Biofilms** and colonies on agar plates
Picioreanu et al. (1998a, b, 1999); Picioreanu (1999)	C	C	C	C	D	D	—	—	Yes	Cell clusters	**Biofilms** and colonies in gel beads
Hermanowicz (1998)	—	—	D	D	D	D	—	—	No	Cell clusters	**Biofilms**
Noguera et al. (1999)	—	—	C	C	D	—	—	—	Yes	Cell clusters	**Biofilms**
Eberl et al. (1999, 2000a)	C	C	C	C	—	—	—	—	Yes	Cell clusters	**Biofilms**
Eberl et al. (2000a)	—	—	C	C	C	—	—	—	Yes	Cell clusters	**Biofilms**

Notes: *Processes included in the models presented in the table: F, flow; C, convection; D, diffusion; R, reaction; G, growth; B, breakage (detachment); A, attachment; M, motility. Letters in the table: C, continuous; D, discrete.

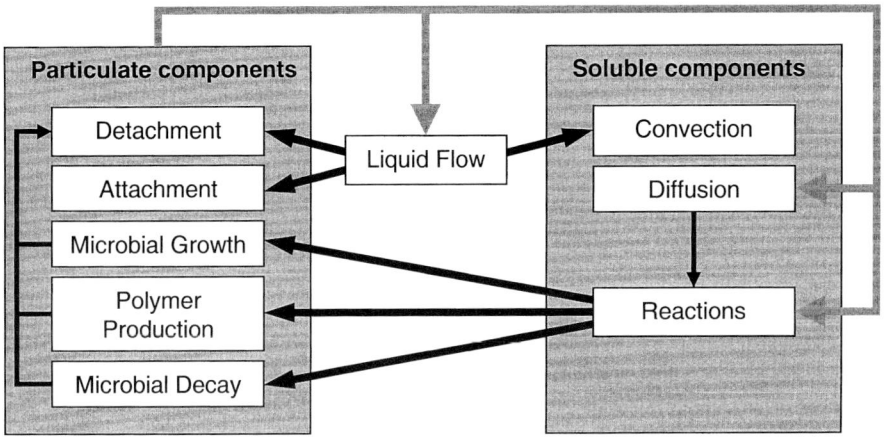

Fig. 2. Interactions between the main processes involved in formation of biofilm structure.

description of the different biofilm processes. Moreover, the mathematical evaluation of parameter significance is essential to define the required level of accuracy of experimental measurements.

BUILDING A MODEL FOR PREDICTION OF BIOFILM STRUCTURE

Biofilms are multiphase systems, comprising solid particles, a liquid phase and, in some cases, a gas phase. From the modelling point of view, we understand the 'biofilm' as being only the solid phase, which includes the extracellular polymeric gel, microbial cells and other entrapped particles of biotic or abiotic origin. *The geometrical structure of a biofilm is the spatial arrangement of particulate biofilm components.* Therefore, the dynamics of solid matter (gel, cells, other solids) produces the biofilm *geometrical heterogeneity.*

Processes in biofilm models

Modelling the structural development of a biofilm is challenging because of the complex interaction between many processes, as shown in Fig. 2. Biofilm development is determined by 'positive' or 'gain' processes, such as cell attachment, cell division and polymer production, which lead to biofilm volume expansion, and 'negative' or 'loss' processes, such as cell detachment and cell death, which contribute to biofilm shrinkage. By changing the balance between these two types of processes, biofilms with different structural properties, such as porosity, compactness or surface roughness, can be formed, as was hypothesized by van Loosdrecht *et al.* (1995) and experimentally shown by Kwok *et al.* (1998). The main biofilm expansion is due to bacterial growth and to EPS production. The nutrients necessary for bacterial growth are dissolved in the liquid flow and to reach the cells they pass first through the boundary layer (external mass transfer) and then through the biofilm matrix (internal mass transfer). The external mass transfer resistance is given by the thickness of the concentration boundary layer (CBL), which is directly correlated to the hydrodynamic boundary layer (HBL) resulting from the flow pattern over the biofilm surface. We can therefore say that on one hand the fluid flow drives biofilm growth by regulating the concentration of substrates and products at the liquid–solid interface, whilst on the other hand, flow shears the biofilm surface and erodes protuberances. While the flow changes the biofilm surface, the interaction is reciprocal because a new biofilm shape leads to different boundary conditions and thus to different flow and concentration fields.

Spatial scales

An essential aspect that must be considered when building any biofilm model is the foreseen spatial scale (spatial resolution). Heterogeneities in systems involving biofilms are characterized by very different length scales. Wood & Whitaker (1998, 1999) identified at least six levels of heterogeneity that might appear in biofilm reactors. Only three spatial levels (Fig. 3) are of interest to us. Level I (macroscale) represents non-uniformities in the solid, liquid and gas particle distribution, which make up the multiphase medium in the reactor. In particle biofilm reactors (e.g. fluidized beds and biofilm airlift reactors), the macroscale level can be characterized by the mean distance between the carrier particles covered with biofilm (typically in the range of millimetres for a fluidized bed). Level II (mesoscale) represents the heterogeneity of the biofilm on the substratum, for which the mean biofilm thickness (usually from 10 to 1000 μm) can be chosen as a characteristic length. This would include irregularities of biofilm surface, and of pores and channels in the biofilm volume. Further down the mesoscale then the constituents of a biofilm might constitute microbial colonies, the non-homogeneous distribution of which can be considered the third spatial level (microscale) of biofilm heterogeneity. A characteristic length at this level can be the distance between two cells (0.1–10 μm). A pictorial representation of these biofilm heterogeneity levels, as they appear in BAS reactors, is shown in Fig. 3.

There are currently models developed for biofilm representation at all three levels. Traditional biofilm models operate at either the reactor scale or the biofilm scale (Wanner & Gujer, 1986; Wanner & Reichert, 1996). For each scale, however, some approaches can be more suitable than others. Kreft *et al.* (1998) work at the microscale and use an individual-based modelling approach. Random variation of cell parameters was chosen in order to simulate heterogeneity of cell size and growth rate in the culture. A 50×50 μm simulation domain is given as an example in Kreft *et al.* (1998), although larger systems can also be used. Other approaches focus mainly on heterogeneity at the

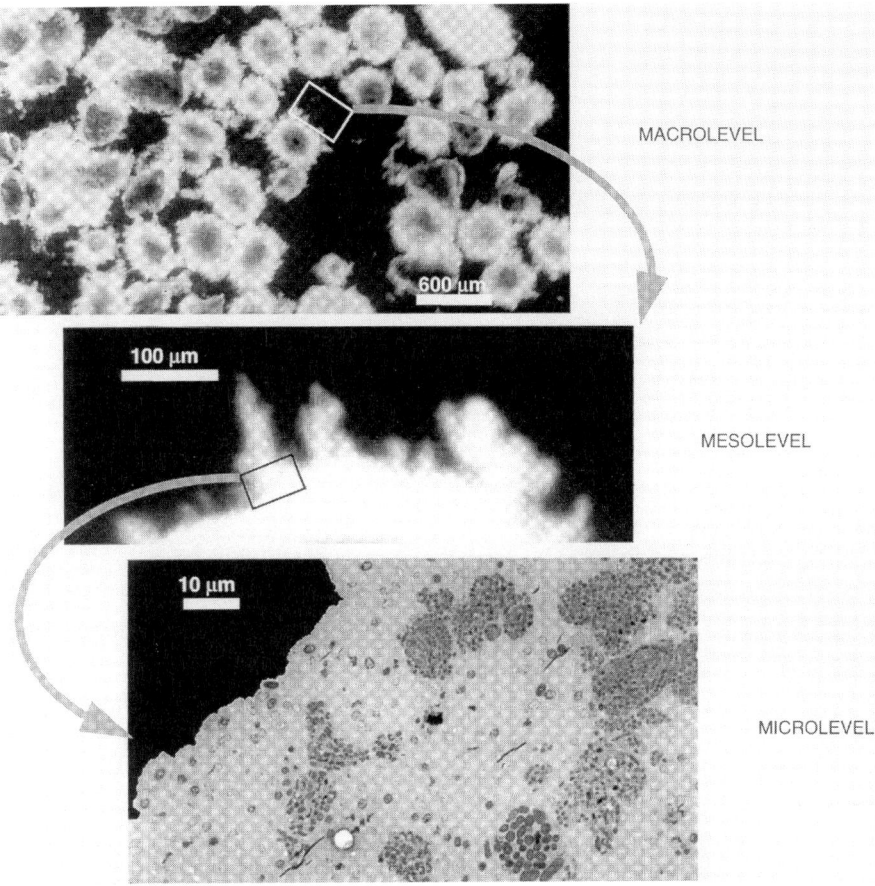

Fig. 3. Three levels of biofilm heterogeneity. A non-uniform distribution of biofilm particle characteristics can be seen at the macrolevel. The mesolevel represents the heterogeneity at the biofilm thickness scale; for example, the biofilm surface roughness, pores and channels in the biofilm volume. The microlevel deals with non-uniform distribution of microbial cells in the colonies.

mesoscale. For example, models of Picioreanu *et al.* (1998a, b, 1999) and Picioreanu (1999) have the goal of describing the formation of geometrical heterogeneity of biofilm surface or biofilm volume. For this purpose, computational domains with a size of at least 1000–2000 μm must be used.

Mathematical modelling of biofilm processes

Biomass growth and spreading. Biomass growth kinetics is dependent on substrate concentrations and, in some cases, simple relationships such as Monod kinetics can be acceptable. In other cases, complications such as substrate/product inhibition, maintenance requirements or biomass decay can be introduced in the model. The result of the microbial growth process, together with production of EPS, is an increase in the biofilm volume, called 'biomass spreading'. Model implementation of such biofilm expansion in all space directions raises quite difficult problems that current modelling approaches must address and which will be discussed.

Several requirements can be postulated for models of biomass spreading: (a) isotropic biomass spreading, i.e. generation of round colonies when there are no growth or space limitations; (b) a sharp biofilm–liquid interface; (c) no biomass mixing in clusters.

Early models for spreading the newly formed amount of biomass (Table 1) made use of an analogy between crystal growth and biofilm accumulation. Both crystallization and biofilm formation are driven by the mass transfer of some essential dissolved compounds from the bulk liquid to a solid surface. A crystal grows by deposition of new layers of material on the existing surface. An external mass transfer process, in which the dissolved matter has to diffuse through boundary layers in order to reach a 'reactive' interface, is coupled to a surface reaction where the soluble matter is transformed in solid phase. Diffusion-limited aggregation (DLA) models can be formulated to describe satisfactorily crystallization and dendritic deposits formed by electrolysis (Vicsek, 1984; for a review see Kaye, 1989). The DLA model has been applied in biology to model the forms of bacterial colonies (Matsushita & Fujikawa, 1990; Fujikawa & Matsushita, 1991; Fujikawa, 1994) and growth of dendritic corals (Nakamori, 1988). The biofilm accumulation process, however, is unlike either crystal or coral growth. The major difference is that the expansion of the solid–liquid biofilm interface is caused by internal pressure generated by the growing biomass. The nutrients diffuse not only across a liquid boundary layer, but also into the biofilm, leading to the appearance of a reaction zone in the bulk biofilm (thus biofilms grow *in volume* and not only *at the surface*). The current difficulty in modelling the spreading of microbial colonies is that a mechanism to release the pressure generated by the growing bacteria must be implemented. Different solutions have been proposed so far, all of them still needing much improvement in order to generate a realistic picture of biomass spreading. According to our knowledge, despite existing modelling efforts, none of the actual biomass-spreading models are realistically suitable for application to multispecies biofilms.

Modelling of biomass growth and spreading is currently done by two main approaches. Usually, the model space is discretized in two- and three-dimensional grids of rectangular elements. Each grid element has four first-order neighbours and another four second-order neighbours in the two-dimensional rectangular space discretization (see Fig. 4). In the pioneering model of Wimpenny & Colasanti (1997), growth

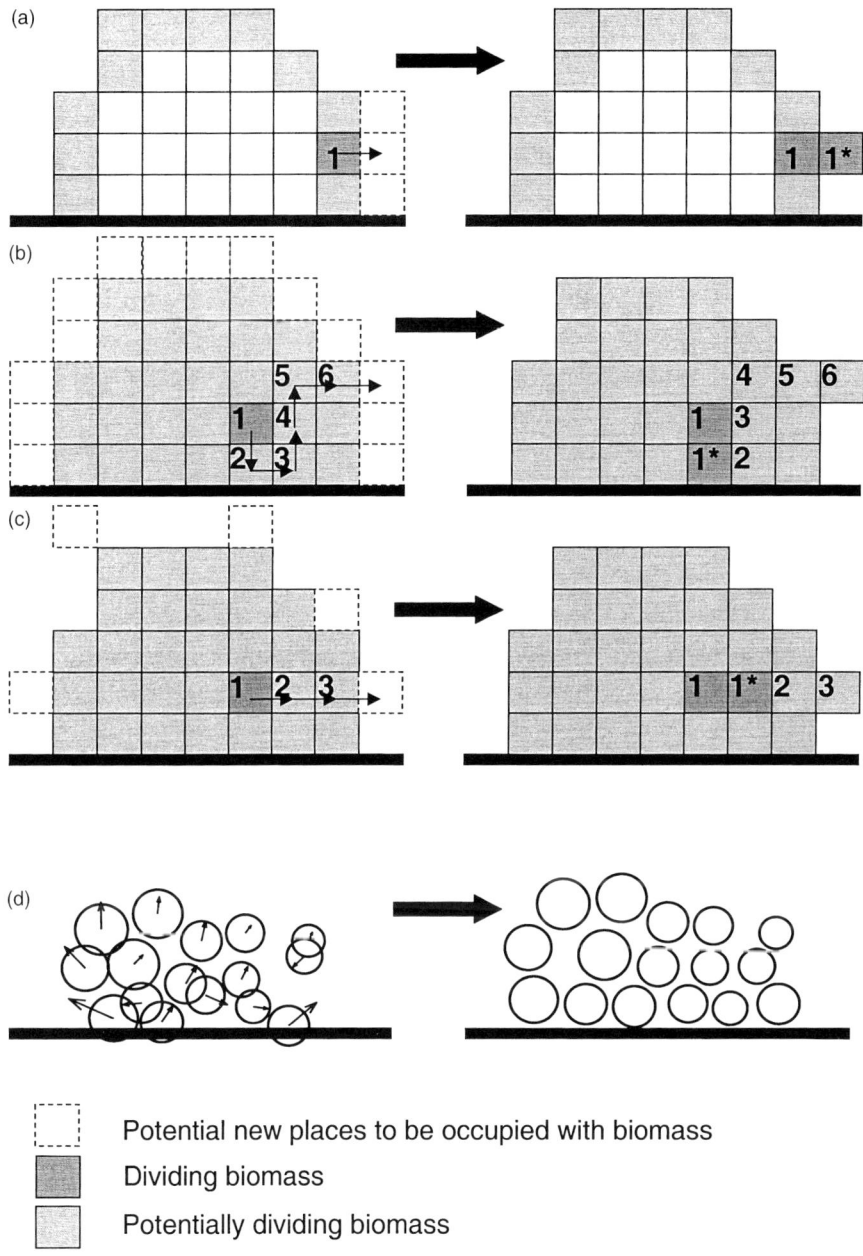

Potential new places to be occupied with biomass

Dividing biomass

Potentially dividing biomass

Fig. 4. Discrete biomass-spreading mechanisms. (a) Surface growth (Wimpenny & Colasanti, 1997); (b) random growth in biofilm volume (Picioreanu et al., 1998a, b); (c) shifting cells in the direction of the biofilm surface (Hermanowicz, 1998); (d) shoving mechanism of individual circular cells (Kreft et al., 1998).

occurred only if there was available free space in the neighbourhood of a cell that can divide (Fig. 4a). This mechanism generated growth only in the outermost cell layer, just as in crystal formation, and the rules do not obey the conservation laws for substrate conversion into biomass. Hermanowicz (1998) proposed a mechanism in which the cell resulting from division will push a whole line of cells in the direction of the nearest biofilm surface, to make a place for itself (Fig. 4c). This is a superior mechanism to Wimpenny's growth automaton (Wimpenny & Colasanti, 1997), but it still has a serious drawback. Without adequate weighting of the four or eight possible pushing directions, the colonies resulting from a single inoculum cell will always show either a rectangular or pyramidal shape. The mechanism proposed independently by Picioreanu *et al.* (1998a, b) generates round cell clusters, while keeping the important feature of growth in bulk biofilm (Fig. 4b). Here, the amount of biomass that has to be redistributed will displace a nearest-neighbour, which, in its turn will displace another neighbour, and so on until a free place is found. One drawback of this mechanism is that this random displacement of neighbouring biomass volumes would produce excessive mixing of a multispecies biofilm. The consequence is the occurrence of a homogeneous species distribution in big biofilm clusters, instead of keeping the identity of monospecies colonies. Noguera *et al.* (1999) allow the excess biomass formed in a time-step to be distributed also in volume elements non-adjacent to the source element. Although these researchers reported a model that aims at modelling multispecies biofilms, this biomass-spreading feature makes the model unsuitable for multispecies. Excess cells belonging to one species can jump over a few neighbouring elements containing other species. By doing this, a more pronounced mixing of species in the biofilm clusters is produced than required by the necessary condition of keeping the cells of one type (as much as possible) together. The mechanism in Noguera *et al.* (1999) shows similarities with the colony-spreading algorithm proposed by Schindler & Rataj (1992) and Schindler & Rovensky (1994). In these colony models, the newly formed cell performs a random walk, starting with the 'mother's' neighbourhood and stopping when a free place is found.

Many of the drawbacks encountered in biofilm-spreading models with discrete spreading directions and discrete displacement distances can be surmounted by allowing cells movement on a continuous set of directions and distances. A realistic model of this kind was proposed by Kreft *et al.* (1998). In this model, the pressure build-up due to biomass increase is relaxed by minimizing the overlap of cells. For each cell, the position is shifted by the vectorial sum of all overlap radii (Fig. 4d). The current disadvantage of this spreading mechanism, when applied together with a finite volume scheme for solution of substrate field, is that cells are assumed to be circular whilst the finite volumes are rectangular. Averaging of reaction rates of cells occupying a certain grid element must therefore be done prior to solution of substrate mass balances. This time-consuming operation makes this, otherwise promising, approach currently applicable only to microscale biofilm representations, where individual cells need to be treated separately.

Substrate transport and conversion. The transport of chemical compounds to cells within a biofilm is important since the concentration of both nutrients and products determines the rates of microbial reactions. Major processes such as microbial growth and EPS production that contribute to biofilm volume increase are driven by nutrient availability. Decay processes are also affected by the concentration level of certain compounds.

Dissolved compounds (solutes) can be transported by molecular *diffusion* and *convection* (sometimes also called *advection*). In convective mechanisms, movement of solute is driven by the flow of solvent, usually water. The convective flux of solute in any point in space is proportional to the liquid velocity, **u**, and to the concentration of the transported compound, *c*. The molecular mass transport mechanism, diffusion, is driven by a concentration gradient (i.e. a difference in solute concentration between two points in space). According to Fick's law, the diffusive flux of mass is proportional to the concentration gradient; the proportionality constant is D (diffusion coefficient). This gradient of concentration from bulk liquid to the solid substratum occurs because of substrate consumption or product generation within the biofilm.

Before reaching the reaction site (i.e. cells, where conversions take place), substrates must be transported (1) through the liquid to the biofilm surface – *external transport* – and (2) from the biofilm–liquid interface through the biofilm to the cells – *internal transport*. Products released in the biofilm travel the opposite way, from the cells to the bulk liquid. Convection dominates the mass transport of substrates in bulk liquid, where high velocities and maximum concentrations are present. Diffusion is important in a thin layer of liquid adjacent to the biofilm surface – the *boundary layer* – and in the biofilm matrix itself, where no convection is possible. For such reasons, the internal diffusion transport of nutrients must be included in any biofilm model that is to generate structure. Because the external resistance to mass transfer is concentrated in this boundary layer, a simplified model for substrate transport can consider only diffusion through the boundary layer. To be able to compute mass transport rates with a simplified approach, we must know at least the *thickness* of the boundary layer and its *shape*. A mean thickness can be estimated from engineering correlations for mass transfer coefficients, as a function of liquid velocity and liquid and solute physical properties (viscosity, density, diffusivity). The shape of the boundary layer is usually approximated in two limit cases. At low liquid velocities, the boundary layer is more or less parallel with the substratum (case considered in the model of Picioreanu *et al.*,

1998b), whereas at higher velocities it tends to follow the biofilm shape (case considered by the models of Hermanowicz, 1998 and Noguera *et al.*, 1999). An experimental determination of the two-dimensional CBLs around biofilms in these two cases was done by using oxygen microelectrodes (De Beer *et al.*, 1994). In many cases, approximation of CBL is sufficient to produce good qualitative representations of the influence of mass transport on biofilm growth. For complicated biofilm surface geometries, or for intermediate flow regimes, however, the external mass transfer should result only after flow calculations have been made (Picioreanu *et al.*, 2000). This is because convection can also significantly contribute to mass transport in biofilms. Correct calculation of convective mass transport poses more difficult problems, because the liquid velocity field must also be known.

Qualitative pictures of substrate concentration field caused by diffusion and reaction can also be obtained by *discrete* methods, e.g. by cellular automata. One example is the work of Wimpenny & Colasanti (1997), where 'quanta' of nutrient have a Brownian motion that simulates the diffusion process. Cellular automata models have also been used to simulate diffusion in biofilms (Barker & Grimson, 1993). Superior results are obtained by the *continuum* methods, because here the powerful apparatus of differential calculus can be used. The results of differential models for substrate transport are quantitative and can be directly compared with data measured in real systems. In any case, finding the field of substrate concentration requires solution of the partial differential equations of mass balance, including diffusion, reaction and, where needed, convection terms. The disadvantage of continuum methods is a higher computational expense compared to discrete methods (cellular automata). These tasks can, however, be easily achieved by modern personal computers. These differential methods were included in multidimensional models for biofilm structure by Picioreanu *et al.* (1998b, 1999), Noguera *et al.* (1999) and Eberl *et al.* (1999, 2000b).

Liquid flow. Due to the inherent computational complexity, accurate hydrodynamics has rarely been considered in biofilm modelling. A few isolated cases are the models of Dillon *et al.* (1996), Picioreanu *et al.* (1999, 2000) and Eberl *et al.* (1999, 2000a) (see Table 1). Knowledge of liquid flow pattern (hydrodynamics) past the biofilm is important because it affects several other essential processes governing biofilm development:

(1) Computation of substrate transport by convection requires knowledge of the liquid velocity field, **u**.
(2) In the attachment of suspended cells, the liquid flow determines cell transport to the solid surface. Together with the substratum geometry, hydrodynamics affects the initial colonization process.
(3) The mechanical effect that liquid exerts on the biofilm is multiple. Liquid flowing

past the biofilm induces stress in its structure, which can lead to further detachment. Forces acting on the biofilm gel can affect its rheology, stretch or compression.

The correct approach in modelling hydrodynamics is to solve the flow equations (liquid momentum balance and liquid mass continuity equations). The solution of such flow equations finds the distribution in space of two state variables: liquid pressure, p, and vector velocity, $\mathbf{u}$. From these state variables can be calculated thereafter the forces that the fluid exerts on a unit of biofilm surface area, called stresses. These stresses are later used to compute biofilm deformations, necessary in the detachment models.

Numerical approximations of the flow field past and through the biofilm structure can be calculated by using traditional computational fluid dynamics (CFD) methods (e.g. finite difference, finite volumes, finite elements). To be included in a biofilm model, the CFD methods should satisfy some specific requirements. First, they should be able to cope easily with a very irregular geometry of the boundaries. Second, the numerical methods used must show flexibility and a rapid mesh generation, since the shape of the biofilm–liquid interface continuously changes in time. The lattice Boltzmann method is one new approach which meets these requirements and, moreover, works very well on modern parallel computers and has a good physical soundness. The first implementation of the lattice Boltzmann method in a multidimensional biofilm model can be found in Picioreanu *et al.* (1999) and more details are given by Picioreanu (1999).

Biomass detachment. Detachment is essential for development of biofilm structure, because it is the main process by which biomass is lost. Within the frame of a biofilm model aimed at structure prediction, only two mechanisms for detachment have been proposed. The qualitative model of Hermanowicz (1998) allows cells to detach randomly with a probability that increases in proportion to the biofilm thickness. This mimics an observation that the biofilm parts which are the most exposed to liquid shear (i.e. outer biofilm layers) have a larger contribution to the total amount of detached biofilm than do parts that are sheltered from shear forces. A quantitative microscopical approach to biofilm detachment was also developed by Picioreanu (1999). This approach is based on the hypothesis that biofilm breaks at points where the mechanical stress exceeds the biofilm mechanical strength. The mechanical stress in the biofilm builds up due to forces acting on the biofilm surface as a result of liquid flow. This model requires knowledge of the mechanical properties of biofilms, such as tensile strength (Ohashi & Harada, 1994, 1996; Ohashi *et al.*, 1999) and elasticity modulus (Stoodley *et al.*, 1999a). These have been only scarcely measured until now.

Attachment. 'Attachment' is a generic name for a whole class of sub-processes that lead to an immobilization of cells suspended in the liquid phase, by the biofilm or by the substratum. Cell attachment involves transport to the solid surface, a reversible adhesion step, and finally firm attachment due to the production of polymer. The attachment process is important in biofilm structure generation because (1) it determines not only the initial colonization phase of the substratum, but also the recolonization that occurs after sloughing events, and (2) it affects species distribution in the biofilm. By attachment, new species existing in the liquid phase can be introduced in an already developing biofilm structure.

The cell attachment process is influenced by both physical and biological factors. The liquid flow pattern and geometry of the substratum determine the places where cells can settle, adhere and, if they stay long enough, form polymer to strengthen their position in the structure.

Attachment of particles (e.g. microbial cells) was introduced in a multispecies biofilm model by Wanner & Reichert (1996), by a variable macroscopic attachment rate. In a multidimensional model that must predict the spatial distribution of species, this is not sufficient, because the location where a certain microbe sticks to the biofilm structure is also important.

In a simple approach, biomass attachment can be regarded as the opposite of the detachment process. Thus new cells can be added to the solid structure in random places on the surface, with a frequency proportional to the desired attachment rate. In addition, the sticking probability of these 'biomass particles' could increase from substratum to the biofilm surface. This approach would constitute, in fact, the opposite of the detachment mechanism proposed in the model of Hermanowicz (1998). Other simple approach (Picioreanu, 1996) simulated attachment by a diffusion-limited-aggregation mechanism (Kaye, 1989). This implicitly creates a higher attachment probability at the top of the biofilm than in the less accessible channels. The attachment rate could be tuned by varying the number and species distribution of cells introduced in the computational space, and also their adhesion probability. A more elaborate approach considered the influence that flow pattern has on cellular attachment. Dillon *et al.* (1996) introduced, in their two-dimensional model, attachment of motile single cells travelling in a liquid stream. Liquid forces pushing these cells are also calculated. When the cells encountered a solid surface (other cells or just solid walls), they stuck. In time, cell–cell and cell–wall bonds were produced as a function of substrate availability. This constitutes in the model of Dillon and co-workers a representation of the cohesive function of EPS.

Timescales – coupling the processes in time

Processes in biofilms occur at very different timescales. One major problem that is encountered is how to accommodate, in the same biofilm model, all the fast and slow physical, chemical and biological processes. The solution comes from a timescale analysis. The order of magnitude of the characteristic times for convective and molecular transport, biomass growth, biomass decay and detachment can be easily calculated by writing the model equations in a non-dimensional form. It was shown in Picioreanu *et al.* (1999) that processes changing the biofilm volume (biomass growth, decay and detachment) are all much slower than processes involved in substrate mass balance (diffusion, convection and reaction). In addition, momentum transport (by convection or viscous dissipation) is much faster than the slowest step (i.e. diffusion) of substrate mass transfer. Usually, we are interested in biofilm evolution during only a few weeks of reactor operation. If the dynamic (non-steady-state) balance equations for momentum, mass and biomass growth were all solved with the smallest time-step, given in this case by the momentum equations, then the computational effort needed to obtain the solution of substrate and biomass balances is far too large. To achieve a realistic calculation speed, the natural timescale separation in biofilm systems is very useful. It is therefore justified to work at three timescales: (1) biomass growth, in the order of hours or days; (2) mass transport of solutes, in the order of minutes; and (3) hydrodynamic processes, in the order of seconds. In other words, while solving the mass balance equation, the flow pattern can be considered at pseudo-equilibrium for a given biofilm shape, whilst at the same time the biomass growth, decay and detachment are in a frozen state.

The above considerations led Picioreanu *et al.* (1999) and Picioreanu (1999) to the next strategy in following the biofilm development with time (Fig. 5).

(1) A *hydrodynamic* step is performed each time the biofilm–liquid interface is modified. Flow equations are solved to find the flow field variables pressure (p), velocities (**u**) and the normal (σ) and tangential (τ) stresses acting on the biofilm surface. This step is needed each time that the geometry of the system has changed, for instance by growth (a positive development) or detachment (negative development).

(2) The stresses at the biofilm surface (σ and τ) constitute boundary conditions for the equations of mechanical equilibrium in the biofilm volume. The *mechanical stresses* and displacements are calculated at different points in the biofilm structure.

(3) The biofilm is allowed to break at those points where the developed stress exceeds the biofilm strength. After biomass *detachment*, a recalculation of the flow field is necessary.

(4) The calculated flow velocities, **u**, are used in solving the *convective-diffusive mass transfer* of soluble components (substrates, products), including also the *transformation* processes. The mass balance equation is solved towards a pseudo-steady-state concentration of substrate, c_S. The substrate concentration in its turn

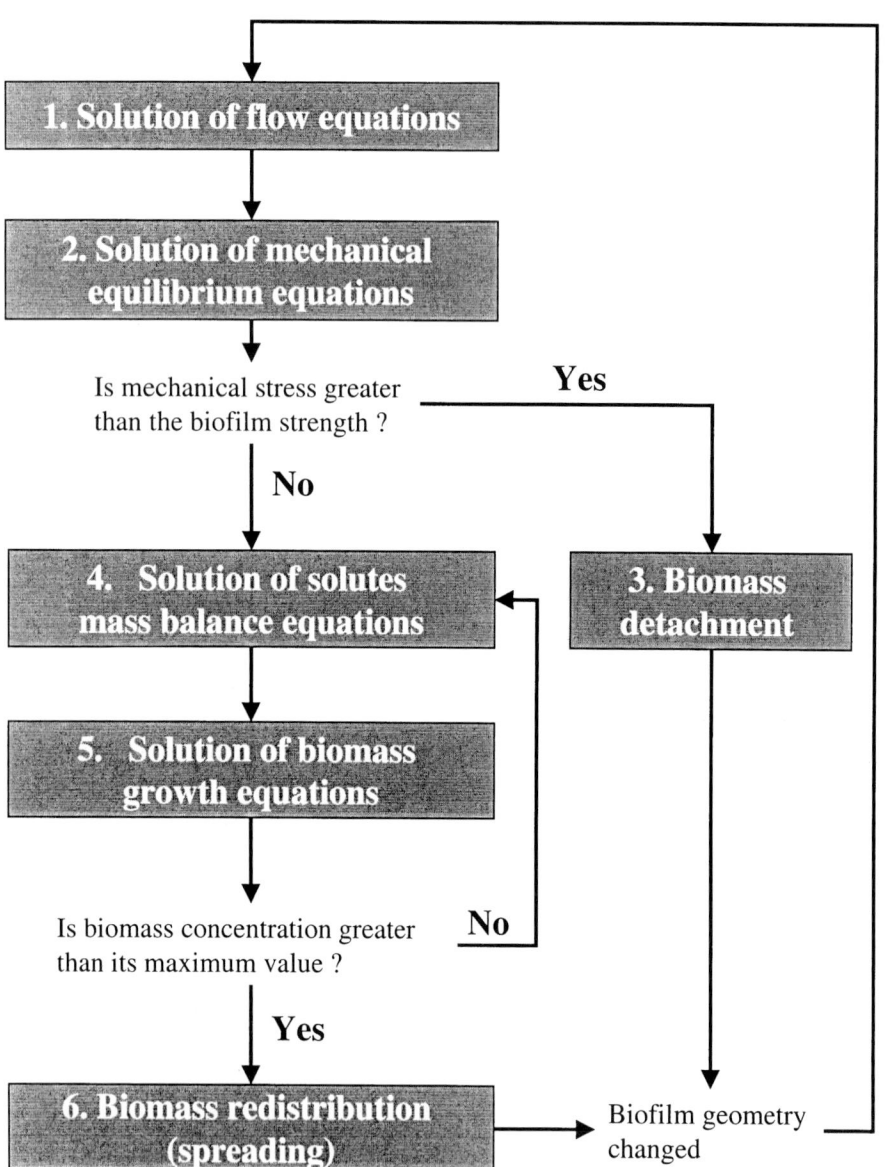

Fig. 5. Algorithm steps for the biofilm model, including substrate convection, substrate diffusion, substrate reaction, biomass growth and biofilm detachment.

will be used in the biomass growth kinetic equation. Both the Navier–Stokes equations and substrate mass balance were solved by Picioreanu (1999) in a dimensionless form by using a lattice Boltzmann algorithm (Ponce Dawson *et al.*, 1993; Chen *et al.*, 1995, 1997; Chen & Doolen, 1998; Picioreanu, 1999).

(5) The *new biomass* content of each grid element, c_X, is calculated by using biomass growth kinetics, including the substrate concentration at steady state calculated before. Typical time-steps used in solving the growth process were between $\Delta t_g = 1000$ s for fast-growing biofilms and 10 000 s for slow-growing biofilms.

(6) As biomass grows, it has to be redistributed in space according to discrete (cellular automata) rules as used in Picioreanu *et al.* (1998a, b), Hermanowicz (1998) or Noguera *et al.* (1999). The biomass content in the grid element in which it has grown above the maximum biomass density, c_{Xm}, is split in two equal parts. Then, empty grid elements are occupied by pushing the neighbours to reach the free space. *Biomass spreading* generates the biofilm structure. However, the geometry of this structure determines the further development of the biofilm by changing the flow behaviour and the nutrient/product transport. After the biofilm volume expands, the flow boundaries change and a new hydrodynamic step (step 1) is necessary. If biomass density did not exceed c_{Xm} in any grid element, the mass balance (step 4) is solved using the same flow field.

Practically, the timescale separation of flow, mass transfer and growth processes is complete. By exploiting the natural timescale separation in biofilms, the step by which the whole algorithm advances in time is the one necessary for the slowest process, namely growth, Δt_g. The loop through steps 1–6 is repeated a few hundred times to simulate biofilm life of a few weeks.

APPLICATIONS OF MODELS FOR BIOFILM STRUCTURE

Diffusion–reaction–growth biofilm models

One-species–one-substrate biofilms. Three processes are essential in any model that studies the influence of nutrient availability on formation of biofilm structure. First is the transport of nutrients, only by diffusion in this simplified model, through the CBL and further into the biofilm matrix. The second process is conversion of these substrates into biomass at a cellular level (the 'reaction step'). Finally, the newly formed biomass must be redistributed in space, leading to an increase in the biofilm volume. A hybrid discrete-differential approach of biofilm structure evolution incorporating these three steps was described by Picioreanu *et al.* (1998a, b). In what follows, this will be called the DRG (diffusion–reaction–growth) model. This quantitative model generates various two- (Fig. 6) and three-dimensional (Fig. 7) biofilm structures, as the result of

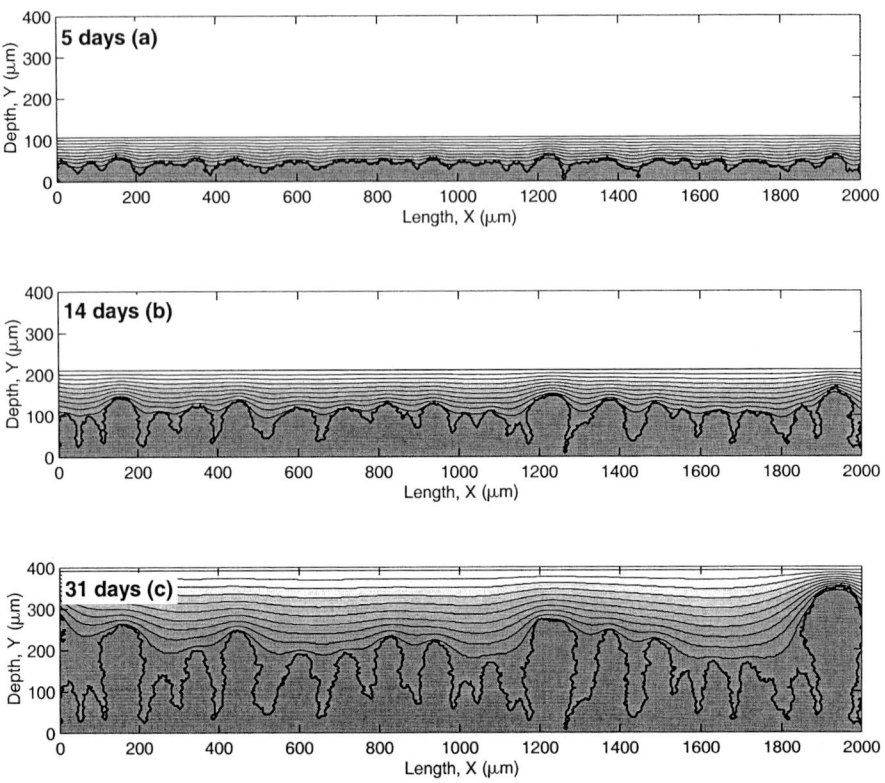

Fig. 6. Evolution in time of a biofilm simulated with the two-dimensional DRG model. Dissolved oxygen concentration was 2 mg l^{-1}, and the rest of the parameters are as in Picioreanu *et al.* (1998b). The thick continuous lines indicate the biofilm surface. Iso-concentration lines show the decrease of substrate concentration from the maximum value in the bulk liquid (white patches) to zero in the biofilm (dark-grey patches), with a variation of 10 % between lines.

different substrate transport rate/growth rate ratios. The model assumes only microbial growth in a static liquid environment, or in any case, only when the CBL becomes parallel to the substratum (case which, according to the microelectrode measurements of De Beer *et al.*, 1994, occurs at low liquid velocities).

Fig. 6 presents the time evolution of a simulated two-dimensional biofilm. Parameters used here are those in Picioreanu *et al.* (1998b). It is clear from Fig. 6 that biofilm complexity increases continuously over time. Initially, the patchy biofilm colonies are completely penetrated by the substrate. They tend to grow in all directions, filling the space existing initially due to random surface inoculation. As biofilm thickness increases, some colonies get the chance to be closer to the substrate source than do

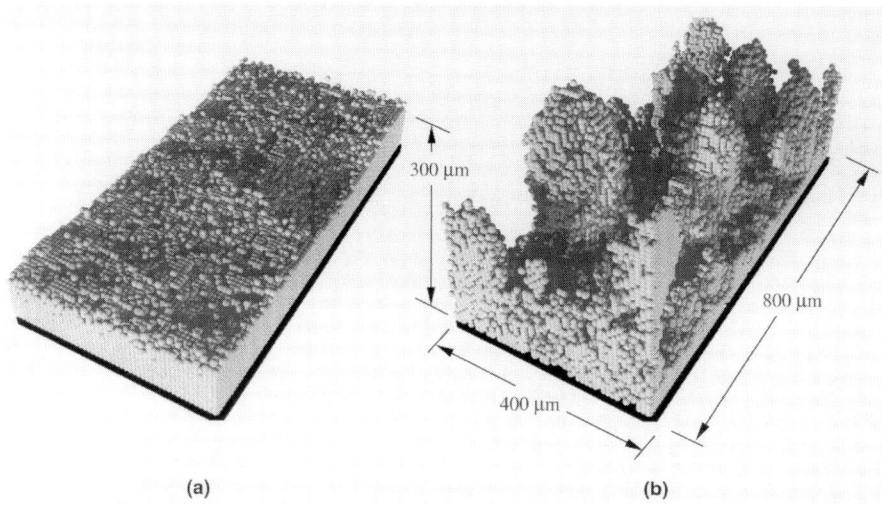

Fig. 7. Spatial biomass distribution in a three-dimensional biofilm simulation with the DRG model, in (a) biomass growth-limited regime after 5 days and (b) substrate-transfer-limited regime after 50 days. Both simulations use the model and parameters as in Picioreanu *et al*. (1998b). Each sphere in this visualization represents a bacterial cluster with a diameter of 4 μm (the grid size used in simulations).

others. These top parts of the biofilm experience a higher concentration and a higher flux of nutrients. New biomass will therefore be formed more quickly, whereas the colonies in the 'valleys' grow in an environment that is depleted of substrate. The voids between the colonies cannot be filled with new biomass any longer. An obvious consequence is that a rough, 'finger-like' biofilm will develop on the top of a compact basal layer.

Porous biofilms, with many channels and voids between the 'finger-like' colonies, were obtained in a substrate-transfer-limited regime (Fig. 7b). Compact and dense biofilms resulted at higher substrate transfer rates, when biofilm development was limited only by microbial metabolism (Fig. 7a). Similar approaches, which can generate biofilm structures with qualitative features resembling the real ones, have also been independently reported by Wimpenny & Colasanti (1997) and Hermanowicz (1998). These models, however, work in abstract time and space, with no direct link between their model parameters and the real values of some widely accepted parameters like substrate diffusivities, reaction rate constants and stoichiometric coefficients. The model presented in Picioreanu *et al*. (1998b) shows that even in this simple case, with one substrate and one bacterial species, a quantitative multidimensional model leads to a better and more rational understanding of biofilm behaviour in response to different environmental conditions.

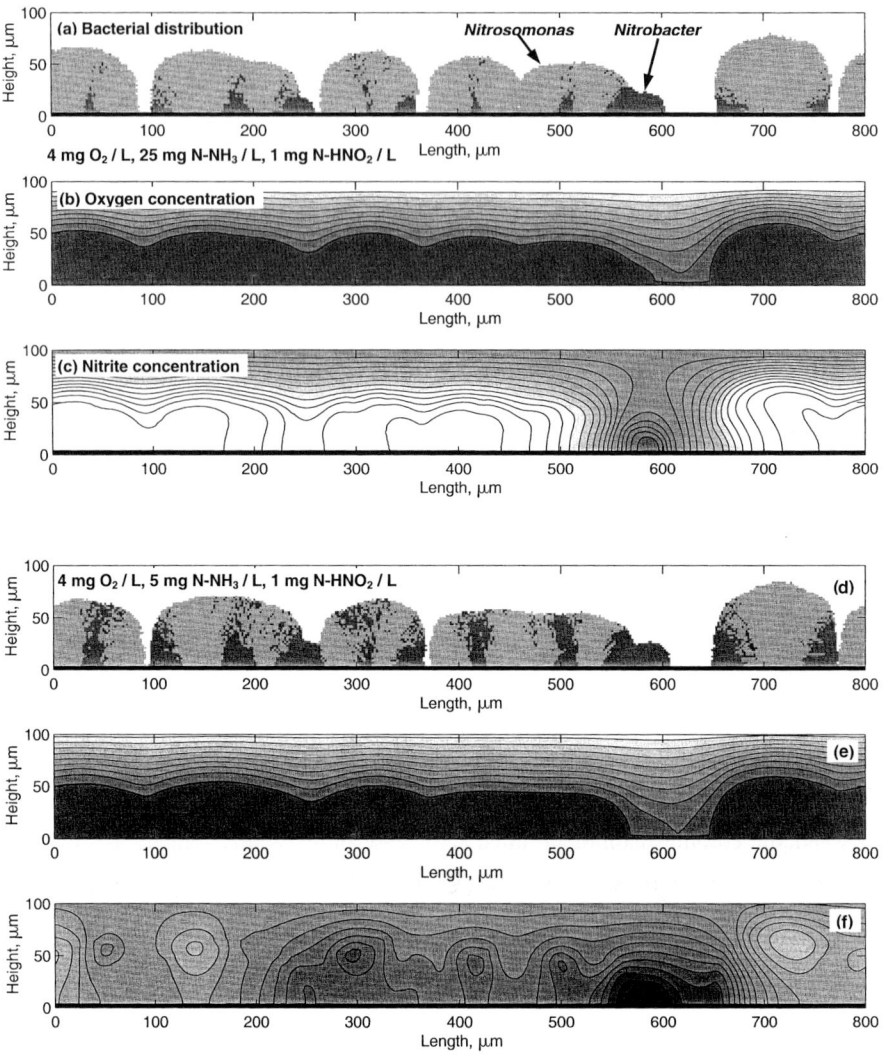

Fig. 8. Two-dimensional simulations of a two-species–three-substrates biofilm at three different sets of ammonia, nitrite and oxygen concentrations in the bulk liquid. The biomass distribution maps (a), (d) and (g) present the ammonia oxidizers (*Nitrosomonas*) in the light-grey regions, and the nitrite oxidizers (*Nitrobacter*) in the dark-grey areas. On the dissolved substrate concentration contour plots, iso-concentration lines delimit regions of 10 % variation in concentration. Dissolved oxygen contour plots (b), (e) and (h) show on the greyscale a variation in concentration from 100 % (white) to 0 % (dark grey) from the bulk liquid concentration. Nitrite concentration maps (c), (f) and (i) show variations between 200 % (white) and 0 % (dark grey) from the bulk nitrite concentration. Ammonia concentration is not shown because this substrate was in neither case growth limiting.

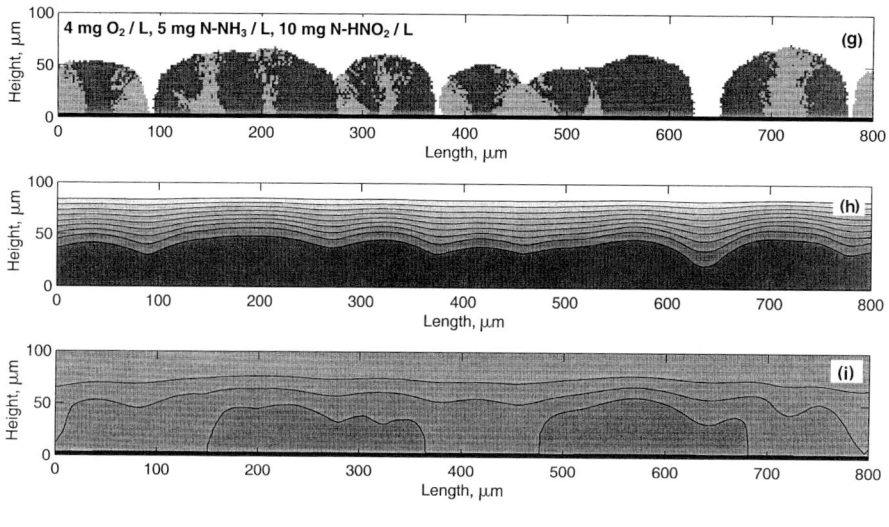

Multispecies–multisubstrate biofilms. The DRG model presented in Picioreanu *et al.* (1998b) was later easily extended to simulate, for example, the evolution of a nitrifying biofilm. Two bacterial species, ammonia oxidizers (e.g. *Nitrosomonas*) and nitrite oxidizers (e.g. *Nitrobacter*), were considered, each having the growth and substrate consumption rates, as well as the kinetic and stoichiometric parameters, that were reported by the review of Wiesmann (1994). Three substrates, oxygen, ammonia and nitrite, with mass transport properties as in Picioreanu *et al.* (1997) were incorporated in the model. In order to minimize unwanted mixing of species in the biofilm clusters, a modified version of the discrete biomass-spreading mechanism from Picioreanu *et al.* (1998b) was used. Results of the two-dimensional simulations of this two-species–three-substrates biofilm, at three different sets of ammonia, nitrite and oxygen concentrations in the bulk liquid, are presented in Fig. 8. At high ammonia concentrations in bulk liquid, high levels of nitrite accumulated inside the biofilm because of a limitation in its transport out of the biofilm (Fig. 8c). The ammonia oxidizers clearly dominated the population distribution in the biofilm (Fig. 8a). As ammonia concentration was decreased, while that of nitrite was increased, then a gradual shift in population distribution towards a dominance of nitrite oxidizers can be seen from Fig. 8(a, d, g). At the same time, the concentration of the intermediate nitrite decreased in the biofilm, relative to the bulk liquid. In Fig. 8(f), there can be seen areas where the nitrite concentration is higher than in the bulk liquid (white areas, corresponding to the presence of *Nitrosomonas*), and other regions where nitrite is completely depleted by the *Nitrobacter* population (the dark regions on that figure).

When nitrite oxidizers dominate (Fig. 8g), the nitrite concentration everywhere in the biofilm is lower than in the bulk liquid (Fig. 8i).

Results obtained using a three-dimensional version of the same DRG model showed the same qualitative trends.

Convection–diffusion–reaction–growth biofilm models

An extension of the two-dimensional DRG model was elaborated in Picioreanu *et al.* (1999) and Picioreanu (1999) for biofilm development, including biomass growth, diffusive and convective transport and transformation of substrates and flow around the biofilm structure. The model (called CDRG: convection–diffusion–reaction–growth) is fully quantitative, being based on first principles as Navier–Stokes equations, substrate mass balances and kinetic laws for biomass growth. The main results of this study are presented briefly in what follows.

The CBL largely follows the substratum surface, when the substratum coverage with biofilm is high; that is, there is a continuous bacterial layer formed on the substratum, or separate biofilm clusters are relatively close to each other, so that convection in the channels is strongly impeded (Fig. 9a–f). This is the 'skimming flow' case suggested by Stoodley *et al.* (1999b). If the CBL were approximately parallel with the substratum, then simplification of the model by considering only a diffusion boundary layer, as in the previous DRG model, would be acceptable. This would have the obvious advantage of skipping the computationally costly steps required for solution of the flow equations and convective mass transport.

The CBL follows closer the biofilm surface than the substratum surface only when the biofilm colonies are isolated *and* at high flow velocity (Fig. 9h, i). Only in this case of 'patchy' biofilm growth can the convective transport in valleys contribute significantly to the increase of the total conversion rate. These observations fall in one of the cases called 'isolated roughness flow' and 'wake interaction flow' by Stoodley *et al.* (1999b). However, when the flow velocity is low, in the entrance region the CBL thickness continuously increases, following neither the shape of the biofilm nor that of the substratum (Fig. 9g).

The main result of studies done with this model was that increased biofilm roughness does not necessarily lead to an enhancement of either substrate conversion rates or external mass transfer (Picioreanu *et al.*, 2000). If there is poor convection in the biofilm channels or valleys, then the main transport mechanism for substrate is only by diffusion, driven by gradients of concentration perpendicular to the carrier. Notably, the overall mass transfer rate decreases in rough biofilms because of an increase in the diffusional path of substrate. Although the total biofilm area is increased by roughness, the effective mass transfer area is in fact decreased. Only a very limited fraction of all biofilm area receives nutrient because only the peaks are accessible for substrate. An obvious limitation of the CDRG model is that biofilm structure is considered immobile. If model biofilm filaments vibrated, induced by liquid movement as Stoodley *et al.* (1998) demonstrated experimentally, external mass transfer could be enhanced.

Our studies demonstrate that roughness alone is insufficient to characterize the influence of surface irregularity on mass transport and conversion. If biofilm peaks are not far enough from each other, convective transport in valleys cannot contribute significantly to the increase of the total conversion rate. A study of mass transfer in biofilms with more complicated surface shape should therefore not lump roughness and cluster frequency into a single measure, but rather consider them separately.

Using this model, three possible factors influencing biofilm geometrical heterogeneity were identified: (a) the ratio between biomass growth rate and internal substrate transfer rate; (b) external mass transfer limitations as a result of different liquid flow rates; (c) initial substratum surface coverage. Model predictions show that in a biofilm growth regime limited by the rate of substrate transport (internal, as well as external), structures with a high degree of surface irregularity develop. Biofilms grow in these conditions as 'finger-like' or filamentous structures with high surface roughness, high surface area, high porosity and low degree of compactness. As the nutrient availability increases, there is a gradual shift towards compact and smooth biofilms.

A smaller fraction of the substratum surface colonized led to a rougher biofilm. Surface irregularity and vertical channels deep in the biofilm are in this case caused by it being impossible for the colonies to spread over the whole surface.

Convection–diffusion–reaction–growth–detachment biofilm models

A two-dimensional model for detachment, based on internal stress created by moving liquid past the biofilm, is able to generate a variety of biofilm formation patterns (Picioreanu, 1999). The two known biofilm detachment mechanisms, *erosion* (loss of small biofilm parts – eventually only cells – mainly from the biofilm surface) and *sloughing* (loss of massive biofilm chunks, often broken from the substratum surface), can be surprisingly modelled in a unitary way. The model assumes that detachment is caused by stress developed in the biofilm structure. From a single breakage criterion, both erosion and sloughing are found in the model.

Analysis of stress build-up in biofilms is important because it determines the places in the structure where fracture most probably will occur. An example of equivalent stress

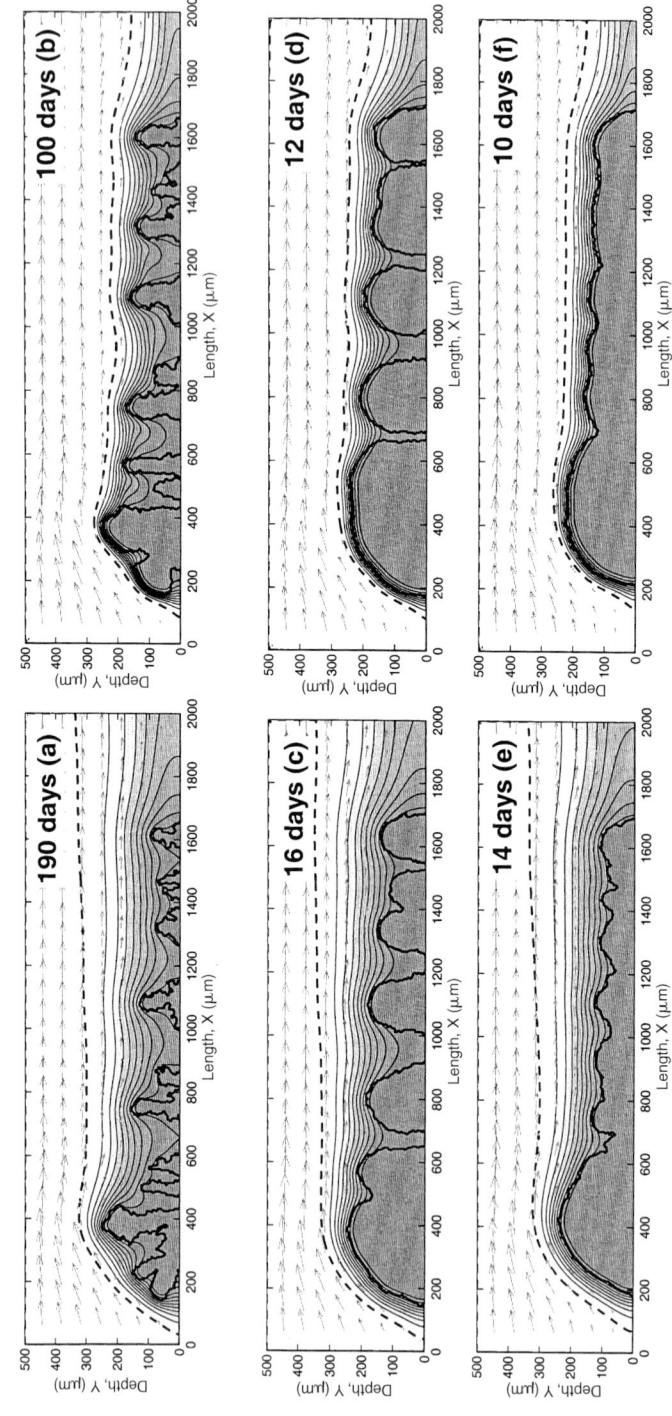

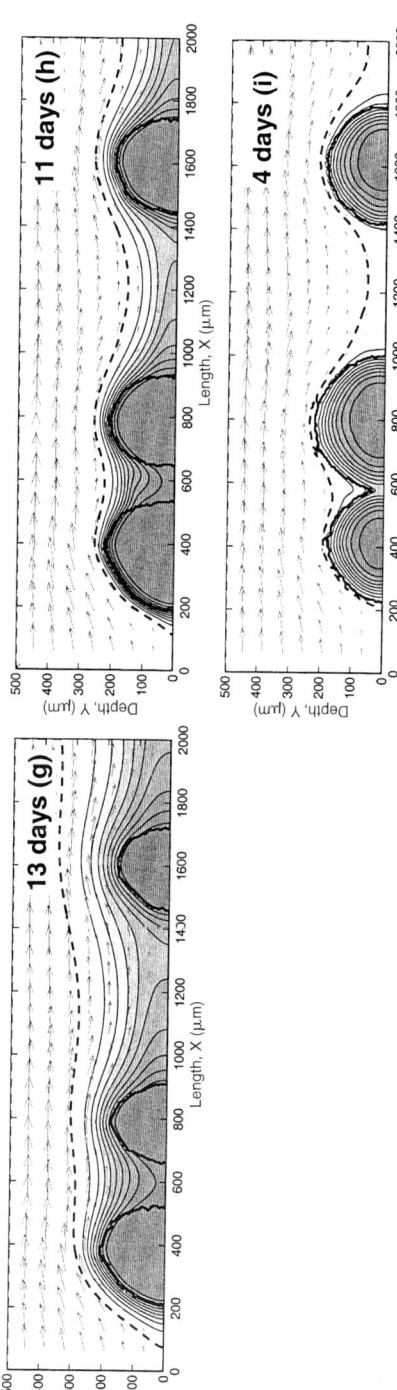

Fig. 9. Two-dimensional simulations of biofilm development at the moment when the maximum biofilm thickness reached approximately 220–250 μm. The environmental conditions were: substrate (oxygen) concentration in inlet (c_{S0}), 0.4 mg l^{-1} (a, b), 4 mg l^{-1} (c–h) and 20 mg l^{-1} (i); the initial substratum surface coverage with biomass, 1 % (g–i), 5 % (a–d) and 40 % (e, f); the liquid velocity on the top boundary, 0.01 m s^{-1} (b, d, f, h, i) and 0.00125 m s^{-1} (a, c, e, g). The grey arrows represent the vector velocity at intervals of 16 grid nodes. The thick continuous lines indicate the biofilm surface. Iso-concentration lines show the decrease of substrate concentration from the maximum value in the bulk liquid (white patches) to zero in the biofilm (dark-grey patches), with a variation of 10 % between lines. The thick dashed contour lines indicate the limit of the CBL, at 98 % from the bulk concentration of substrate. All model equations and parameters used in these simulations can be found in Picioreanu (1999).

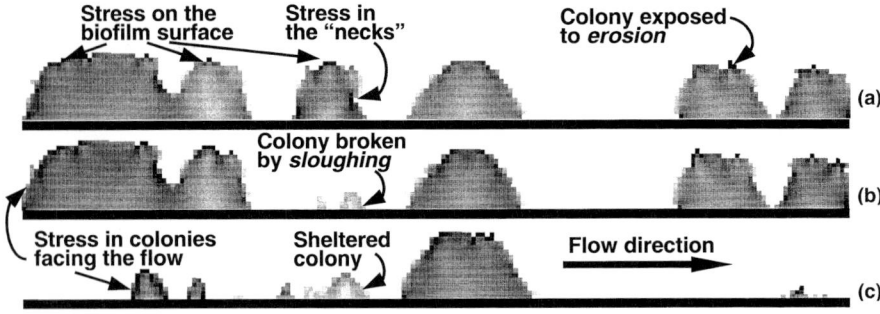

Fig. 10. Distribution of mechanical stress in some simulated biofilm structures. In each of the three configurations, the equivalent stress was normalized relative to the maximum value existing at that moment in the structure. On the greyscale, black represents finite elements where the mechanical stress is maximum, and white where it is minimum. The stress fields (a)–(c) presented here correspond to the biofilm structures (e)–(g) in Fig. 11.

distribution in the biofilm structure is shown in Fig. 10. A few typical situations when the biofilm breaks were identified.

(a) In compact biofilm clusters (like the middle hemispherical one in Fig. 10), the highest stress develops near the biofilm surface. This is especially caused by the shear component of stress. Biofilm *erosion*, defined as continuous loss of small amounts of biomass from the biofilm surface, can be the result of these stresses.

(b) In finger-like clusters, the stress builds up also near the biofilm–carrier interface, due to a high bending momentum. 'Necks' especially concentrate stress (Fig. 10a). In general, cracks form in the narrowest part (the smallest cross-sectional area) of the biofilm filaments. This leads to breakage of a whole biofilm patch, a phenomenon known as *sloughing*.

(c) Usually, the isolated colonies or simply the first one that faces the flow are less protected from the flow shearing effect. Although it grows faster than other colonies, because it receives more substrate, stress builds-up more easily in the first structure from the left in Fig. 10(a, b). Consequently, these colonies will also be eroded and sheared at a higher rate. This coincides with experimental observations by Gjaltema *et al.* (1994), who reported that sloughing always started at the upstream side of the slides in a RotoTorque bioreactor.

These findings are in agreement with the experimental results of Ohashi & Harada (1994), showing that convex parts of the biofilm surface were more subject to detachment than concave parts.

The flow regime has a double influence on biofilm formation (Fig. 11). As expected, at higher flow velocities the substrate flux towards the biofilm is increased because of the diminished external resistance in the boundary layer. This intensifies the biofilm growth rate. However, high flow velocity means a high shear stress at the biofilm surface, which generates a greater detachment rate. At Reynolds number $Re = 13.3$ (Fig. 11a–d), the biofilms reached smaller amounts of biomass per carrier area, and they were thinner than those grown at $Re = 6.7$ (lower liquid velocity; Fig. 11e–h). These results are very normal, because the higher the Re the stronger the forces that biofilm structure must withstand. Continuous erosion shaped at $Re = 13.3$ a more compact and thus mechanically more stable biofilm (Fig. 11a–d). The biofilm grown at lower liquid velocity is more heterogeneous and mechanically weaker, thus it is more susceptible to massive biomass loss (e.g. by sloughing). The causes for sloughing must therefore be sought not only in the biofilm strength, but also in its shape. The model simulations shown above predict, as we already expected, that erosion makes the biofilm surface smoother, whereas sloughing leads to an increased biofilm surface roughness.

An avalanche effect was observed several times in biomass loss, in simulations of biofilm development. Breakage of some biofilm structures left other colonies highly exposed to strong liquid shear. When these also break, others will follow, leading to massive sloughing in a short period. This model behaviour is in complete accordance with Stoodley's suggestion that 'some sloughing events are triggered' when 'the surface cover is reduced by the detachment of a few cell clusters' and, as a result, 'increasing drag on neighbouring downstream clusters may then cause them to detach too' (Stoodley *et al.*, 1999b).

Our simulation studies also showed that faster-growing biofilms have a faster detachment rate than slow-growing biofilms, under similar hydrodynamic conditions. This is in agreement with the experimental evidence showing that detachment is dependent on both shear and growth rates. Massive sloughing can be avoided by high liquid shear, combined with low biomass growth rates. This fact was also experimentally observed in the biofilm airlift reactor, when slow-growing micro-organisms like nitrifiers formed much stronger and compact biofilms than fast-growing heterotrophic populations.

To study the effect of carrier surface texture on initial biofilm development stages, the same experimental conditions as in Fig. 11(a–d) were kept in the simulation presented in Fig. 12(a–d), but an irregular substratum (carrier) geometry was created. Sheltered in the carrier valleys, only a few bacterial colonies developed a patchy biofilm (Fig. 12). Finally, the biofilm entered a steady state, with all the newly formed biomass being sheared off by the liquid (Fig. 12c, d). The clusters completely filled the cavity in which

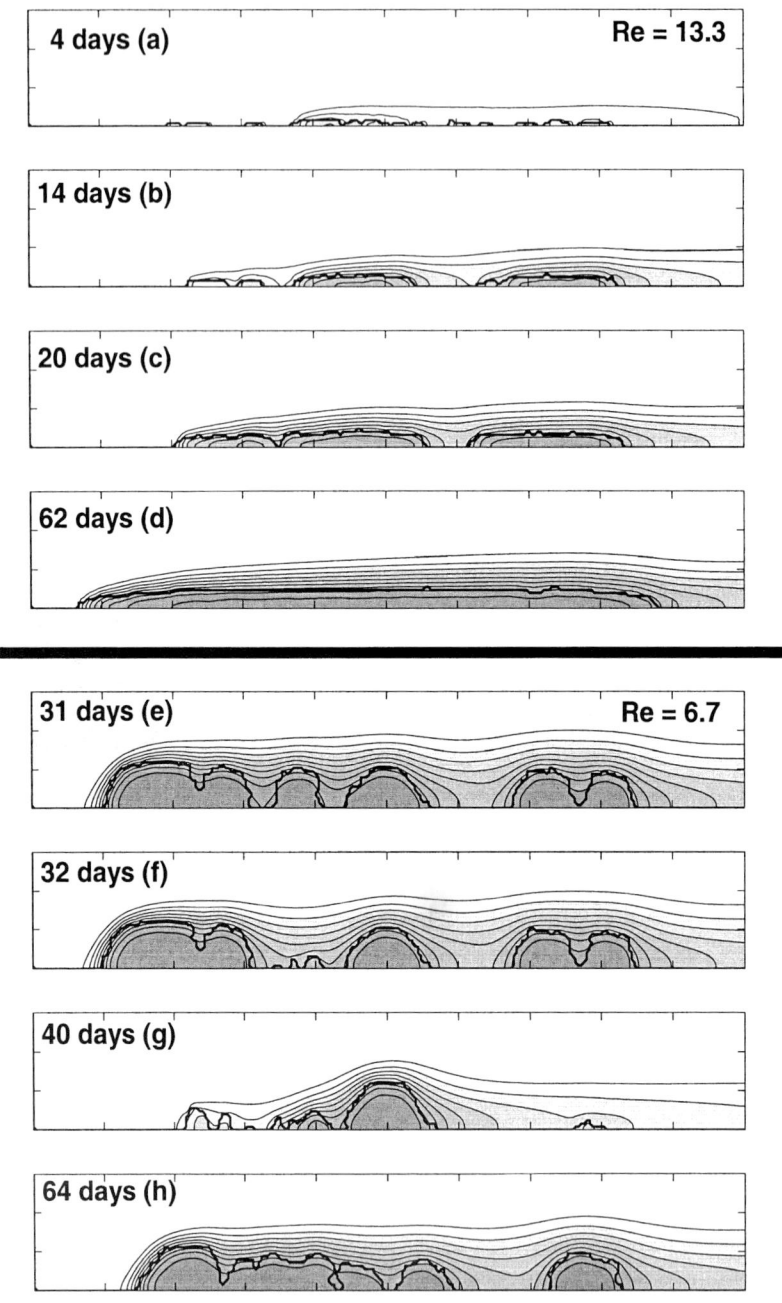

Fig. 11. Simulation of biofilm development at two liquid velocities: in cases (a)–(d), the liquid velocity is double that in cases (e)–(h). Signification of lines and greyscale is as in Fig. 9. Part of the computational system shown here has the dimensions 1000 × 150 μm. The complete model and parameters used in these simulations are given in Picioreanu (1999).

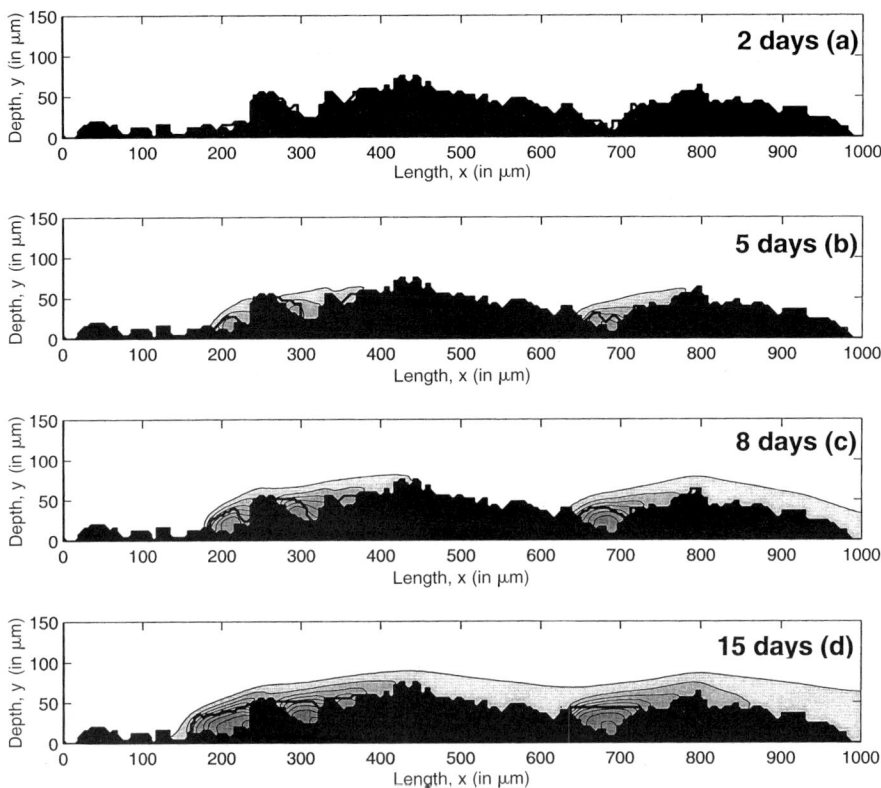

Fig. 12. Simulated biofilm development on a rough substratum. Signification of lines and greyscale is as in Fig. 9. The black area is the substratum irregular relief.

they started to grow, but they were unable to colonize the substratum further. Any attempt to accumulate more biofilm results in its exposure to high shear stress, which produces a high loss rate. The only mechanism that could lead to a colonization of neighbouring valleys is reattachment, because none of the existing colonies is able to transport biomass over the carrier peaks in the empty neighbouring valleys. This numerical experiment thus reveals again the necessity to introduce in the model processes of cell attachment and reattachment.

CONCLUSIONS

This review hopefully demonstrated that results of recent studies can give us confidence in the possibility of modelling and predicting biofilm geometrical complexity. It is not only biological factors, such as the type of micro-organism that makes up the biofilm, that have to be taken into account to explain biofilm complexity. Equally, or even more important, are the environmental conditions in which biofilms form, such as the hydrodynamic regime, substrate/product transport rates, and the substratum geometry. All modelling approaches (quantitative or only qualitative), starting from basic principles, have confirmed the hypothesis that porous biofilms can form as a result of nutrient depletion in biofilm depth. On the other hand, a deterministic detachment model has shown how liquid forces can shape the biofilm surface, to generate either compact or patchy structures.

Despite their recent successes, however, it can be clearly stated that models aiming at prediction of biofilm structural heterogeneity are still in their infancy. Many possibly important processes have been neglected. Future work is needed to include the role of EPS in biofilm formation, and to take into account variations in structural properties such as biofilm density and strength. Another possible mechanism that regulates biofilm structure, cell–cell signalling, has to be critically studied as well. The hypothesis of cell–cell signalling was suggested as a result of the impossibility to explain, by other mechanisms, pore formation in biofilms. We have shown, however, that only by using laws of physics that are generally valid, regardless of the biological system studied, can a great variety of heterogeneous biofilm structures form. Because physical mechanisms govern the world at a deeper level than biological phenomena, we prefer to explain formation of geometrical heterogeneity in biofilms by mechanisms included in the models presented here, rather than by 'bacterial talking'. Implementation of cell–cell communication and EPS production in the existing modelling frames would not be a problem, but these mechanisms are still poorly understood. Nevertheless, further experiments must be conducted in parallel with development of biofilm models, because testing a model is always essential for its credibility.

REFERENCES

Allison, D. G., Heys, S. J. D., Willcock, L., Holah, J. & Gilbert, P. (1999). Cellular detachment and dispersal from bacterial biofilms: a role for quorum sensing? In *Biofilms: the Good, the Bad and the Ugly*, pp. 279–293. Edited by J. W. T. Wimpenny, P. Gilbert, J. Walker, M. Brading & R. Bayston. Cardiff: BioLine.

Alphenaar, A. (1995). *Anaerobic granular sludge*. PhD thesis, Wageningen Agricultural University.

Atkinson, B. & Davies, I. J. (1974). The overall rate of substrate uptake (reaction) by microbial films. Part I – A biological rate equation. *Trans Inst Chem Eng* **52**, 248–259.

Barker, G. C. & Grimson, M. J. (1993). A cellular automaton model of microbial growth. *Binary* **5**, 132–137.

Beeftink, H. H. (1987). *Anaerobic bacterial aggregates. Variety and variation*. PhD thesis, University of Amsterdam.

Ben-Jacob, E., Schochet, O., Tenenbaum, A., Cohen, I., Czirók, A. & Vicsek, T. (1994). Generic modelling of cooperative growth patterns in bacterial colonies. *Nature* **368**, 46–49.

Bishop, P. L. & Rittmann, B. E. (1995). Modelling heterogeneity in biofilms: report of the discussion session. *Water Sci Technol* **32**, 263–265.

Bishop, P. L., Gibbs, J. T. & Cunningham, B. E. (1997). Relationship between concentration and hydrodynamic boundary layers over biofilms. *Environ Technol* **18**, 375–386.

Caldwell, D. E., Korber, J. R. & Lawrence, D. R. (1993). Analysis of biofilm formation using 2D vs 3D digital imaging. *J Appl Bacteriol* **74**, S52–S66.

Casey, E. (1999). *The membrane-aerated biofilm reactor: fundamental reaction engineering aspects governing performance*. PhD thesis, National University of Ireland.

Chang, H. T., Rittmann, B. E., Amar, D., Heim, R., Ehrlinger, O. & Lesty, Y. (1991). Biofilm detachment mechanisms in a liquid fluidized bed. *Biotechnol Bioeng* **38**, 499–506.

Characklis, W. G. (1981). Fouling biofilm development: a process analysis. *Biotechnol Bioeng* **23**, 1923–1960.

Characklis, W. G., Trulear, M. G., Bryers, J. D. & Zelver, N. (1982). Dynamics of biofilm processes: methods. *Water Res* **16**, 1207–1216.

Chen, S. & Doolen, G. D. (1998). Lattice Boltzmann method for fluid flows. *Annu Rev Fluid Mech* **30**, 329–364.

Chen, S., Dawson, S. P., Doolen, G. D., Janecky, D. R. & Lawniczak, A. (1995). Lattice methods and their applications to reacting systems. *Comput Chem Eng* **19**, 617–646.

Chen, Y., Ohashi, H. & Akiyama, M. (1997). Simulation of laminar flow over a backward-facing step using the lattice BGK method. *JSME Int J Ser B* **40**, 25–32.

Christensen, B. E. & Characklis, W. G. (1990). Physical and chemical properties of biofilms. In *Biofilms*, pp. 93–130. Edited by W. G. Characklis & K. C. Marshall. New York: Wiley.

Costerton, J. W., Lewandowski, Z., De Beer, D., Caldwell, D., Korber, D. & James, G. (1994). Minireview: biofilms, the customized microniche. *J Bacteriol* **176**, 2137–2142.

Davies, D. G., Chakrabarty, A. M. & Geesey, G. G. (1994). Exopolysaccharide production in biofilms: substratum activation of alginate gene expression by *Pseudomonas aeruginosa* during biofilm development in continuous culture. *Appl Environ Microbiol* **59**, 1181–1186.

Davies, D. G., Parsek, M. R., Pearson, J. P., Iglewski, B. H., Costerton, J. W. & Greenberg, E. P. (1998). The involvement of cell-to-cell signals in the development of a bacterial biofilm. *Science* **280**, 295–298.

De Beer, D. & Stoodley, P. (1995). Relation between the structure of an aerobic biofilm and transport phenomena. *Water Sci Technol* **32**, 11–18.

De Beer, D., Stoodley, P., Roe, F. & Lewandowski, Z. (1994). Effects of biofilm structures on oxygen distribution and mass transport. *Biotechnol Bioeng* **43**, 1131–1138.

De Beer, D., Stoodley, P. & Lewandowski, Z. (1996). Liquid flow and mass transport in heterogeneous biofilms. *Water Res* **30**, 2761–2765.

Dillon, R., Fauci, L., Fogelson, A. & Gaver, D. (1996). Modeling biofilm processes using the immersed boundary method. *J Comput Phys* **129**, 57–73.

Eberl, H., Picioreanu, C. & van Loosdrecht, M. C. M. (1999). Modelling geometrical heterogeneity in biofilms. In *Proceedings of the 13th International Conference of High Performance Computing Systems & Applications*, June 1999, Kingston, Canada.

Eberl, H., Picioreanu, C., Heijnen, J. J. & van Loosdrecht, M. C. M. (2000a). A three-dimensional numerical study on the correlation of spatial structure, hydrodynamic conditions, and mass transfer and conversion in biofilms. *Chem Eng Sci* (in press).

Eberl, H., Parker, D. & van Loosdrecht, M. C. M. (2000b). A new deterministic spatio-temporal continuum model for biofilm development. *J Theor Med* (submitted).

Fox, P., Suidan, M. T. & Bandy, J. T. (1990). A comparison of media types in acetate fed expanded-bed anaerobic reactors. *Water Res* **24**, 827–835.

Fujikawa, H. (1994). Diversity of the growth patterns of *Bacillus subtilis* colonies on agar plates. *FEMS Microbiol Ecol* **13**, 159–168.

Fujikawa, H. & Matsushita, M. (1991). Bacterial fractal growth in the concentration field of nutrient. *J Physiol Soc Jpn* **60**, 88–94.

Garrido, J. M., van Benthum, W. A. J., van Loosdrecht, M. C. M. & Heijnen, J. J. (1997). Influence of dissolved oxygen concentration on nitrite accumulation in a biofilm airlift suspension reactor. *Biotechnol Bioeng* **53**, 168–178.

Gibbs, J. T. & Bishop, P. L. (1995). A method for describing biofilm surface roughness using geostatistical techniques. *Water Sci Technol* **32**, 91–98.

Gjaltema, A., Arts, P. A. M., van Loosdrecht, M. C. M., Kuenen, J. G. & Heijnen, J. J. (1994). Heterogeneity of biofilms in rotating annular reactor: occurrence, structure and consequences. *Biotechnol Bioeng* **44**, 194–204.

Gjaltema, A., van der Marel, N., van Loosdrecht, M. C. M. & Heijnen, J. J. (1997). Adhesion and biofilm development on suspended carriers in airlift reactors: hydrodynamic conditions versus surface characteristics. *Biotechnol Bioeng* **55**, 880–889.

Hermanowicz, S. W. (1998). A model of two-dimensional biofilm morphology. *Water Sci Technol* **37**, 219–222.

Hermanowicz, S. W., Schindler, U. & Wilderer, P. (1995). Fractal structure of biofilms: new tools for investigation of morphology. *Water Sci Technol* **32**, 99–105.

Hermanowicz, S. W., Schindler, U. & Wilderer, P. (1996). Anisotropic morphology and fractal dimensions of biofilms. *Water Res* **30**, 753–755.

Kaye, B. H. (1989). *A Random-Walk Through Fractal Dimensions*. Weinheim: VCH.

Kreft, J.-U., Booth, G. & Wimpenny, J. W. T. (1998). BacSim, a simulator for individual-based modelling of bacterial colony growth. *Microbiology* **144**, 3275–3287.

Kreft, J.-U., Picioreanu, C., Wimpenny, J. W. T. & van Loosdrecht, M. C. M. (1999). Individual-based modelling of biofilms: why? In *Biofilms: the Good, the Bad and the Ugly.* Edited by J. W. T. Wimpenny, P. Gilbert, J. Walker, M. Brading & R. Bayston. Cardiff: BioLine.

Kwok, W. K., Picioreanu, C., Ong, S. L., van Loosdrecht, M. C. M., Ng, W. J. & Heijnen, J. J. (1998). Influence of biomass production and detachment forces on biofilm structures in a biofilm airlift suspension reactor. *Biotechnol Bioeng* **58**, 400–407.

Lawrence, J. R., Korber, D. R., Hoyle, B. D., Costerton, J. W. & Caldwell, D. E. (1991). Optical sectioning of microbial biofilms. *J Bacteriol* **173**, 6558–6567.

Lewandowski, Z., Webb, D., Hamilton, M. & Harkin, G. (1999). Quantifying biofilm structure. *Water Sci Technol* **39**, 71–76.

Matsushita, M. & Fujikawa, H. (1990). Diffusion-limited growth in bacterial colony formation. *Physica A* **168**, 498–506.

Murga, R., Stewart, P. S. & Daly, D. (1995). Quantitative analysis of biofilm thickness variability. *Biotechnol Bioeng* **45**, 503–510.

Nakamori, T. (1988). Skeletal growth model of the dendritic hermatypic corals limited by light shelter effect. In *Proceedings of the 6th International Coral Reef Symposium* **3**, pp. 113–118.

Noguera, D. R., Pizarro, G., Stahl, D. A. & Rittmann, B. E. (1999). Simulation of multispecies biofilm development in three dimensions. *Water Sci Technol* **39**, 123–130.

Nowell, A. R. M. & Church, M. (1979). Turbulent flow in a depth limited boundary layer. *J Geophys Res C: Ocean Atmos* **84**, 4816–4824.

Ohashi, A. & Harada, H. (1994). Adhesion strength of biofilm developed in an attached growth reactor. *Water Sci Technol* **29**, 281–288.

Ohashi, A. & Harada, H. (1996). A novel concept for evaluation of biofilm adhesion strength by applying tensile force and shear force. *Water Sci Technol* **34**, 201–211.

Ohashi, A., Koyama, T., Syutsubo, K. & Harada, H. (1999). A novel method for evaluation of biofilm tensile strength resisting to erosion. *Water Sci Technol* **39**, 261–268.

Picioreanu, C. (1996). Modelling biofilms with cellular automata. *Final Report to the European Environmental Research Organisation*, April 1996, Wageningen, The Netherlands.

Picioreanu, C. (1999). *Multidimensional modeling of biofilm structure*. PhD thesis, Delft University of Technology.

Picioreanu, C., van Loosdrecht, M. C. M. & Heijnen, J. J. (1997). Modelling the effect of oxygen concentration on nitrite accumulation in a biofilm airlift suspension reactor. *Water Sci Technol* **36**, 147–156.

Picioreanu, C., van Loosdrecht, M. C. M. & Heijnen, J. J. (1998a). A new combined differential-discrete cellular automaton approach for biofilm modeling: application for growth in gel beads. *Biotechnol Bioeng* **57**, 718–731.

Picioreanu, C., van Loosdrecht, M. C. M. & Heijnen, J. J. (1998b). Mathematical modeling of biofilm structure with a hybrid differential-discrete cellular automaton approach. *Biotechnol Bioeng* **58**, 101–116.

Picioreanu, C., van Loosdrecht, M. C. M. & Heijnen, J. J. (1999). Discrete-differential modelling of biofilm structure. *Water Sci Technol* **39**, 115–122.

Picioreanu, C., van Loosdrecht, M. C. M. & Heijnen, J. J. (2000). A theoretical study on the effect of surface roughness on mass transport and transformation in biofilms. *Biotechnol Bioeng* **68**, 355–369.

Ponce Dawson, S., Chen, S. & Doolen, G. D. (1993). Lattice Boltzmann computations for reaction-diffusion equations. *J Chem Phys* **98**, 1514–1523.

Rittmann, B. E. & McCarty, P. L. (1980). Model of steady-state-biofilm kinetics. *Biotechnol Bioeng* **22**, 2343–2357.

Schindler, J. & Rataj, T. (1992). Fractal geometry and growth models of a *Bacillus* colony. *Binary* **4**, 66–72.

Schindler, J. & Rovensky, L. (1994). A model of intrinsic growth of a *Bacillus subtilis* colony. *Binary* **6**, 105–108.

Schramm, A., Larsen, L. H., Revsbech, N. P., Ramsing, N. B., Amann, R. & Schleifer, K.-H. (1996). Structure and function of a nitrifying biofilm as determined by *in situ* hybridization and use of microelectrodes. *Appl Environ Microbiol* **62**, 4641–4647.

Stewart, P. S. (1998). A review of experimental measurements of effective diffusive permeabilities and effective diffusion coefficients in biofilms. *Biotechnol Bioeng* **59**, 261–272.

Stoodley, P., Boyle, J. D., Dodds, I. & Lappin-Scott, H. M. (1997). Consensus model of biofilm structure. In *Biofilms: Community Interactions and Control*, pp. 1–9. Edited by J. W. T. Wimpenny, P. Handley, P. Gilbert, H. Lappin-Scott & M. Jones. Cardiff: BioLine.

Stoodley, P., Lewandowski, Z., Boyle, J. D. & Lappin-Scott, H. M. (1998). Oscillation characteristics of biofilm streamers in turbulent flowing water as related to drag and pressure drop. *Biotechnol Bioeng* **57**, 536–544.

Stoodley, P., Lewandowski, Z., Boyle, J. D. & Lappin-Scott, H. M. (1999a). Structural deformation of bacterial biofilms caused by short-term fluctuations in fluid shear: an in situ investigation of biofilm rheology. *Biotechnol Bioeng* **65**, 83–92.

Stoodley, P., Boyle, J. D., De Beer, D. & Lappin-Scott, H. M. (1999b). Evolving perspectives of biofilm structure. *Biofouling* **14**, 75–90.

Stoodley, P., Jørgensen, F., Williams, P. & Lappin-Scott, H. M. (1999c). The role of hydrodynamics and AHL signaling molecules as determinants of the structure of *Pseudomonas aeruginosa* biofilms. In *Biofilms: the Good, the Bad and the Ugly*, pp. 279–293. Edited by J. W. T. Wimpenny, P. Gilbert, J. Walker, M. Brading & R. Bayston. Cardiff: BioLine.

Takács, I. & Fleit, E. (1995). Modelling of the micromorphology of the activated sludge floc: low DO, low F/M bulking. *Water Sci Technol* **31**, 235–243.

Tijhuis, L., van Loosdrecht, M. C. M. & Heijnen, J. J. (1994). Formation and growth of heterotrophic aerobic biofilms on small suspended particles in airlift reactors. *Biotechnol Bioeng* **44**, 595–608.

Tijhuis, L., Hijman, B., van Loosdrecht, M. C. M. & Heijnen, J. J. (1995). Influence of detachment, substrate loading rate and reactor scale on the formation of biofilms in airlift reactors. *Appl Microbiol Biotechnol* **45**, 7–17.

Van Benthum, W. A. J., Garrido, J. M., Tijhuis, L., Van Loosdrecht, M. C. M. & Heijnen, J. J. (1996). Formation and detachment of biofilms and granules in a nitrifying biofilm airlift suspension reactor. *Biotechnol Prog* **12**, 764–772.

Van Loosdrecht, M. C. M., Eikelboom, D., Gjaltema, A., Mulder, A., Tijhuis, L. & Heijnen, J. J. (1995). Biofilm structures. *Water Sci Technol* **32**, 235–243.

Verran, J., Lees, G. & Shakespeare, A. P. (1991). The effect of surface roughness on the adhesion of *Candida albicans* to acrylic. *Biofouling* **3**, 183–192.

Vicsek, T. (1984). Pattern formation in diffusion-limited aggregation. *Phys Rev Lett* **53**, 2281–2284.

Wanner, O. & Gujer, W. (1986). A multispecies biofilm model. *Biotechnol Bioeng* **28**, 314–328.

Wanner, O. & Reichert, P. (1996). Mathematical modeling of mixed-culture biofilms. *Biotechnol Bioeng* **49**, 172–184.

Wiesmann, U. (1994). Biological nitrogen removal from wastewater. *Adv Biochem Eng Biotechnol* **51**, 113–154.

Wimpenny, J. W. T. & Colasanti, R. (1997). A unifying hypothesis for the structure of microbial biofilms based on cellular automaton models. *FEMS Microbiol Ecol* **22**, 1–16.

Wood, B. D. & Whitaker, S. (1998). Diffusion and reaction in biofilms. *Chem Eng Sci* **53**, 397–425.

Wood, B. D. & Whitaker, S. (1999). Cellular growth in biofilms. *Biotechnol Bioeng* **64**, 656–670.

Yang, S. & Lewandowski, Z. (1995). Measurement of local mass transfer coefficient in biofilms. *Biotechnol Bioeng* **48**, 737–744.

Zhang, T. C. & Bishop, P. L. (1994a). Density, porosity and pore structure of biofilms. *Water Res* **28**, 2267–2277.

Zhang, T. C. & Bishop, P. L. (1994b). Evaluation of tortuosity factors and effective diffusivities in biofilms. *Water Res* **28**, 2279–2287.

Zhang, T. C., Bishop, P. L. & Gibbs, J. T. (1994). Effect of roughness and thickness of biofilms on external mass transfer resistance. In *Critical Issues in Water and Wastewater Treatment, National Conference in Environmental Engineering*, pp. 593–600. New York: ASCE.

Microbial community interactions in biofilms

Philip D. Marsh[1,2] and George H. W. Bowden[3]

[1] Research Division, CAMR, Salisbury SP4 0JG, UK
[2] Leeds Dental Institute, University of Leeds, LS2 9LU, UK
[3] Department of Oral Biology, Faculty of Dentistry, The University of Manitoba, Winnipeg, Canada

INTRODUCTION

In nature, colonization of habitats by mixtures of bacterial populations is the rule rather than the exception. Diverse groups of micro-organisms are invariably isolated from samples from environmentally exposed habitats. Evidence is accumulating that such mixtures of organisms are not merely passive neighbours but that they are involved in a wide range of dynamic physical and metabolic interactions. Indeed, these interactions appear to be essential for the attachment, growth and survival of species at a site, and also enable organisms to persist in what often appear to be overtly hostile environments.

Such interacting mixtures of micro-organisms are termed microbial communities, and are generally found on a surface, spatially organized as a biofilm (see other chapters in this volume). Microbial communities have been described from habitats ranging from aquatic environments, anaerobic digesters, plant surfaces and soil particles, to the digestive tract of humans and animals. Although the nomenclature of the species from these diverse habitats is different, this chapter will demonstrate that the function (or *niche*; Alexander, 1971) of the community members is often similar.

Existence within a microbial community can have profound consequences for the component populations. These include (a) a broader habitat range for colonization, (b) an increased metabolic diversity and efficiency, so that substrates normally recalcitrant to catabolism by individual organisms can often be broken down by consortia to simpler products, (c) increased resistance to environmental stress and to host defence

SGM symposium 59: Community structure and co-operation in biofilms. Editors D. Allison, P. Gilbert, H. Lappin-Scott, M. Wilson. Cambridge University Press. ISBN 0 521 79302 5 ©SGM 2000.

factors (Caldwell *et al.*, 1997a, b; Shapiro, 1998) and, in some cases, (d) an increased ability to cause disease (pathogenic synergism; van Steenbergen *et al.*, 1984; Brook, 1987a). Thus microbial communities display emergent properties (Odum, 1986), that is, the properties of the community are more than the sum of those of its component populations. Some of the mechanisms that contribute, or may contribute, to these emergent properties will be discussed in this chapter.

Early studies of microbial communities tended to focus primarily on isolating species from clinical or environmental samples, and characterizing the properties of these organisms in pure culture under 'artificial' conditions in the laboratory. The characteristics of the organism determined in this way are then extrapolated back to their activities in the natural habitat, often with little recognition or insight into how the component species might interact and modify each other's behaviour. Contemporary microbiology is beginning to address this issue, and is recognizing the importance of the 'community lifestyle' on the overall capabilities of the microflora at a site. Indeed, it has been suggested recently that such communities could be viewed holistically as a 'multicellular organism' in which there is a cellular division of labour that may be facilitated by both specific spatial organization and sophisticated signalling networks among the member populations (Caldwell *et al.*, 1997b; Shapiro, 1998). This theme will be continued in this review but, where possible, examples will be chosen from human microbiology, and extended to include the role of microbial communities in disease, since these aspects have often been neglected in the past.

DEVELOPMENT OF MICROBIAL COMMUNITIES

Microbial communities usually develop on environmentally exposed surfaces as a biofilm. The physical, chemical and biological properties of each surface will govern the composition and activities of the colonizing bacteria (Alexander, 1971). Numerous micro-organisms can be transported to a surface (either passively, or actively if motile), but the properties of each site will determine those strains which (a) attach, (b) grow, and possibly dominate the community, (c) remain minor components, and (d) will be unable to establish. In this way, the habitat is selective, and influences the development and final composition of the community; similarly, the habitat can be hostile to broad groups of bacteria. For example, faecal lactobacilli derived from the mother during birth can be found in the mouth of babies, but they are unable to persist as they are not adapted to growth at this site (Carlsson & Gothefors, 1975; Carlsson *et al.*, 1975).

Bacteria which first colonize a habitat have been termed pioneer species (Alexander, 1971). The metabolism of these organisms can modify the habitat and local environment to such an extent as to make conditions more suitable for the growth of secondary colonizers with more demanding requirements. These, and subsequent colonizers, continue to modify the environment, providing additional niches, which can be filled by different species. The accessibility of the mouth for sampling has permitted the detailed identification of such pioneer species. Streptococci, such as *Streptococcus mitis* biovar 1 and *Streptococcus oralis*, colonize the tongue and oral mucosa of newborn infants (Pearce *et al.*, 1995; Fitzsimmons *et al.*, 1996). Significantly, the majority of these pioneer streptococci produce an IgA1 protease (Cole *et al.*, 1994) that can partially degrade secretory IgA (sIgA), a key component in the protection of mucosal surfaces (Kilian *et al.*, 1996). Later, after the eruption of teeth, different bacteria that carry adhesins for the saliva-based conditioning films on enamel adhere to the tooth surface (Marsh & Martin, 1999). The early colonizers of tooth enamel again include streptococci (especially *Streptococcus sanguis* and *S. oralis*; Nyvad, 1993) and *Neisseria* spp. that consume oxygen, and liberate CO_2 and H_2, creating micro-environments suitable for microaerophiles and eventually obligate anaerobes (Könönen *et al.*, 1994). The early colonizers on the teeth, whether alive or dead, also provide novel receptors for attachment by later colonizers that form physical associations with cells of the pioneer strains by a process termed coaggregation or coadhesion (see the chapter by Kolenbrander and others, this volume).

After the initial and later phases of colonization, several bacterial populations will coexist within the habitat, and interactions can occur (Alexander, 1971; Margulis, 1991). At this point, the nature and availability of nutrient plays an important role in community interactions such as protocooperation and competition (Table 1). Nutrient modifies the flora of dental plaque (Bowden & Li, 1997), and may introduce changes that can be associated with oral disease (Marsh & Martin, 1999). The diet of the host is also particularly important in influencing the bacterial communities in the large intestine of humans (Ballongue *et al.*, 1997; Macfarlane & Macfarlane, 1997; Macfarlane *et al.*, 1997; Salminen & Salminen, 1997) and in the rumen (Wallace, 1985; Weimer, 1998). In these habitats, changes introduced in the bacterial community by host diet can also have a positive or negative impact on the health of the host (Flint, 1997; Fuller & Gibson, 1997; McBain & Macfarlane, 1998). Diet is also an important determinant of the intestinal flora in human infants, where breast, compared to bottle, feeding may produce differences in the species and proportions of bifidobacteria in the intestine (Fuller & Gibson, 1997). Generally, breast-fed infants are more resistant to gastrointestinal upsets.

The process of changes in bacterial species and their proportions in a habitat during colonization is termed succession (Alexander, 1971). A number of different types of microbial interaction occur during succession, and examples are listed in Table 1 (Alexander, 1971; Margulis, 1991). Some types, such as competition, protocooperation and amensalism, may be more widespread or at least more easily detected among

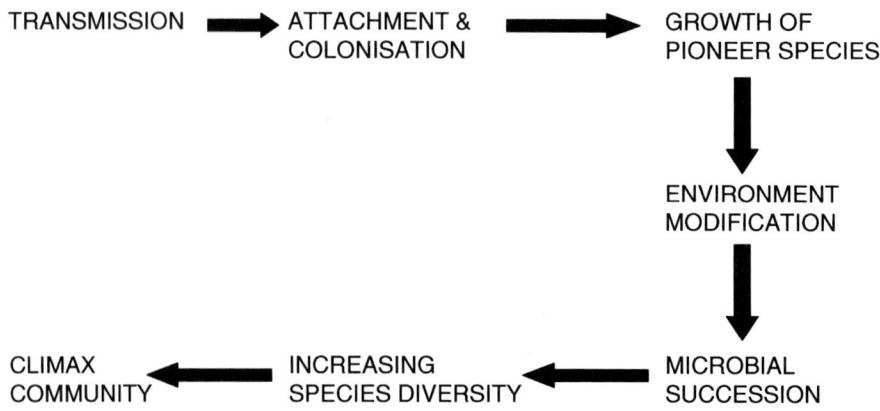

Fig. 1. Ecological stages in the establishment of microbial communities (Marsh & Martin, 1999; reproduced with permission).

populations in bacterial communities, while others are less common. Microbial succession is a key process in community development, and the possible linking of coadhesion and metabolism may facilitate a favourable spatial organization of interacting species (Bradshaw *et al.*, 1998), and ensure that organisms are most likely to attach when conditions are most conducive to their growth (Fig. 1). The trend to increasing diversity during succession reflects the relatively few niches in the uncolonized site, and the multitude of potential niches in the fully populated habitat (Alexander, 1971). Eventually, the composition of the microbial community at a site becomes stable over time (the 'climax community'; Alexander, 1971). This ability to maintain stability in a variable environment has been termed 'microbial homeostasis' (Alexander, 1971). Homeostasis is not due to any metabolic indifference among the component species, but is a result of a dynamic tension, which is dependent on a series of intermicrobial interactions (both synergistic and antagonistic; Table 1) (Fig. 2). It has been proposed that communities with a high species diversity and metabolic complexity are better able to withstand environmental perturbation (Alexander, 1971).

The relationship between the habitat and the microflora, however, is reciprocal, so that a substantial change in key environmental factors can result in a reorganization of the community to reflect these changes (Alexander, 1971). For example, eutrophication can occur when nitrogenous fertilizer leaches into aquatic habitats such as rivers and lakes, resulting in the overgrowth of algae followed by a loss of microbial and plant life by oxygen depletion. Some examples from humans and animals have been cited above, and are often associated with changes in diet, such as the increase in acidogenic and acid-tolerating species in dental plaque following elevated frequencies of sugar

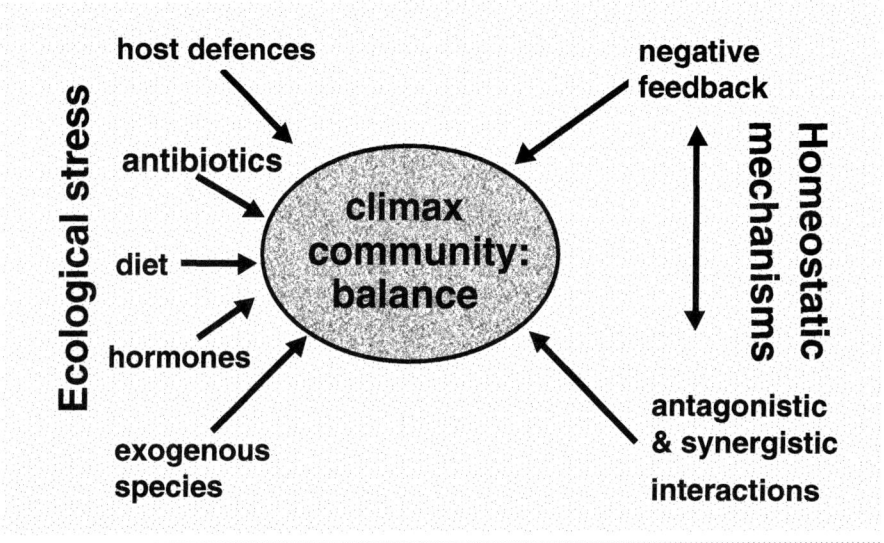

Fig. 2. Factors involved in the maintenance of homeostasis in microbial communities (Marsh & Martin, 1999; reproduced with permission).

consumption (Minah *et al.*, 1985). Others include the increase of *Propionibacterium acnes* on the skin following hormonal changes during puberty. Thus it may be possible to control the composition or activity of microbial communities during development by changing the nutritional status or local environmental conditions at a site (Marsh & Bradshaw, 1997).

CHARACTERISTICS OF MICROBIAL COMMUNITIES

Although different populations of bacteria can be isolated from specific habitats, it cannot be assumed that collectively they should all be described as, or be included into, a single community. Criteria that define microbial communities relative to mixtures of bacteria have been described (Caldwell *et al.*, 1997b). It has been argued that true bacterial communities express 'communality', which involves synergy (i.e. a community should function more effectively than the component populations), autopoiesis (the ability of populations to self-organize themselves) and homeostasis (the ability to maintain community stability within a variable environment) (Caldwell *et al.*, 1997b). The concept of symbiosis among living organisms, including bacteria, has been addressed by Margulis (1991), who provides definitions of the features associated with symbiosis, covering a variety of potential interactions among members of a community. Also, Alexander (1971) has defined and given examples of the possible interactions among populations in microbial communities. Although definitions of symbiosis and some terminology may vary, there is general agreement on the types of interactions that occur within microbial communities (Table 1). Central to the microbial community lifestyle is the concept that its members will interact, usually to the benefit of the survival of the community. However, as stated above, it cannot be assumed that all populations present in a habitat are active members of the community. It is possible that a community will be made up of combinations of different consortia and individuals (Caldwell *et al.*, 1997b). Some populations of bacteria may be quiescent (Koch, 1997), perhaps playing a sporadic role in the community, while others may be only active locally in a spatially separate sub-community. Localized microcolonies are known to exist, for example, in dental plaque (Listgarten, 1999), which, while it shows evidence of community activity, including food webs (Marsh & Bradshaw, 1999), may have spatially isolated populations and communities. Also, it could be proposed that other populations present never interact as a member of the community. This latter possibility is supported to some extent by the observation that the numbers of cells of two populations of oral bacteria accumulating on a surface *in vivo* (Beckers & van der Hoeven, 1984) or *in vitro* (Bowden & Li, 1997) were similar in mixed or monoculture. In this case, the biculture may be an example of a mixture, rather than a community, but the point can be made that bacteria can be in close proximity but occupy distinct niches, without obvious interaction, i.e. 'neutralism' (Alexander, 1971). The observations listed above mean that within an overall framework of communality, any potential interactions may not operate continually, may be limited to localized areas in structured communities, or, in the case of individual populations, they may not occur. Significantly, a population may also find a 'refuge', by being spatially separated from other antagonistic populations in the community (Alexander, 1971). Communities probably retain their flexibility and enhance their survival through adapting their population interactions to specific environments. It seems most likely that populations within a community respond to both external environmental cues and signals internal to the community, which operate between and among component populations. Consequently, the response of one population could initiate a cascade effect, involving several populations, including those that, when they are examined in isolation, may seem to be physiologically remote from the initiating population.

STRUCTURE OF MICROBIAL COMMUNITIES

In nature, the majority of bacteria, and consequently bacterial communities, are found on solid surfaces, often as biofilms, in habitats with an associated fluid phase (Hamilton, 1987; Costerton *et al.*, 1994). In humans and animals, examples of such communities are found in the mouth (Marsh & Martin, 1999), large intestine (Macfarlane & Macfarlane, 1997; Macfarlane *et al.*, 1997) and rumen (Flint, 1997; Hobson & Wallace, 1982). However, it must be recognized that these major habitats do not support a single community of consistent composition. Rather, the habitat is colonized by a wide range of communities, each associated with a specific microenvironment within the major habitat. In the oral cavity, the biofilms on different areas of the teeth vary in microbial composition (Marsh & Martin, 1999) and differ from those on the tongue (Milnes *et al.*, 1993) and the oral mucosa (Frandsen *et al.*, 1991). Similar variations occur on the mucosa of the intestine (Macfarlane & Macfarlane, 1997) and the rumen (Cheng *et al.*, 1995). In addition, the mouth, large intestine and rumen each has a fluid phase that contains populations of bacteria. Whether the populations in saliva constitute a community is not known, but this seems unlikely, given the transient nature of the organisms. However, in the intestine and rumen, the populations in the fluid in the lumen may be more stable. Importantly, masses of particulate matter, serving as surfaces for growth, are found within the fluid phase of these two latter habitats (Macfarlane & Macfarlane, 1997; Macfarlane *et al.*, 1997; Mackie & White, 1990). Therefore, bacterial communities associated with humans and animals are commonly found on tissue surfaces, but they may also be present within a fluid phase on suspended particulate matter.

Biofilms are the most structurally complex of these communities, being composed of cells and extracellular matrix. Their structure is influenced by a variety of physical and nutritional factors (van Loosdrecht *et al.*, 1995; Bowden & Li, 1997; Wimpenny & Colasanti, 1997). Significant structural features of biofilms include localized microcolonies of micro-organisms embedded within an extracellular matrix. The matrix may contain 'water channels' (Costerton *et al.*, 1994, 1995), implying the possibility of fluid-mediated transfer of molecules within the structure. Biofilms would seem to provide an almost optimum organization for: (a) efficient signalling among cells, (b) transfer of genetic material, (c) collaborative degradation of substrates and host defence molecules and (d) adaptation to, and survival from, environmental stress (Bowden & Hamilton, 1998). In addition, one outcome of spatially organized populations could be localized microenvironments, increasing the diversity of niches and consequently enhancing the diversity among populations. All of the characteristics mentioned above can be suggested to support and promote bacterial communities.

BACTERIAL DIVERSITY WITHIN COMMUNITIES

A common feature of bacterial communities is the diversity among their component populations. Several hundred species can be cultured from relatively small samples from sites within habitats, but this figure hides the true level of diversity since it has been estimated that between 50 and 98% of microbial populations cannot yet be cultivated. Molecular approaches (e.g. sequencing of 16S rRNA genes from community-extracted DNA) are now being used to identify these 'non-culturable' organisms (Welling *et al.*, 1997; Ward *et al.*, 1998; Wade, 1999). The diversity of genera and species forming bacterial communities has long been known, but the diversity among strains of a species is now being recognized following the application of biological, physiological, serological and multilocus enzyme typing methods (Selander *et al.*, 1986; Selander & Musser, 1990; Goodfellow & O'Donnell, 1993; Musser, 1996). Indeed, as a result of analysis of bacterial genomes, it has become apparent that strains of a species may be highly diverse. This approach is typified by the work on bacterial population genetics, particularly with significant human pathogens (Musser, 1996). Much less is known of the population genetics and its impact on the resident or commensal microflora of humans and animals, although these organisms often exist as closely associated populations within a community. Some data on the relative diversity of resident bacterial populations associated with humans are available (Bowden & Hamilton, 1998). For example, several clonal types of oral streptococci and *Actinomyces* may be associated with a single host, and a host may be colonized by unique strains (Hohwy & Kilian, 1995; Fitzsimmons *et al.*, 1996; Bowden *et al.*, 1999). Another characteristic of bacterial diversity within a community is that clonal types may be transient. Although a given species may be a constant component, the strains representing that species may undergo clonal variation through mutation or exchange of genetic material, or clonal replacement (Musser, 1996; Bowden, 1999). Therefore, when the diversity of bacteria in communities is considered, the possibility of clonal variation and replacement has to be added to the already well recognized diversity of genera, species, and biochemical, physiological and antigenic types. Clonal variation and replacement may influence community interactions if the clonal variant either expresses a novel character or loses a character common among strains of the species. Loss or gain of a character significant to a community may act as a stimulus for community modification.

POSSIBLE INTERACTIONS AMONG MEMBERS OF BACTERIAL COMMUNITIES

Symbiosis and antagonism

A range of potential interactions among populations in microbial communities has been described by Alexander (1971), while Margulis (1991) has discussed symbiosis and the relationships among populations (Table 1). These interactions can be beneficial to one or more of the participating populations, and enable organisms to modify their local environment, and make it more favourable for growth, and collaborate where necessary in the catabolism of complex substrates. Some of these concerted and co-ordinated activities are described in more detail in later sections.

In addition, antagonism (amensalism) among micro-organisms is also a major contributing factor in determining the composition of microbial communities. The production of antagonistic compounds (such as bacteriocins or lantibiotics) can give an organism a competitive advantage when interacting with other microbes. In studies

Table 1. Terms used to describe bacterial interactions

Term	Definition
Symbiosis	Interactive association between two or sometimes more dissimilar bacteria (species) for a significant portion of their life history
Interactions in symbiosis	
Mutualism	A beneficial interaction between two specific partners that are interdependent
Commensalism	One partner gains benefit from another, while the second is unaffected
Protocooperation	Both partners co-operate to their mutual benefit but the relationship is non-specific and each may survive alone in a different environment
Competition	Competition between two bacteria of the same or different species for an essential resource
Amensalism	One organism produces a product which is toxic to other bacteria in the habitat
Parasitism	One organism grows at the expense of the other
Population relationships	
Spatial	
Obligate	One organism requires physical contact with another throughout most or all its life history
Facultative	One organism can complete its life history apart from the other
Temporal	
Allelochemical	Chemical compounds produced by one partner evoke a behavioural or growth response in the other
Behavioural	The behaviour of each partner is required for establishment or maintenance of the association
Cyclical	Physical association is periodically interrupted and reformed
Permanent	Physical association is required throughout the life history of each partner
Metabolic	
Metabolite	A product of metabolism (amino acid, carbohydrate, nucleotide) of one partner is a semiochemical (invokes a response) for the other
Biotrophy	One partner requires a nutrient that is a metabolic product of the other
Symbiotrophy, necrotrophy	One partner's nutritional needs are completely met by the other partner, which remains alive (parasitism), or is weakened or killed (parasitism, pathogenesis)
Genetic	
Gene-product transfer	Protein or RNA synthesized off the genome of one partner is used by the other partner
Gene transfer	Genes of one partner are transferred to the genome of the other

using either gnotobiotic animals (van der Hoeven & Rogers, 1979) or human volunteers (Hillman *et al.*, 1987), the degree of colonization of teeth by oral streptococci was proportional to their level of *in vitro* bacteriocin activity.

The production of antagonistic factors will also be a mechanism whereby exogenous species are prevented from colonizing pre-established microbial communities, and will contribute to the phenomenon of 'colonization resistance' (van der Waaij *et al.*, 1971). Some strains of *Streptococcus salivarius* produce a lantibiotic inhibitor [enocin, salivaricin or BLIS (bacteriocin-like inhibitory substances); James & Tagg, 1991; Tagg & Ragland, 1991] active against Lancefield group A streptococci. It has been claimed that *S. salivarius* is more frequently isolated from the throats of children who do not become colonized following exposure to group A streptococci than from those who do become infected (Sanders & Sanders, 1982). Similar interference against group A streptococci may also be a property of other members of the pharyngeal microflora (Brook & Gober, 1999).

Despite the production of inhibitory factors, the existence of discrete microhabitats within a biofilm results in organisms being spatially organized, and therefore bacteria can coexist in a community with species that would be incompatible with one another in a homogeneous environment.

Gene transfer

Evidence for horizontal gene exchange among bacteria in microbial communities has come from the detection of microbial DNA persisting in the environment, the ability to induce competence in recipient cells under conditions resembling those encountered *in vivo*, and the existence of (a) mosaic genes and (b) identical genes (or parts of genes) among evolutionarily unrelated organisms (Lorenz & Wakernagel, 1994). Such transfer can have obvious significance when virulence factors such as toxins or antibiotic resistance are transferred within or among strains of bacterial species (Maynard Smith *et al.*, 1991; Poulsen *et al.*, 1994, 1998; Bennett, 1995; Maiden & Feavers, 1995; Lee, 1996; Haubek *et al.*, 1997; Leng *et al.*, 1997; Morelli *et al.*, 1997; Reichmann *et al.*, 1997; Hakenbeck *et al.*, 1998). The mechanisms, limitations on transfer, and the outcome of transfer among bacteria have been characterized (Rainey *et al.*, 1993; Lorenz & Wakernagel, 1994; Wilkins, 1995). Conclusive evidence for gene transfer has been derived from controlled *in vitro* and microcosm studies (Angles *et al.*, 1993; Goodman *et al.*, 1993; Beaudoin *et al.*, 1998; Christensen *et al.*, 1998). In addition, plasmid DNA has been shown to be transferred to naturally competent bacteria in river water (Williams *et al.*, 1996). Although there are less conclusive data on gene transfer in bacterial communities on human surfaces, the presence of genes with a mosaic structure among strains of *S. mitis* and *Streptococcus pneumoniae* encoding low-affinity penicillin-binding proteins is evidence of recombination events rather than mutations (Coffey *et al.*, 1991; Hakenbeck *et al.*, 1998). It is reasonable to assume that such transfer of antibiotic resistance must have taken place when the organisms were in close proximity on a surface in a common host. Natural competence in *S. pneumoniae* is induced by a secreted peptide pheromone (the 'competence-stimulating peptide', CSP) (Håvarstein *et al.*, 1995). This pheromone induces competence when its concentration reaches a critical level, ensuring that competence develops only at particular cell densities; such circumstances would be readily achieved in biofilms found on mucosal and dental surfaces in the oropharynx. A range of species of the *S. mitis* group inhabiting these surfaces produce CSPs which display some sequence heterogeneity (Håvarstein *et al.*, 1997). Generally, each pheromone type (pherotype) induces competence only in strains producing identical molecules, implying that there could be specific 'cross-talk' among members of a community. However, species of streptococci from the *Streptococcus anginosus* group produce and respond to identical CSPs. There is also evidence that natural transformation can occur in *Neisseria meningitidis* (Bowler *et al.*, 1994) and in streptococci (Poulsen *et al.*, 1998) with respect to penicillin-binding proteins and sIgA proteases, respectively. Similarly, intact 'pathogenicity islands' appear to transfer between distinct bacterial populations, and confer complex virulence properties to the recipients (Lee, 1996). One aspect of CSPs that may be of significance to biofilms is the observation that competency declines in stationary phase cells (Håvarstein & Morrison, 1999), suggesting that cells during the later stages of biofilm development may not respond so well to CSPs. However, sudden changes in the nutrient environment of the biofilm encouraging growth of specific biofilm populations might facilitate competency and responses to CSPs.

CONSEQUENCES OF A COMMUNITY LIFESTYLE

Extended habitat range

Many of the micro-organisms recovered from habitats persist and grow successfully under macroenvironmental conditions that appear to be overtly hostile. For example, the mouth is essentially an overtly aerobic environment and yet many of the predominant bacteria in dental plaque are obligately anaerobic (Marsh & Martin, 1999). Similarly, several bacterial species in plaque are sensitive to low pH and yet survive repeated exposure to pH values below 5.0 following the intake of fermentable carbohydrates in the diet (Marsh & Bradshaw, 1999). Studies of the properties of microbial communities have shown that bacterial interactions play a critical role in the persistence of species under such hostile conditions (Hamilton, 1987). Some key examples of these interactions will now be described.

Oxygen. As stated above, obligate anaerobes are the dominant groups of bacteria in many habitats, including those that appear overtly aerobic. Also, surfaces prior to colonization can often be aerobic, and yet the climax community can include high numbers of anaerobes (Hamilton, 1987). Anaerobic organisms cope with the toxic effects of oxygen by two main strategies. They interact with oxygen-consuming species which reduce the environmental levels of oxygen sufficiently to enable them to detoxify the residual low levels with a range of protective enzyme systems (Marquis, 1995). For example, the rumen of herbivores contains large numbers of anaerobic micro-organisms including bacteria and ciliate protozoa (approx. 10^{10} and 10^6, respectively). These organisms, together with resident fungi, degrade plant polymers to volatile fatty acids (the main source of carbon and energy for the animal), H_2 and CO_2. The H_2 and CO_2 are used by methanogenic bacteria to produce methane and energy. Studies have shown that rumen methanogens have a symbiotic relationship with ciliate protozoa (Williams & Lloyd, 1993). Some protozoa possess an organelle, the hydrogenosome, that generates hydrogen during the oxidation of pyruvate to acetate and CO_2. This organelle enables the protozoa to scavenge low levels of oxygen, indeed using oxygen as a terminal electron acceptor, providing environmental conditions suitable for methanogenesis. The methanogens can be found on the surface of the ciliates and also, on occasions, intracellularly (Finlay et al., 1994). Laboratory studies have shown that methanogens in co-culture with rumen holotrich protozoa have a 10-fold increase in apparent oxygen tolerance (Hillman et al., 1988). It has also been proposed that the transfer of hydrogen between microbial species is enhanced by close physical association between protozoa and methanogens (Krumholz et al., 1983).

Direct evidence that specific physical associations among members of human microbial communities can provide protection for obligate anaerobes from the toxic effects of oxygen has been obtained from laboratory studies of communities of oral bacteria (Bradshaw et al., 1996, 1997, 1998), using a two-stage, mixed-culture, biofilm model. A stable microbial community was established in the anaerobic first-stage fermenter, and this was then passed continuously into an actively aerated vessel containing surfaces for biofilm formation. Surprisingly, the obligate anaerobes grew in both the biofilm and the planktonic culture, even in the aerated second stage. The predominant species, however, was an oxygen-consuming species, Neisseria subflava (>80% of the total microflora in early communities), which had been present at barely detectable levels in the community grown anaerobically. No dissolved oxygen could be detected in the planktonic phase, and this was attributed to the metabolism of the aerobe N. subflava; the redox potential remained reduced (approx. -250 mV). Separate studies of pure culture biofilms had shown that gradients of oxygen can develop over only a few cell distances (Costerton et al., 1994, 1995), and it was assumed that similar gradients developed in these oral biofilms and in mixed-culture aggregates of cells in the planktonic phase. In a subsequent experiment, therefore, N. subflava was deliberately omitted from the inoculum, and yet the anaerobes still persisted. The loss of the aerobe was compensated for by an increase in the levels of facultatively anaerobic species, especially the streptococci (Bradshaw *et al.*, 1996). The data suggested that close cell–cell contact between oxygen-consuming and oxygen-sensitive species must be occurring, enabling the obligate anaerobes to survive, especially in the planktonic phase. As stated elsewhere in this volume, coaggregation (or coadhesion) is a key process during the formation of biofilms such as dental plaque, facilitating intra- and intergeneric attachment (Kolenbrander, 1988; Kolenbrander *et al.*, 1990; Kolenbrander & London, 1993; Kolenbrander and others, this volume). It was shown that although the oxygen-consumer (*N. subflava*) coaggregated only poorly with the obligate anaerobes, this interaction could be markedly enhanced in the presence of *Fusobacterium nucleatum*, which could act as a bridging organism between otherwise weakly coaggregating pairs of strains (Bradshaw *et al.*, 1998). The key role for this coaggregation was confirmed when consortia lacking *F. nucleatum* were reintroduced into the aerated biofilm vessel. Viable counts of the black-pigmenting anaerobes within the community (*Prevotella nigrescens, Porphyromonas gingivalis*) fell by three orders of magnitude (Bradshaw *et al.*, 1998). These findings suggest that the role of coaggregation is not necessarily restricted to anchoring cells during the establishment of a microbial community, but may also facilitate metabolism among strains that depend for their survival on close physical cell–cell contact (an example of an obligate spatial relationship; Table 1). It is probable that similar physical interactions will occur to ensure that organisms needing to interact for nutritional or other environmental-modifying purposes are appropriately spatially organized.

pH. Many bacterial species have a relatively narrow pH range for growth, and yet, when present as part of a community, survive repeated exposure to apparently extremes of pH that are inhibitory to the same cells growing in pure culture (for examples, see McDermid *et al.*, 1986; Bradshaw *et al.*, 1989). Their survival is probably due to several of the properties of biofilms when they function as surface-associated microbial communities. Gradients develop in key parameters in biofilms (Costerton *et al.*, 1987, 1994, 1995), and this environmental and spatial heterogeneity can enable organisms to grow that would be incompatible with one another in a homogeneous habitat. Recent studies using two-photon excitation microscopy coupled with fluorescent lifetime imaging of pH-sensitive dyes have confirmed conditions of environmental heterogeneity in complex biofilms in terms of pH over short distances (Fig. 3; Vroom *et al.*, 1999). The gradients in pH were not linear in either the x–y- or x–z-axes; zones of discrete pH were observed adjacent to areas of quite differing pH.

Also, bacteria are able to modulate their local pH, especially in biofilms, by up-regulating genes involved with acid or base production. Urease and arginine deiminase activity by streptococci (Rogers *et al.*, 1987; Casiano-Colon & Marquis, 1988; Chen *et*

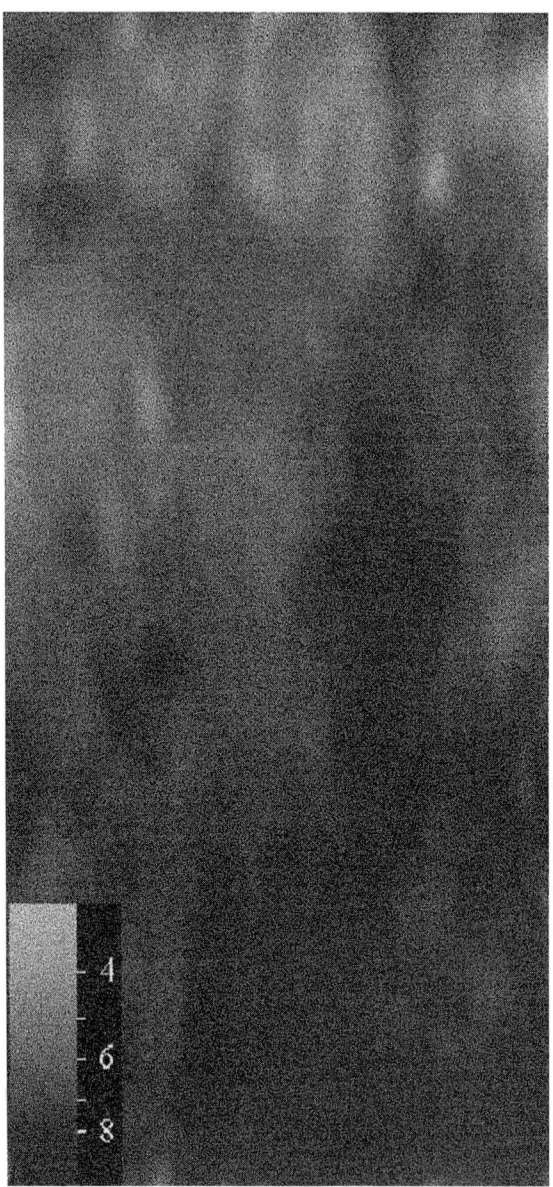

Fig. 3. Demonstration of environmental heterogeneity in terms of pH in a microbial community. A consortium of oral bacteria was grown as a biofilm, and visualized using two-photon excitation microscopy after staining with carboxyfluorescein. Fluorescent lifetime imaging was used in conjunction with the pH-sensitive dye to map pH gradients in the biofilm (Vroom *et al.*, 1999; reproduced with permission of the American Society for Microbiology). The key displays the scale for the range pH 4.0–8.0. The depth of the image is 50 μm.

al., 1996) is maintained and even enhanced at low pH, liberating ammonia and raising the pH. These enzymes can be active at pH values lower than those at which the bacteria can grow. Individual bacteria also possess specific molecular strategies which enable them to adapt rapidly to sudden changes in pH (Foster, 1995; Hall *et al.*, 1995; Bowden & Hamilton, 1998). Thus microbial communities are able to defy the constraints imposed by the external macroenvironment by creating, through their metabolism, a mosaic of microenvironments that enable the survival and growth of the component species.

Increased metabolic diversity and efficiency

In order to proliferate at a site, micro-organisms need to acquire all of their essential nutrients, and modify the local environment to make conditions more favourable for growth, as described above. In many habitats, the primary sources of carbon, nitrogen and energy are complex organic molecules derived from the local environment, such as polysaccharides, lipids, glycoproteins and proteins. A common feature of microbial communities is that the breakdown of such molecules requires the co-operative action of several groups of micro-organisms with complementary enzyme profiles. The catabolism of such molecules also involves both the sequential and the concerted metabolism of the substrate, in which the substrate is broken into simpler products by primary feeders, and these products are further converted into simpler products by secondary feeders until, in complete microbial communities, the terminal end products of metabolism are compounds such as CH_4, CO_2, H_2 and H_2S. Thus the decomposition of plant material involves several groups of bacteria, including cellulolytic bacteria, primary and secondary fermenters, syntrophs and methanogens. Similarly, in the human and animal digestive tract, endogenous nutrients such as proteins and glycoproteins are primary sources of nutrient, and play a major role in maintaining the diversity of the resident microflora. These endogenous nutrients require the concerted action of a range of proteases, peptidases and glycosidases (see Fig. 4). Individual bacterial species have limited capabilities to degrade these molecules, and studies from several habitats have shown that bacteria generally grow only poorly in the presence of these complex molecules in pure culture. Higher cell densities are achieved if species with complementary enzyme profiles are grown in binary or triplet culture (for examples, see Gharbia & Shah, 1989; van der Hoeven & Camp, 1991; Homer & Beighton, 1992), but especially in more complex communities (Hoskins *et al.*, 1985; ter Steeg *et al.*, 1987, 1988; ter Steeg & van der Hoeven, 1989) (Table 2). Three phases of metabolism were recognized when consortia of oral bacteria were grown on human serum. Firstly, saccharolytic bacteria such as *Streptococcus* spp., *Bifidobacterium* spp. and *Eubacterium* spp. predominated, and utilized the oligosaccharide side chains. Secondly, proteolysis occurred and any remaining carbohydrates were consumed. Finally, amino acid fermentations predominated (ter Steeg & van der Hoeven, 1989).

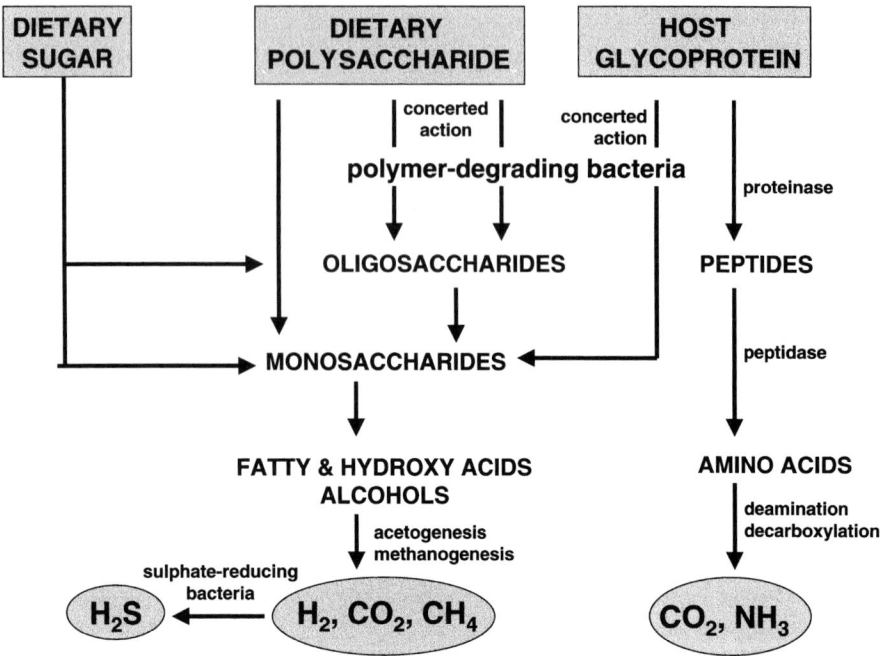

Fig. 4. Schematic diagram to illustrate the interactions involved in the catabolism of complex dietary and host molecules by communities of micro-organisms from the digestive tract. Catabolism involves the concerted action of distinct species, as well as the cycling of complex molecules into simpler end products of metabolism in food chains.

As a consequence of this community action, serum glycoproteins, including immunoglobulins, were extensively degraded. Similar studies were undertaken using defined communities of oral or enteric bacteria, with pre-selected biochemical attributes, growing on glycoproteins (Hoskins *et al.*, 1985; Bradshaw *et al.*, 1994). Sugars released during the microbial catabolism of glycoproteins in the gut can also support the growth of non-mucin-degrading populations of faecal bacteria (Hoskins *et al.*, 1992).

In habitats like the colon, microbial communities also catabolize dietary polysaccharides and other sugars. Complex polysaccharides are broken down in a series of hydrolytic steps by diverse communities of bacteria to short-chain fatty acids (SCFAs) and other organic acids, alcohols, H_2 and CO_2 (Fig. 4). The ultimate fate of these end products depends on the amount of absorption of metabolites from the colon and rectum, and on the balance of the three H_2 disposal mechanisms which operate in the colon (methanogenesis, dissimilatory sulphate reduction and acetogenesis) (Hudson & Marsh, 1995). The SCFAs act as a major source of energy for the colonic mucosa, and

Table 2. Concerted action between two species of enteric bacteria in the degradation of the oligosaccharide side chains of a mucin glycoprotein

Adapted from Hoskins *et al.* (1985).

Bacterium	Enzyme activity		Percentage loss mucin hexoses	Bacterial growth*
	α-Galactosidase	β-*N*-Acetyl hexosaminidase		
Ruminococcus	+	−	43	32
Bifidobacterium	−	+	46	22
Ruminococcus + *Bifidobacterium*	+	+	96	139

Note: *Bacterial growth is expressed as the change in OD_{600} ($\times 10^3$).

influence mucosal cell turnover (Conway, 1995). This is an example of mutualism (Table 1) since the microflora of the colon utilize the mucosal cells and the secreted mucins as nutrients, while the cells benefit from the SCFAs.

The sequential breakdown of complex molecules leads to the establishment of simple food chains, or more complex food webs, whereby the product of one organism (primary feeder) becomes the substrate for another (secondary feeder) (biotrophy; Table 1). Such interactions contribute significantly to homeostasis in microbial communities because the component species become mutually dependent on each other for nutrient acquisition. A potential increase by one species could be held in check by the lack of a response by a codependent organism. Many of these large and complex host molecules can be regarded as providing an array of discrete nutrient niches which can be exploited only by particular species or consortia.

A common food chain involves the production of lactate from the fermentation of a range of sugars, which is then cycled to simpler fermentation products (e.g. propionic and acetic acids) by lactate-utilizing species. Such acids can be further catabolized by other bacteria to CO_2, CH_4, H_2, etc. (Fig. 4). This food chain occurs in microbial communities from a wide range of habitats, although the participating species may be quite distinct. Thus in the mouth, Gram-positive species such as *Streptococcus* and *Actinomyces* spp. convert dietary sugars to lactate, and this is utilized by *Veillonella* spp. (Mikx & van der Hoeven, 1975). In the colon, the ingestion of fructo-oligosaccharides, which exist naturally in many plants, can lead to elevated levels of both lactate-producers (such as *Bifidobacterium* and *Lactobacillus* species) and lactate-utilizers (*Veillonella* spp.) (Hudson & Marsh, 1995). In contrast, in soil communities, lactate utilization is undertaken by *Desulfovibrio* species. Such food chains enhance metabolic efficiency by reducing feed-back inhibition due to end product accumulation. For example, glycolysis by streptococci was increased by almost threefold when cells metabolized monosaccharides in the presence of a lactate-utilizing organism (Hamilton & Ng, 1983).

Increased protection from host/environmental stresses

Organisms can cope with a wider range of environmental stresses when part of a structured microbial community. For example, the strategies employed by bacteria to cope with extremes of oxygen and pH within a microbial community were described in earlier sections. Communities can also preferentially survive exposure to a number of other stresses, including those experienced with the host in health and disease.

Antimicrobial resistance. As early as 1948, it had been shown that *Actinomyces israelii*, present as whole colonies, was more resistant to penicillin than suspended cells (Holm, 1948). *In vivo*, this protection also probably extended to other bacterial populations, such as *Actinobacillus actinomycetemcomitans*, which are often associated with the granules of *A. israelii* found in the lesions of actinomycosis. This finding suggested that growth as colonies or as aggregates (granules) provided physical protection, limiting the penetration of the antibiotic. It is now well accepted that bacteria are more resistant to antimicrobial agents when growing as a biofilm (Costerton *et al.*, 1987; Wilson, 1996; Gilbert *et al.*, 1997). This may be due to the relatively slow growth rate of cells on a surface, or the expression of a novel surface-associated phenotype. In addition, community-mediated effects may also play an important role; cells can be protected by the restricted penetration of an antimicrobial into the biofilm due to selective binding to extracellular polysaccharides, or to the inactivation of the agent by enzymes (e.g. β-lactamases; Brook *et al.*, 1983; Walker *et al.*, 1987; Brook, 1989, 1996; van Winkelhoff *et al.*, 1997) concentrated within the biofilm, and produced by other community members. Similarly, organisms could be protected from components of the host defences by being physically close to neighbours producing a neutralizing enzyme, such as sIgA protease (Kilian *et al.*, 1996) or other immunoglobulin- and complement-degrading proteases (Sundqvist *et al.*, 1985) (Fig. 5). Microbial communities may also afford protection from phagocytosis for cells deep within a spatially organized consortium (Costerton *et al.*, 1981, 1987). Similar strategies may help protect organisms from bacterially generated inhibitors. Many bacteria in communities in the gut or mouth produce hydrogen peroxide at concentrations capable of inhibiting sensitive species. Studies have shown that catalase production by high-density cultures of *Escherichia coli* can provide 'group-protection', whereas isolated organisms remain susceptible (Ma & Eaton, 1992). Similarly, *Salmonella typhimurium* can be protected from environmental stress by the presence of competing organisms, perhaps by reducing the impact of oxidative damage by consuming oxygen (Aldsworth *et al.*, 1998).

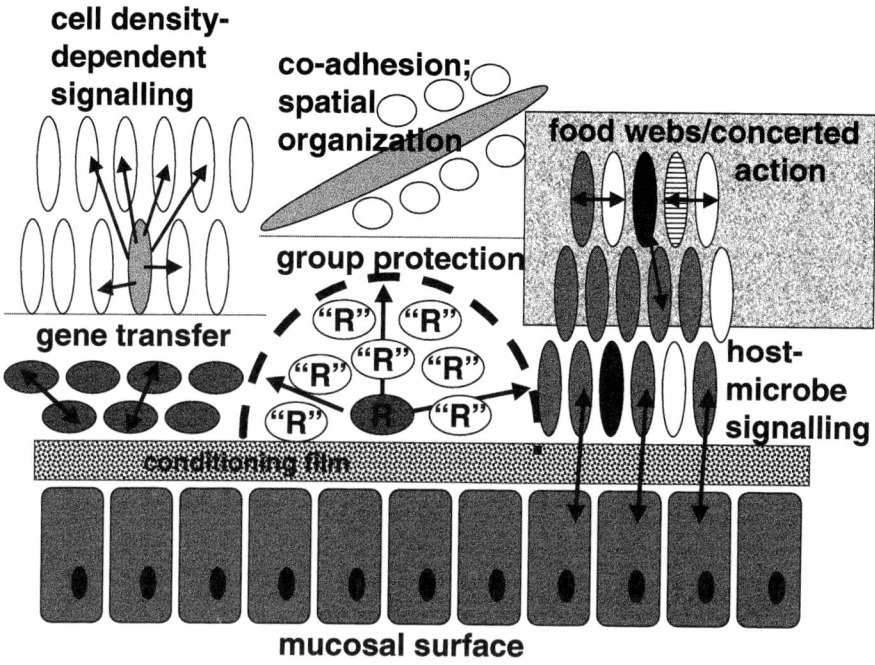

Fig. 5. Schematic representation of a microbial community illustrating the types of interaction that can take place both among the component populations (e.g. metabolic interactions, gene transfer, quorum sensing, inactivation of inhibitors and/or host defences) and between the micro-organisms and the host surface. Group protection can be achieved when a truly resistant population (R) produces, for example, an enzyme that neutralizes an inhibitor and affords protection ("R") to neighbouring cells that are intrinsically sensitive to that agent.

Pathogenic synergism

Some species of bacteria have the ability to colonize, overcome the defences of the host, and initiate disease when in pure culture (overt pathogens). However, in other situations, disease is the result of the action of consortia of microbes. Thus abscesses at various body sites and periodontal disease (diseases of the gum and peridontium; Fig. 6) are examples of synergistic infections, whereby organisms that are individually unable to satisfy all of the requirements necessary to cause disease combine forces to do so (Roberts, 1967; Fabricius *et al.*, 1982; van Steenbergen *et al.*, 1984; Brook, 1987a, b; Isenberg, 1988; Onderdonk *et al.*, 1990; Baumgartner *et al.*, 1992) (Table 3).

In a variety of animal models, various combinations of anaerobic and aerobic bacteria produced levels of sepsis or disease that could not be induced by individual species (see Brook, 1987a). Synergy has also been reported between '*Bacteroides melaninogenicus*'

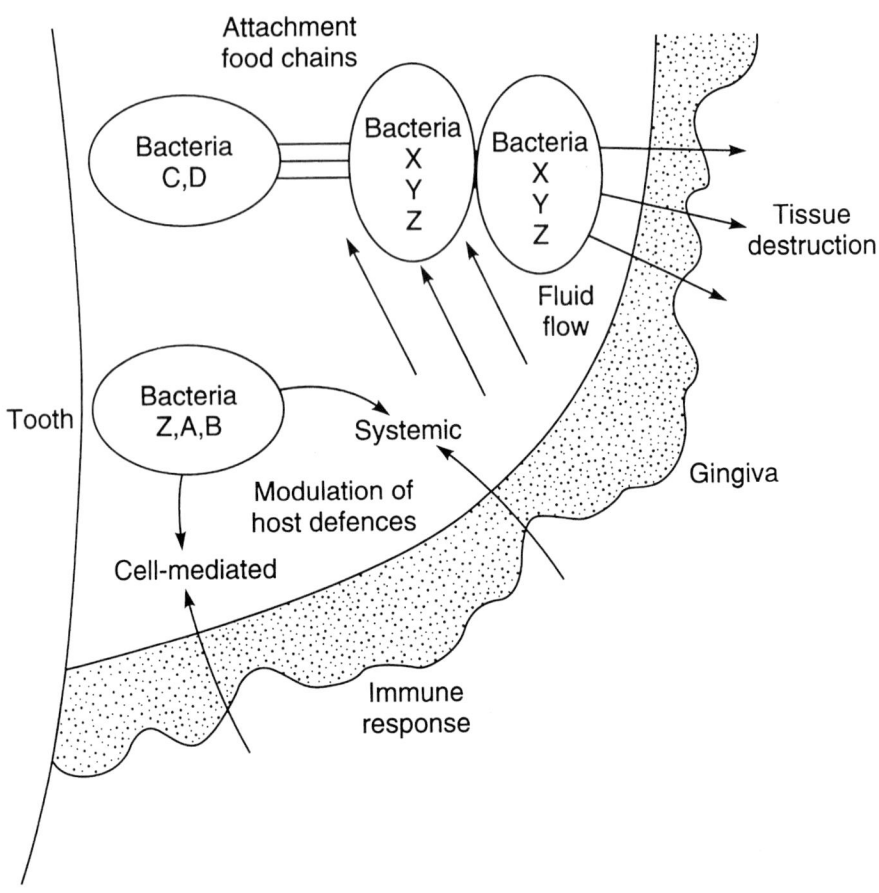

Fig. 6. Pathogenic synergy and the aetiology of periodontal diseases. Bacteria capable of causing tissue damage directly (e.g. species X, Y and Z) may be dependent on the presence of other cells (e.g. organisms C and D) for nutrient acquisition (e.g. via food chains or through concerted action) or for coadhesion. Similarly, both groups of bacteria may be reliant on other organisms (e.g. Z, A and B) to modulate the host defences. Individual bacteria may have more than one role (e.g. organism Z) in the aetiology of disease, and different species may play the same role in distinct sites (Marsh & Martin, 1999; reproduced with permission).

(now reclassified as *Prevotella melaninogenica*), '*Bacteroides asaccharolyticus*' (now *Porphyromonas asaccharolytica*) and *Peptostreptococcus micros* (Sundqvist *et al.*, 1979); '*Bacteroides asaccharolyticus*' and *Klebsiella pneumoniae*, where the latter organism provided succinate for '*B. asaccharolyticus*' (Mayrand & McBride, 1980); *Porphyromonas gingivalis*, *F. nucleatum*, *Eubacterium saburreum* and *Capnocytophaga ochracea* (Grenier & Mayrand, 1983); and *E. coli* and *Bacteroides fragilis* (Rotstein & Kao, 1988). Also, the β-haemolysin of *E. coli* has been highlighted as one

Table 3. Pathogenic synergism of *Fusobacterium nucleatum* with other anaerobic bacteria

Each group had seven mice; each mouse was scored from 0 (no observed pathology) to 4 (death). Data adapted from Baumgartner *et al.* (1992).

Bacterium	Pathogenic score	
	Pure culture	Mixed culture
Fusobacterium nucleatum	12–14	—
Porphyromonas gingivalis	0–5	15–21*
Prevotella intermedia	0	18–26*
Prevotella melaninogenica	0	14
Peptostreptococcus micros	2–7	10–14

Note: *Bacteria in mixed culture were significantly more pathogenic than *F. nucleatum* in pure culture.

factor that may potentiate pathogenic synergy among colonic bacteria (Ushijima *et al.*, 1990).

Mechanisms to explain such pathogenic synergy include communal protection from phagocytosis and intracellular killing, production of essential growth factors (such as succinate for *Porphyromonas*), modification of the local environment (e.g. lowering of redox potential) (Brook, 1987a), and the protection of sensitive species by the inactivation of inhibitors, such as by the production of β-lactamases (see above). This protection has also been termed 'indirect pathogenicity' (Brook, 1989), and is a significant clinical problem in ear, nose and throat infections where some of the more obvious pathogens are shown to be antibiotic sensitive, but are rendered 'resistant' by other members of the mixed infection. In these cases, treatment has to be targeted at all of the component populations in a mixed infection (Brook, 1989, 1996).

Although interactions between populations of bacteria from mixed infections can be demonstrated, often one species appears to play a relatively significant role in the infection. Thus *P. gingivalis* and *P. asaccharolytica* are important contributors to infections in a guinea pig abscess model (Sundqvist *et al.*, 1979; Mayrand & McBride, 1980; Grenier & Mayrand, 1983). In addition, the significance of capsulated strains in mixed infections has been studied by Brook and his collaborators (Brook, 1987a, b, 1988; Brook & Walker, 1983; Brook *et al.*, 1983). Capsulated strains can be more significant in mixed infections and there is evidence of emergence of capsulated strains of *Bacteroides* spp. in mixed infections, possibly as a result of genetic transfer *in situ*

(Brook, 1988). This latter observation may be a key example indicating transfer of genes, or switching on of genes *in vivo*, in mixed infections or communities.

Opportunistic pathogens are often components of the resident bacterial communities associated with humans and animals. Such organisms become pathogenic by gaining access to sites not normally accessible to them, or by exploiting changes in local environmental conditions, and becoming a more prominent member of a community. These areas of microbial pathogenicity (mixed infections, emergence of opportunistic pathogens) warrant reassessment (Smith, 1995); the increased availability of gene mutants and knowledge of cell–cell signalling strategies among bacteria will permit greater insights into how consortia of organisms co-operate and co-ordinate their activities, to cause disease.

CELL–CELL SIGNALLING

It is known that populations of bacteria sense and respond to their environments (Mekalanos, 1992; Lazazzera & Grossman, 1998; O'Toole & Kolter, 1998), exhibit intercellular signalling (Fuqua *et al.*, 1994; Salmond *et al.*, 1995; Gray, 1997; O'Toole & Kolter, 1998; Dunny & Winans, 1999) and also interact with cells of their hosts (Abraham *et al.*, 1998; Telford *et al.*, 1998; Meyer, 1999; Naumann *et al.*, 1999). These characteristics are also likely to be expressed by individual populations localized within biofilm communities. Signalling molecules in Gram-positive bacteria are generally post-translationally processed peptides (Kleerebezem *et al.*, 1997), while N-acyl homoserine lactones are the key signalling compounds in Gram-negative species (Greenberg, 1998). These molecules are diffusible, and their action is cell-density-dependent, and hence this process has been termed 'quorum sensing'. The effects of such molecules have generally been studied in pure cultures. These molecules, however, also provide the opportunity within microbial communities for both communication between different cell types ('cross-talk'), and also the co-ordination of community actions. An important consideration in this circumstance would be signal specificity, and the effects of the local environment and of other bacteria on this type of communication. The effect of signalling molecules produced within a biofilm might be (a) restricted to the producer population, (b) modified or eliminated by the environment, or (c) recognized by other populations in the community. Some but not all of the CSPs of streptococci can be species-specific (Håvarstein & Morrison, 1999), but interspecies quorum sensing is possible (Gray, 1997; Visick & Ruby, 1999), suggesting that both intra- and interpopulation communication could occur in biofilms (Fig. 5). Moreover, it has been shown that quorum sensing may be important to biofilm development (Davies *et al.*, 1998) and that acylated homoserine lactones are produced by biofilm cells *in situ* in their habitats (Stickler *et al.*, 1998).

The potential for complexity is vast when cell–cell signalling is considered together with the other metabolic and physiological interactions and responses that could occur within diverse biofilm communities. Likewise, it has been proposed that the resident bacteria that colonize the mucosal surfaces of man and animals may also produce signalling molecules that modulate the normal inflammatory response of the host to colonization by micro-organisms (Henderson, 1998; Henderson & Wilson, 1998). These 'microkines' may inhibit the induction of pro-inflammatory cytokine networks, enabling the resident microflora and the mammalian host to communicate and reach a mutually beneficial equilibrium (Henderson & Wilson, 1998) (Fig. 5). Also, when pathogens colonize host tissues, signalling may occur which triggers responses in both the host cells and in the microbe (Abraham *et al.*, 1998); similar events may occur with complex communities.

Given the potential significance of signalling molecules, coupled to the known interdependencies among biofilm populations, modulation of signalling could provide an important means of control of biofilm structure and community physiology. Such modifications could reduce the pathogenic impact of biofilm communities, and reduce or eliminate their resistance to removal and to antibacterial agents. A full understanding of the role of cell–cell signalling in complex microbial communities will be a major goal for research in the future.

CONCLUDING REMARKS

Microbial communities are ubiquitous; samples taken from diverse sites invariably yield consortia of micro-organisms. Evidence suggests that the majority of the component species are actively participating in community functions. Microbes have evolved to exploit the benefits of a community lifestyle, the advantages of which, however, are only just beginning to be recognized by microbiologists. These include potential benefits in terms of increased metabolic efficiency and substrate accessibility, enhanced resistance to environmental stress and inhibitors, a broader habitat range for growth (Caldwell *et al.*, 1997b; Shapiro, 1998), and (on occasions) a raised ability to cause disease. Such properties have led some to propose that microbial communities could be considered as multicellular organisms (Caldwell *et al.*, 1997b; Shapiro & Dworkin, 1997; Andrews, 1998; Shapiro, 1998). Microbial communities are obviously not obligately multicellular, but they clearly display aspects of 'multicellularity' in terms of spatial organization to enhance cellular division of labour, the use of sophisticated communication systems, and metabolic co-operation to fully exploit accessible nutrients. This results in a form of integrated and co-ordinated response by distinct species in order to optimize population survival. In the future, strategies to modulate the activity of communities will be an important target for environmental, industrial and medical microbiologists.

REFERENCES

Abraham, S. N., Jonsson, A.-B. & Normark, S. (1998). Fimbriae-mediated host pathogen cross-talk. *Curr Opin Microbiol* **1**, 75–81.

Aldsworth, T. G., Sharman, R. L., Dodd, C. E. R. & Stewart, G. S. A. B. (1998). A competitive microflora increases the resistance of *Salmonella typhimurium* to inimical processes: evidence for a suicide response. *Appl Environ Microbiol* **64**, 1323–1327.

Alexander, M. (1971). *Microbial Ecology*. New York: Wiley.

Andrews, J. H. (1998). Bacteria as modular organisms. *Annu Rev Microbiol* **52**, 105–126.

Angles, M. L., Marshall, K. C. & Goodman, A. E. (1993). Plasmid transfer between marine bacteria in the aqueous phase and biofilms in reactor microcosms. *Appl Environ Microbiol* **59**, 843–850.

Ballongue, J., Schumann, C. & Quignon, P. (1997). Effects of lactulose and lactitol on colonic microflora and enzyme activity. *Scand J Gastroenterol* **222** (suppl.), 41–44.

Baumgartner, J. C., Falkler, W. A. & Beckerman, T. (1992). Experimentally induced infection by oral anaerobic microorganisms in a mouse model. *Oral Microbiol Immunol* **7**, 253–256.

Beaudoin, D. L., Bryers, J. D., Cunningham, A. B. & Peretti, S. W. (1998). Mobilization of broad host range plasmid from *Pseudomonas putida* to established biofilm of *Bacillus azotoformans*. 1. Experiments. *Biotechnol Bioeng* **57**, 272–279.

Beckers, H. J. A. & van der Hoeven, J. S. (1984). The effects of mutual interaction and host diet on the growth rates of the bacteria *Actinomyces viscosus* and *Streptococcus mutans* during colonisation of tooth surfaces in di-associated gnotobiotic rats. *Arch Oral Biol* **29**, 231–236.

Bennett, P. M. (1995). The spread of drug resistance. In *Population Genetics of Bacteria*, pp. 317–344. Society for General Microbiology Symposium 52. Edited by S. Baumberg, J. P. W. Young, E. M. H. Wellington & J. R. Saunders. Cambridge: Cambridge University Press.

Bowden, G. H. W. (1999). Oral biofilm an archive of past events? In *Dental Plaque Revisited*, pp. 211–235. Edited by H. N. Newman & M. Wilson. Cardiff: BioLine.

Bowden, G. H. W. & Hamilton, I. R. (1998). Survival of oral bacteria. *Crit Rev Oral Biol Med* **9**, 54–85.

Bowden, G. H. & Li, Y. H. (1997). Nutritional influences on biofilm development. *Adv Dent Res* **11**, 81–99.

Bowden, G. H. W., Nolette, N., Ryding, H. A. & Cleghorn, B. M. (1999). The diversity and distribution of the predominant ribotypes of *Actinomyces naeslundii* genospecies 1 and 2 in samples from enamel and healthy and carious root surfaces of teeth. *J Dent Res* **78**, 1800–1809.

Bowler, L. D., Zhang, Q.-Y., Riou, J.-Y. & Spratt, B. G. (1994). Interspecies recombination between the *penA* genes of *Neisseria meningitidis* and commensal *Neisseria* species during the emergence of penicillin resistance in *N. meningitidis*: natural events and laboratory simulation. *J Bacteriol* **176**, 333–337.

Bradshaw, D. J., McKee, A. S. & Marsh, P. D. (1989). Effects of carbohydrate pulses and pH on population shifts within oral microbial communities *in vitro*. *J Dent Res* **68**, 1298–1302.

Bradshaw, D. J., Homer, K. A., Marsh, P. D. & Beighton, D. (1994). Metabolic cooperation in oral microbial communities during growth on mucin. *Microbiology* **140**, 3407–3412.

Bradshaw, D. J., Marsh, P. D., Allison, C. & Schilling, K. M. (1996). Effect of oxygen, inoculum composition and flow rate on development of mixed culture oral biofilms. *Microbiology* **142**, 623–629.

Bradshaw, D. J., Marsh, P. D., Watson, G. K. & Allison, C. (1997). Oral anaerobes cannot survive oxygen stress without interacting with aerobic/facultative species as a microbial community. *Lett Appl Microbiol* **25**, 385–387.

Bradshaw, D. J., Marsh, P. D., Watson, G. K. & Allison, C. (1998). Role of *Fusobacterium nucleatum* and coaggregation in anaerobe survival in planktonic and biofilm oral microbial communities during aeration. *Infect Immun* **66**, 4729–4732.

Brook, I. (1987a). Role of encapsulated anaerobic bacteria in synergistic infections. *CRC Crit Rev Microbiol* **14**, 171–193.

Brook, I. (1987b). Bacteraemia and seeding capsulate *Bacteroides* spp. and anaerobic cocci. *J Med Microbiol* **23**, 61–67.

Brook, I. (1988). Pathogenicity of capsulate and non-capsulate members of *Bacteroides fragilis* and *B. melaninogenicus* groups in mixed infection with *Escherichia coli* and *Streptococcus pyogenes*. *J Med Microbiol* **27**, 191–198.

Brook, I. (1989). Direct and indirect pathogenicity of beta-lactamase-producing bacteria in mixed infections in children. *Crit Rev Microbiol* **16**, 161–180.

Brook, I. (1996). Microbiology and management of sinusitis. *J Otolaryngol* **25**, 249–256.

Brook, I. & Gober, A. E. (1999). Interference by aerobic and anaerobic bacteria in children with recurrent group A beta-hemolytic streptococcal tonsillitis. *Arch Otolaryngol Head Neck Surg* **125**, 552–554.

Brook, I. & Walker, R. I. (1983). Infectivity of organisms recovered from polymicrobial abscesses. *Infect Immun* **42**, 986–989.

Brook, I., Pazzaglia, G., Coolbaugh, J. C. & Walker, R. I. (1983). *In vivo* protection of group A beta-haemolytic streptococci from penicillin by beta-lactamase-producing *Bacteroides* species. *J Antimicrob Chemother* **12**, 599–606.

Caldwell, D. E., Atuku, E., Wilkie, D. C., Wivcharuk, K. P., Karthikeyan, S., Korber, D. R., Schmid, D. F. & Wolfaardt, G. M. (1997a). Germ theory vs. community theory in understanding and controlling the proliferation of biofilms. *Adv Dent Res* **11**, 4–13.

Caldwell, D. E., Wolfaardt, G. M., Korber, D. R. & Lawrence, J. R. (1997b). Do bacterial communities transcend Darwinism? *Adv Microb Ecol* **15**, 105–191.

Carlsson, J. C. & Gothefors, L. (1975). Transmission of *Lactobacillus jensenii* and *Lactobacillus acidophilus* from mother to child at time of delivery. *J Clin Microbiol* **1**, 124–128.

Carlsson, J. C., Grahnen, H. & Jonsson, G. (1975). Lactobacilli and streptococci in the mouth of children. *Caries Res* **9**, 333–339.

Casiano-Colon, A. & Marquis, R. E. (1988). Role of the arginine deiminase system in protecting oral bacteria and an enzymatic basis for acid tolerance. *Appl Environ Microbiol* **54**, 1318–1324.

Chen, Y. Y., Clancy, K. A. & Burne, R. A. (1996). *Streptococcus salivarius* urease: genetic and biochemical characterization and expression in a dental plaque streptococcus. *Infect Immun* **64**, 585–592.

Cheng, K.-J., McAllister, T. A. & Costerton, J. W. (1995). Biofilms of the ruminant digestive tract. In *Microbial Biofilms*, pp. 221–232. Edited by H. M. Lappin-Scott & J. W. Costerton. Cambridge: Cambridge University Press.

Christensen, B. B., Sternberg, C., Andersen, J., Eberl, L., Moller, S., Givskov, M. & Molin, S. (1998). Establishment of new genetic traits in a microbial biofilm community. *Appl Environ Microbiol* **64**, 2247–2255.

Coffey, T. J., Dowson, C. G., Daniels, M., Zhou, J., Martin, C., Spratt, B. G. & Musser,

J. M. (1991). Horizontal transfer of multiple penicillin-binding protein genes, and capsular biosynthetic genes, in natural populations of *Streptococcus pneumoniae*. *Mol Microbiol* **5**, 2255–2260.

Cole, M. F., Evans, M., Fitzsimmons, S., Johnson, J., Pearce, C., Sheridan, M. J., Wientzen, R. & Bowden, G. (1994). Pioneer oral streptococci produce immunoglobulin A1 protease. *Infect Immun* **62**, 2165–2168.

Conway, P. L. (1995). Microbial ecology of the human large intestine. In *Human Colonic Bacteria. Role in Nutrition, Physiology and Pathology*, pp. 1–24. Edited by G. R. Gibson & G. T. Macfarlane. Boca Raton, FL: CRC Press.

Costerton, J. W., Irwin, R. T. & Cheng, K. T. (1981). The bacterial glycocalyx in nature and disease. *Annu Rev Microbiol* **35**, 299–324.

Costerton, J. W., Cheng, K. J., Geesey, G. G., Ladd, T. I., Nickel, J. C., Dasgupta, M. & Marrie, T. J. (1987). Bacterial biofilms in nature and disease. *Annu Rev Microbiol* **41**, 435–464.

Costerton, J. W., Lewandowski, Z., DeBeer, D., Caldwell, D. E., Korber, D. R. & James, G. (1994). Biofilms, the customized microniche. *J Bacteriol* **176**, 2137–2142.

Costerton, J. W., Lewandowski, Z., Caldwell, D. E., Korber, D. R. & Lappin-Scott, H. M. (1995). Microbial biofilms. *Annu Rev Microbiol* **49**, 711–745.

Davies, D. G., Parsek, M. R., Pearson, J. P., Iglewski, B. H., Costerton, J. W. & Greenberg, E. P. (1998). The involvement of cell-to-cell signals in the development of a bacterial film. *Science* **280**, 295–298.

Dunny, G. M. & Winans, S. C. (1999). *Cell-cell Signalling in Bacteria*. Washington, DC: American Society for Microbiology.

Fabricius, L., Dahlen, G., Holm, S. E. & Moller, A. J. R. (1982). Influence of combinations of oral bacteria on periapical tissues of monkeys. *Scand J Dent Res* **90**, 200–206.

Finlay, B. J., Esteban, G., Clarke, K. J., Williams, A. G., Embley, T. M. & Hirt, R. P. (1994). Some rumen ciliates have endosymbiotic methanogens. *FEMS Microbiol Lett* **117**, 157–162.

Fitzsimmons, S., Evans, M., Pearce, C., Sheridan, M. J., Wientzen, R., Bowden, G. & Cole, M. F. (1996). Clonal diversity of *Streptococcus mitis* biovar 1 isolates from the oral cavity of human neonates. *Clin Diagn Lab Immunol* **3**, 517–522.

Flint, H. J. (1997). The rumen microbial ecosystem – some recent developments. *Trends Microbiol* **5**, 483–488.

Foster, J. W. (1995). Low pH adaptation and the acid tolerance response in *Salmonella typhimurium*. *Crit Rev Microbiol* **21**, 215–237.

Frandsen, E. V., Pedrazzoli, V. & Kilian, M. (1991). Ecology of viridans streptococci in the oral cavity and pharynx. *Oral Microbiol Immunol* **6**, 129–133.

Fuller, R. & Gibson, G. R. (1997). Modification of the intestinal microflora using probiotics and prebiotics. *Scand J Gastroenterol* **222** (suppl.), 28–31.

Fuqua, W. C., Winans, S. C. & Greenberg, E. P. (1994). Quorum sensing in bacteria: the LuxR-LuxI family of cell density-responsive transcriptional regulators. *J Bacteriol* **176**, 269–275.

Gharbia, S. E. & Shah, H. N. (1989). The influence of peptides on the uptake of amino acids in *Fusobacterium*; predicted interactions with *Porphyromonas gingivalis*. *Curr Microbiol* **19**, 231–235.

Gilbert, P., Das, J. & Foley, I. (1997). Biofilm susceptibility to antimicrobials. *Adv Dent Res* **11**, 160–167.

Goodfellow, M. & O'Donnell, A. G. (1993). *Handbook of New Bacterial Systematics*. London: Academic Press.

Goodman, A. E., Hild, E., Marshall, K. C. & Hermansson, M. (1993). Conjugative plasmid transfer between bacteria under simulated marine oligotrophic conditions. *Appl Environ Microbiol* **59**, 1035–1040.

Gray, K. M. (1997). Intercellular communication and group behaviour in bacteria. *Trends Microbiol* **5**, 184–186.

Greenberg, E. P. (1998). Quorum sensing in Gram-negative bacteria: acylhomoserine lactone signalling and cell-cell communication. In *Microbial Pathogenesis: Current and Emerging Issues*, pp. 17–26. Edited by D. J. LeBlanc, M. S. Lantz & L. M. Switalski. Indiana: Indiana University.

Grenier, D. & Mayrand, D. (1983). Studies of mixed anaerobic infections involving *Bacteroides gingivalis. Can J Microbiol* **29**, 612–618.

Hakenbeck, R., Konig, A., Kern, I., van der Linden, M., Keck, W., Billot-Klein, D., Legrand, R., Schoot, B. & Gutmann, L. (1998). Acquisition of five high-M_r penicillin-binding protein variants during transfer of high-level β-lactam resistance from *Streptococcus mitis* to *Streptococcus pneumoniae. J Bacteriol* **180**, 1831–1840.

Hall, H. K., Karem, K. L. & Foster, J. W. (1995). Molecular responses of microbes to environmental pH stress. *Adv Microb Physiol* **37**, 229–272.

Hamilton, I. R. & Ng, S. K. C. (1983). Stimulation of glycolysis through lactate consumption in a resting cell mixture of *Streptococcus salivarius* and *Veillonella parvula. FEMS Microbiol Lett* **20**, 61–65.

Hamilton, W. A. (1987). Biofilms: microbial interactions and metabolic activities. In *Ecology of Microbial Communities*, pp. 361–385. Society for General Microbiology Symposium 41. Edited by M. Fletcher, T. R. G. Gray & J. G. Jones. Cambridge: Cambridge University Press.

Haubek, D., Direnzo, J. M., Tinoco, E. M. B., Westergaard, J., Lopez, N. J., Chung, C.-P., Poulsen, K. & Kilian, M. (1997). Racial tropism of a highly toxic clone of *Actinobacillus actinomycetemcomitans* associated with juvenile periodontitis. *J Clin Microbiol* **35**, 3037–3042.

Håvarstein, L. S. & Morrison, D. A. (1999). Quorum sensing and peptide pheromones in streptococcal competence for genetic transformation. In *Cell-cell Signalling in Bacteria*, pp. 9–26. Edited by G. M. Dunny & S. C. Winans. Washington, DC: American Society for Microbiology.

Håvarstein, L. S., Coomaraswamy, G. & Morrison, D. A. (1995). An unmodified heptadecapeptide pheromone induces competence for genetic transformation in *Streptococcus pneumoniae. Proc Natl Acad Sci USA* **92**, 11140–11144.

Håvarstein, L. S., Hakenbeck, R. & Gaustad, P. (1997). Natural competence in the genus *Streptococcus*: evidence that streptococci can change pherotype by interspecies recombinational exchanges. *J Bacteriol* **179**, 6589–6594.

Henderson, B. (1998). Bacteria/host interactions in the periodontal diseases: clues to the development of novel therapeutics for the periodontal diseases. In *Dental Plaque Revisited*, pp. 443–456. Edited by H. N. Newman & M. Wilson. Cardiff: BioLine.

Henderson, B. & Wilson, M. (1998). Commensal communism and the oral cavity. *J Dent Res* **77**, 1674–1683.

Hillman, J. D., Dzuback, A. L. & Andrews, S. W. (1987). Colonization of the human oral cavity by a *Streptococcus mutans* mutant producing increased bacteriocin. *J Dent Res* **66**, 1092–1094.

Hillman, K., Lloyd, D. & Williams, A. G. (1988). Interactions between the methanogen *Methanosarcina barkeri* and rumen holotrich ciliate protozoa. *Lett Appl Microbiol* **7**, 49–53.

Hobson, P. N. & Wallace, R. J. (1982). Microbial ecology and activities in the rumen: part 1. *Crit Rev Microbiol* **9**, 165–225.

Hohwy, J. & Kilian, M. (1995). Clonal diversity of the *Streptococcus mitis* biovar 1 population in the human oral cavity and pharynx. *Oral Microbiol Immunol* **10**, 19–25.

Holm, P. (1948). Some investigations into the penicillin sensitivity of human-pathogenic Actinomycetes and some comments on penicillin treatment of actinomycosis. *Acta Pathol Microbiol Scand* **45**, 376–404.

Homer, K. A. & Beighton, D. (1992). Synergistic degradation of bovine serum albumin by mutans streptococci and other dental plaque bacteria. *FEMS Microbiol Lett* **90**, 259–262.

Hoskins, L. C., Agustines, M., McKee, W. B., Boulding, E. T., Kriaris, M. & Niedermeyer, G. (1985). Mucin degradation in human colon ecosystems. Isolation and properties of fecal strains that degrade ABH blood group antigens and oligosaccharides from mucin glycoproteins. *J Clin Invest* **75**, 944–953.

Hoskins, L. C., Boulding, E. T., Gerker, T. A., Harouny, V. R. & Kriaris, M. S. (1992). Mucin glycoprotein degradation by mucin oligosaccharide-degrading strains of human faecal bacteria. Characterisation of saccharide cleavage products and their potential role in the nutritional support of larger faecal populations. *Microb Ecol Health Dis* **5**, 193–207.

Hudson, M. J. & Marsh, P. D. (1995). Carbohydrate metabolism in the colon. In *Human Colonic Bacteria. Role in Nutrition, Physiology and Pathology*, pp. 61–73. Edited by G. R. Gibson & G. T. Macfarlane. Boca Raton, FL: CRC Press.

Isenberg, H. D. (1988). Pathogenicity and virulence: another view. *Clin Microbiol Rev* **1**, 40–53.

James, S. M. & Tagg, J. R. (1991). The prevention of dental caries by BLIS-mediated inhibition of mutans streptococci. *N Z Dent J* **87**, 80–83.

Kilian, M., Reinholdt, J., Lomholt, H., Poulsen, K. & Frandsen, E. V. G. (1996). Biological significance of IgA1 proteases in bacterial colonisation and pathogenesis: critical evaluation of experimental evidence. *APMIS* **104**, 321–338.

Kleerebezem, M., Quadri, L. E., Kuipers, O. P. & de Vos, W. M. (1997). Quorum sensing by peptide pheromones and two-component signal-transduction systems in Gram-positive bacteria. *Mol Microbiol* **24**, 895–904.

Koch, A. L. (1997). Microbial physiology and ecology at slow growth. *Microbiol Mol Biol Rev* **61**, 305–308.

Kolenbrander, P. E. (1988). Intergeneric coaggregation among human oral bacteria and ecology of dental plaque. *Annu Rev Microbiol* **42**, 627–656.

Kolenbrander, P. E. & London, J. (1993). Adhere today, here tomorrow: oral bacterial adherence. *J Bacteriol* **175**, 3247–3252.

Kolenbrander, P. E., Andersen, R. N. & Moore, L. V. (1990). Intrageneric coaggregation among strains of human oral bacteria: potential role in primary colonization of the tooth surface. *Appl Environ Microbiol* **56**, 3890–3894.

Könönen, E., Asikainen, S., Saarela, M., Karjalainen, J. & Jousimies-Somer, H. (1994). The oral Gram-negative anaerobic microflora in young children: longitudinal changes from edentulous to dentate mouth. *Oral Microbiol Immunol* **9**, 136–141.

Krumholz, L. R., Forsberg, C. W. & Veira, D. M. (1983). Association of methanogenic bacteria with rumen protozoa. *Can J Microbiol* **29**, 676–680.

Lazazzera, B. A. & Grossman, A. D. (1998). The ins and outs of peptide signalling. *Trends Microbiol* **6**, 288–294.

Lee, C. A. (1996). Pathogenicity islands and the evolution of bacterial pathogens. *Infect Agents Dis* **5**, 1–7.

Leng, Z., Riley, D. E., Berger, R. E., Krieger, J. N. & Roberts, C. R. (1997). Distribution and mobility of the tetracycline resistance determinant *tet*Q. *J Antimicrob Chemother* **40**, 551–559.

Listgarten, M. (1999). Formation of dental plaque and other biofilms. In *Dental Plaque Revisited*, pp. 187–210. Edited by H. N. Newman & M. Wilson. Cardiff: BioLine.

Lorenz, M. G. & Wakernagel, W. (1994). Bacterial gene transfer by natural genetic transformation in the environment. *Microbiol Rev* **58**, 563–602.

Ma, M. & Eaton, J. W. (1992). Multicellular oxidant defense in unicellular organisms. *Proc Natl Acad Sci USA* **89**, 7924–7928.

McBain, A. J. & Macfarlane, G. T. (1998). Ecological and physiological studies on large intestinal bacteria in relation to production of hydrolytic and reductive enzymes involved in formation of genotoxic metabolites. *J Med Microbiol* **47**, 407–416.

McDermid, A. S., McKee, A. S., Ellwood, D. C. & Marsh, P. D. (1986). The effect of lowering the pH on the composition and metabolism of a community of nine oral bacteria grown in a chemostat. *J Gen Microbiol* **132**, 1205–1214.

Macfarlane, G. T. & Macfarlane, S. (1997). Human colonic microbiota: physiology and metabolic potential of intestinal bacteria. *Scand J Gastroenterol* **222** (suppl.), 3–9.

Macfarlane, S., McBain, A. J. & Macfarlane, G. T. (1997). Consequences of biofilm and sessile growth in the large intestine. *Adv Dent Res* **11**, 59–68.

Mackie, R. I. & White, B. A. (1990). Recent advances in rumen microbial ecology: potential impact on nutrient output. *J Dairy Sci* **73**, 2971–2995.

Maiden, M. C. J. & Feavers, I. M. (1995). Population genetics and global epidemiology of the human pathogen *Neisseria meningitidis*. In *Population Genetics of Bacteria*, pp. 267–293. Society for General Microbiology Symposium 52. Edited by S. Baumberg, J. P. W. Young, E. M. H. Wellington & J. R. Saunders. Cambridge: Cambridge University Press.

Margulis, L. (1991). Symbiogenesis and symbionticism. In *Symbiosis as a Source of Evolutionary Innovation*, pp. 1–14. Edited by M. Margulis & R. Fester. Cambridge, MA: MIT Press.

Marquis, R. E. (1995). Oxygen metabolism, oxidative stress and acid-base physiology of dental plaque biofilms. *J Ind Microbiol* **15**, 198–207.

Marsh, P. D. & Bradshaw, D. J. (1997). Physiological approaches to plaque control. *Adv Dent Res* **11**, 176–185.

Marsh, P. D. & Bradshaw, D. J. (1999). Microbial community aspects of dental plaque. In *Dental Plaque Revisited*, pp. 237–253. Edited by H. N. Newman & M. Wilson. Cardiff: BioLine.

Marsh, P. D. & Martin, M. V. (1999). *Oral Microbiology*, 4th edn. Oxford: Wright.

Maynard Smith, J., Dowson, C. G. & Spratt, B. G. (1991). Localized sex in bacteria. *Nature* **349**, 29–31.

Mayrand, D. & McBride, B. C. (1980). Ecological relationships of bacteria involved in a simple, mixed anaerobic infection. *Infect Immun* **27**, 44–50.

Mekalanos, J. J. (1992). Environmental signals controlling expression of virulence determinants in bacteria. *J Bacteriol* **174**, 1–7.

Meyer, T. F. (1999). Pathogenic *Neisseriae*: complexity of pathogen-host interplay. *Clin Infect Dis* **28**, 433–441.

Mikx, F. H. M. & van der Hoeven, J. S. (1975). Symbiosis of *Streptococcus mutans* and *Veillonella alcalescens* in mixed continuous culture. *Arch Oral Biol* **20**, 407–410.

Milnes, A. R., Bowden, G. H., Gates, D. & Tate, R. (1993). Predominant cultivable microorganisms on the tongue of pre-school children. *Microb Ecol Health Dis* **6**, 229–235.

Minah, G. E., Solomon, E. S. & Chu, K. (1985). The association between dietary sucrose consumption and microbial population shifts at six oral sites in man. *Arch Oral Biol* **30**, 397–401.

Morelli, G., Malorny, B., Muller, K., Seiler, A., Wang, J.-F., de Valle, J. & Achtman, M. (1997). Clonal descent and microevolution of *Neisseria meningitidis* during 30 years of epidemic spread. *Mol Microbiol* **25**, 1047–1064.

Musser, J. M. (1996). Molecular population genetic analysis of emerged bacterial pathogens: selected insights. *Emerg Infect Dis* **2**, 1–17.

Naumann, M., Rudel, T. & Meyer, T. F. (1999). Host cell interactions and signalling with *Neisseria gonorrhoeae*. *Curr Opin Microbiol* **2**, 62–70.

Nyvad, B. (1993). Microbial colonization of human tooth surfaces. *APMIS* **101**, 7–45.

Odum, E. P. (1986). Introductory review: perspective of ecosystem theory and application. In *Ecosystem Theory and Application*, pp. 1–11. Edited by N. Polunin. Chichester: Wiley.

Onderdonk, A. B., Cisneros, R. C., Finberg, R., Crabb, J. H. & Kasper, D. L. (1990). Animal model system for studying virulence of and host responses to *Bacteroides fragilis*. *Rev Infect Dis* **12**, S169–S177.

O'Toole, G. A. & Kolter, R. (1998). Initiation of biofilm formation in *Pseudomonas fluorescens* WCS365 proceeds via multiple, convergent signalling pathways: a genetic analysis. *Mol Microbiol* **28**, 449–461.

Pearce, C., Bowden, G. H., Evans, M., Fitzsimmons, S. P., Johnson, J., Sheridan, M. J., Wientzen, R. & Cole, M. F. (1995). Identification of pioneer viridans streptococci in the oral cavity of human neonates. *J Med Microbiol* **42**, 67–72.

Poulsen, K., Theilade, E., Lally, E. T., Demuth, D. R. & Kilian, M. (1994). Population structure of *Actinobacillus actinomycetemcomitans*: a framework for studies of disease associated properties. *Microbiology* **140**, 2049–2060.

Poulsen, K., Reinholdt, J., Jespersgaard, C., Boye, K., Brown, T. A., Hauge, M. & Kilian, M. (1998). A comprehensive genetic study of streptococcal immunoglobulin A1 proteases: evidence for recombination within and between species. *Infect Immun* **66**, 181–190.

Rainey, P. B., Moxon, E. R. & Thompson, I. P. (1993). Intraclonal polymorphism in bacteria. *Adv Microb Ecol* **13**, 263–300.

Reichmann, P., Konig, A., Linares, J., Alcaide, F., Tenover, F. C., McDougal, L., Swidsinski, S. & Hakenbeck, R. (1997). A global gene pool for high-level cephalosporin resistance in commensal *Streptococcus* species and *Streptococcus pneumoniae*. *J Infect Dis* **176**, 1001–1012.

Roberts, D. S. (1967). The pathogenic synergy of *Fusiformis necrophorus* and *Corynebacterium pyogenes*. II. The response of *F. necrophorus* to a filterable product of *C. pyogenes*. *Br J Exp Pathol* **48**, 674–679.

Rogers, A. H., Zilm, P. S. & Gully, N. J. (1987). Influence of arginine on the coexistence of *Streptococcus mutans* and *S. milleri* in glucose-limited continuous culture. *Microb Ecol* **14**, 193–202.

Rotstein, O. D. & Kao, J. (1988). The spectrum of *Escherichia coli-Bacteroides fragilis* pathogenic synergy in an intraabdominal infection model. *Can J Microbiol* **34**, 352–357.

Salminen, S. & Salminen, E. (1997). Lactulose, lactic acid bacteria, intestinal microecology and mucosal protection. *Scand J Gastroenterol* **222** (suppl.), 45–48.

Salmond, G. P. C., Bycroft, B. W., Stewart, G. S. A. B. & Williams, P. (1995). The bacterial enigma: cracking the code of cell-cell communication. *Mol Microbiol* **16**, 615–624.

Sanders, C. C. & Sanders, W. E. (1982). Enocin: an antibiotic produced by *Streptococcus salivarius* that may contribute to protection against infections due to group A streptococci. *J Infect Dis* **146**, 683–690.

Selander, R. K. & Musser, J. M. (1990). Population genetics of bacterial pathogenesis. In *The Bacteria. Volume XI, Molecular Basis of Bacterial Pathogenicity*, pp. 11–36. Edited by B. H. Inglewski & V. L. Clark. New York: Academic Press.

Selander, R. K., Caugant, D. A., Ochman, H., Musser, J. M., Gilmour, M. N. & Whittam, S. (1986). Methods of multilocus enzyme electrophoresis for bacteria population genetics and systematics. *Appl Environ Microbiol* **51**, 873–884.

Shapiro, J. A. (1998). Thinking about bacterial populations as multicellular organisms. *Annu Rev Microbiol* **52**, 81–104.

Shapiro, J. A. & Dworkin, M. (1997). *Bacteria as Multicellular Organisms*. Oxford: Oxford University Press.

Smith, H. (1995). The revival of interest in mechanisms of bacterial pathogenicity. *Biol Rev Camb Philos Soc* **70**, 277–316.

Stickler, D. J., Morris, N. S., McLean, R. J. C. & Fuqua, C. (1998). Biofilms on indwelling urethral catheters produce quorum-sensing signal molecules in situ and in vitro. *Appl Environ Microbiol* **64**, 3486–3490.

Sundqvist, G. K., Eckerbom, M. I., Larsson, A. P. & Sjogren, U. T. (1979). Capacity of anaerobic bacteria from necrotic pulps to induce purulent infections. *Infect Immun* **25**, 685–693.

Sundqvist, G., Carlsson, J., Herrmann, B. & Tarnvik, A. (1985). Degradation of human immunoglobulins G and M and complement factors C3 and C5 by black-pigmented *Bacteroides*. *J Med Microbiol* **19**, 85–94.

Tagg, J. R. & Ragland, N. L. (1991). Applications of BLIS typing to studies of the survival on surfaces of salivary streptococci and staphylococci. *J Appl Bacteriol* **71**, 339–342.

Telford, G., Wheeler, D., Williams, P., Tomkins, P. T., Appleby, P., Sewell, H., Stewart, G. S. A. B., Bycroft, B. W. & Pritchard, D. I. (1998). The *Pseudomonas aeruginosa* quorum-sensing molecule *N*-(3oxododecanoyl) L-homoserine lactone has immunomodulatory activity. *Infect Immun* **66**, 36–42.

ter Steeg, P. F. & van der Hoeven, J. S. (1989). Development of periodontal microflora on human serum. *Microb Ecol Health Dis* **2**, 1–10.

ter Steeg, P. F., van der Hoeven, J. S., de Jong, M. H., van Munster, P. J. J. & Jansen, M. J. H. (1987). Enrichment of subgingival microflora on human serum leading to accumulation of *Bacteroides* species, peptostreptococci and fusobacteria. *Antonie Leeuwenhoek* **53**, 261–272.

ter Steeg, P. F., van der Hoeven, J. S., de Jong, M. H., van Munster, P. J. J. & Jansen, M. J. H. (1988). Modelling the gingival pocket by enrichment of subgingival microflora in human serum in chemostats. *Microb Ecol Health Dis* **1**, 73–84.

Ushijima, T., Takahashi, M. & Seto, A. (1990). The role of *Escherichia coli* haemolysin in the pathogenic synergy of colonic bacteria in sub-cutaneous abscess formation in mice. *J Med Microbiol* **33**, 17–22.

van der Hoeven, J. S. & Camp, P. J. M. (1991). Synergistic degradation of mucin by *Streptococcus oralis* and *Streptococcus sanguis* in mixed chemostat cultures. *J Dent Res* **68**, 1041–1044.

van der Hoeven, J. S. & Rogers, A. H. (1979). Stability of the resident microflora and bacteriocinogeny of *Streptococcus mutans* as factors affecting its establishment in specific pathogen free rats. *Infect Immun* **23**, 206–212.

van der Waaij, D., Berghuis de Vries, J. M. & Lekker-Kerk van der Wees, J. E. C. (1971). Colonisation resistance of the digestive tract in conventional and antibiotic-treated mice. *J Hyg* **69**, 405–411.

van Loosdrecht, M. C. M., Eikelboom, D., Gjaltema, A., Mulder, A., Tijhuis, I. & Heijnen, J. J. (1995). Biofilm structures. *Water Sci Technol* **32**, 35–43.

van Steenbergen, T. J. M., van Winkelhoff, A. J. & de Graaff, J. (1984). Pathogenic synergy: mixed infections in the oral cavity. *Antonie Leeuwenhoek* **50**, 789–798.

van Winkelhoff, A. J., Winkel, E. G., Barendregt, D., Dellemijn-Kippuw, N., Stijne, A. & van der Velden, U. (1997). Beta-lactamase producing bacteria in adult periodontitis. *J Clin Periodontol* **24**, 538–543.

Visick, K. L. & Ruby, E. G. (1999). The emergent properties of quorum sensing: consequences to bacteria of autoinducer signalling in their natural environment. In *Cell-cell Signalling in Bacteria*, pp. 333–352. Edited by G. M. Dunny & S. C. Winans. Washington, DC: American Society for Microbiology.

Vroom, J. M., de Grauw, K. J., Gerritsen, H. C., Bradshaw, D. J., Marsh, P. D., Watson, G. K., Allison, C. & Birmingham, J. J. (1999). Depth penetration and detection of pH gradients in biofilms using two-photon excitation microscopy. *Appl Environ Microbiol* **65**, 3502–3511.

Wade, W. (1999). Unculturable bacteria in oral biofilms. In *Dental Plaque Revisited*, pp. 313–322. Edited by H. N. Newman & M. Wilson. Cardiff: BioLine.

Walker, C. B., Tyler, K. T., Low, S. B. & King, C. J. (1987). Penicillin-degrading enzymes in sites associated with adult periodontitis. *Oral Microbiol Immunol* **2**, 129–131.

Wallace, R. J. (1985). Synergism between different species of proteolytic rumen bacteria. *Curr Microbiol* **12**, 59–64.

Ward, D. M., Ferris, M. J., Nold, S. C. & Bateson, M. M. (1998). A natural view of microbial diversity within hot spring cyanobacterial mat communities. *Microbiol Mol Biol Rev* **62**, 1353–1370.

Weimer, P. J. (1998). Manipulating ruminal fermentation: a microbial ecological perspective. *J Anim Sci* **76**, 3114–3122.

Welling, G. W., Elfferich, P., Raangs, G. C., Wildeboer-Veloo, A. C. M., Jansen, G. J. & Degener, J. E. (1997). 16S ribosomal RNA-targeted oligonucleotide probes for monitoring of intestinal tract bacteria. *Scand J Gastroenterol* **222** (suppl.), 17–19.

Wilkins, B. M. (1995). Gene transfer by bacterial conjugation: diversity of systems and functional specializations. In *Population Genetics of Bacteria*, pp. 59–88. Society for General Microbiology Symposium 52. Edited by S. Baumberg, J. P. W. Young, E. M. H. Wellington & J. R. Saunders. Cambridge: Cambridge University Press.

Williams, A. G. & Lloyd, D. (1993). Biological activities of symbiotic and parasitic protozoa and fungi in low-oxygen environments. *Adv Microb Ecol* **13**, 211–262.

Williams, H. G., Day, M. J., Fry, J. C. & Stewart, G. J. (1996). Natural transformation in river epilithon. *Appl Environ Microbiol* **62**, 2994–2998.

Wilson, M. (1996). Susceptibility of oral bacterial biofilms to antimicrobial agents. *J Med Microbiol* **44**, 79–87.

Wimpenny, J. W. T. & Colasanti, R. (1997). A unifying hypothesis for the structure of microbial biofilms based on cellular automaton models. *FEMS Microbiol Ecol* **22**, 1–16.

Microbial communities: aggregates of individuals or co-ordinated systems

Søren Molin, Janus A. J. Haagensen, Kim B. Barken and
Claus Sternberg

Molecular Microbial Ecology Group, Department of Microbiology, Technical University of
Denmark, DK-2800 Lyngby, Denmark

INTRODUCTION

Less than 50 years of molecular biology have documented that a few thousand genes
suffice to construct prokaryotic life forms, which could be argued to be the most
successful of all life forms throughout all biological evolution. These were the first to
appear, and most likely the last to sustain as living organisms on earth. Molecular
biology has also taught us that one explanation for the success of bacteria is the
continuous evolution of not only new traits, but also of new regulatory elements,
making these organisms very adaptable to a range of environments and to shifts
between these. Expression of a large fraction of the genes in modern bacterial genomes
is controlled at several levels – from the very specific metabolite-directed derepression
scheme to global regulatory loops controlling the overall state of the cell and its
interactions with the environment. Through complex webs of signal-transduction
events bacteria sense their environment, and react to a large number of factors by fine-
tuning expression of a large number of genes in apparently 'intelligent' ways.

It is therefore an almost logical extrapolation to assume that these simple, but highly
regulated, organisms may also possess the capacity to engage in multicellular system
constructions, in which features of co-ordinated organizational designs develop. The
embryonic development of compartmentalized organisms with different organs and
tissues from the undifferentiated fertilized egg-cell may have its parallel in prokaryotic
organisms forming highly structured communities with distributed functionality based
on a programmed, progressive differentiation process. Several recent findings in
microbiology seem to support this line of thinking, and it is now frequently assumed

SGM symposium 59: Community structure and co-operation in biofilms. Editors D. Allison, P. Gilbert, H. Lappin-Scott, M. Wilson.
Cambridge University Press. ISBN 0 521 79302 5 ©SGM 2000.

that bacteria are capable of two distinct life modes: the 'primitive' unorganized planktonic state (one-dimensional, mechanic, individualistic), and the 'complex' organized sessile state (three-dimensional, interactive, social).

Two phenomena discovered in recent years have provided support for this picture. The first is, in fact, a feature of bacterial gene regulation characterized as a control element from cultures of planktonic cells, quorum sensing (Fuqua *et al.*, 1994). Secretion of secondary metabolites exerting gene regulatory properties may accumulate as a function of population density, and when a threshold level is eventually reached, synchronous gene expression is induced in all individual cells. This type of gene regulation is seen as a cell-to-cell communication activity, since the inducing compound is extracellular (Fuqua *et al.*, 1994). The second phenomenon relates to the sessile life mode of bacteria (Costerton *et al.*, 1995). There has been a growing interest among microbiologists in studying surface-attached microbial communities, mainly because bacteria normally live and develop in surface communities in natural environments, but certainly also because recent technological breakthrough developments now permit detailed investigations of complex consortia of bacteria located on different substrata (Caldwell *et al.*, 1993; Costerton *et al.*, 1987, 1995). The 'marriage' of molecular technology with light microscopy (especially confocal scanning laser microscopy) has been instrumental in the high-resolution imaging of microbial assemblies as highly structured and apparently organized communities (Caldwell *et al.*, 1992). Observations of cell clusters separated by water-filled channels, which supply nutrients and remove waste products (Massol-Deyá *et al.*, 1995), make the parallel to the animal body or the plant almost self-evident.

Recently, these two phenomena have become unified. Investigations of biofilm formation with different variants of *Pseudomonas aeruginosa* showed that wild-type cells developed structured communities with stable and robust properties resisting treatments with detergents and other chemicals. In contrast, mutants affected in the cell-to-cell communication machinery formed loose, fluffy structures in the biofilms, which were highly sensitive to treatments with detergents (Davies *et al.*, 1998). In other words, microbial community development seems to be controlled by extracellular signals, which appear to be directly or indirectly responsible for a co-ordinated community performance. Thus the construction of a multicellular bacterial organism (the biofilm consortia) can be considered to occur along similar lines to that of animals and plants.

In this article, the emerging picture of co-ordinated microbial communities as multicellular organisms will be confronted with a series of experimental results obtained from laboratory-based biofilms, analysed within the framework of a recently presented investigation model (Molin & Molin, 1997).

THE SCIO MODEL

The development of an analytical model for investigations of microbial communities was considered important as soon as it was realized that bacteria may cover surfaces in many different ways. In a first attempt, a model (SCIO: S for Structure, C for Co-ordination, I for Interactions, O for Organisms) was designed which on one hand is descriptive, but which also takes into account the principal features of complex systems, allowing the analysis to eventually transcend from one type of system to another. Currently, a quantitative analysis model (COMSTAT) is under development, which aims at parameter-based descriptions of surface communities based on statistical data processing of structural pattern information. The two models complement each other. In the present context, where the aim is to assess the assumptions concerning the multicellular nature of microbial communities, it is mainly the SCIO model which offers interpretative power.

The SCIO model is rooted in social sciences and systems theory, and the details of the model and its development have been described previously (Molin & Molin, 1997). The important notion to emphasize is that the model assumes microbial communities to be considered as open systems, in which 'the individual elements (the bacteria) of the system are only the tip of the iceberg constituted by complex, dynamic patterns of interdependencies'. In fact, 'the essential characteristic of the open system is its organization, which refers to the structural set-up that is controlled by information and fueled by energy' (quotations from Molin & Molin, 1997). Thus when applying the model to studies of microbial communities, it becomes important to search specifically for interdependencies and for evidence of controlled organization based on intercellular information exchange. In the first presentation of SCIO, the four systemic features forming the empirical basis for the model will be briefly introduced and correlated with experimental data collections.

Organisms

Both in laboratory-based and in natural communities, organism compositions are frequently quite complex. Hence, *in situ* determination of the different species present is a necessary prerequisite for subsequent investigations and interpretations. In Fig. 1, two alternative approaches to the identification of the relevant organisms are presented. *In situ* rRNA hybridization (Fig. 1a) is independent of the organisms (at least in principle) and also of prior cultivation (Amann *et al.*, 1995). A drawback is the requirement for fixation, which implies killing of the entire community (still pictures). Tagging the relevant cells with marker genes (Fig. 1b) requires genetic modification and hence prior cultivation of the cells (Prosser, 1994), but the advantage is a non-destructive identification allowing on-line monitoring. This approach is gaining increasing attention as the number of reporter genes with different colours increases. In very complex communities, the investigator must choose which organisms to monitor, since

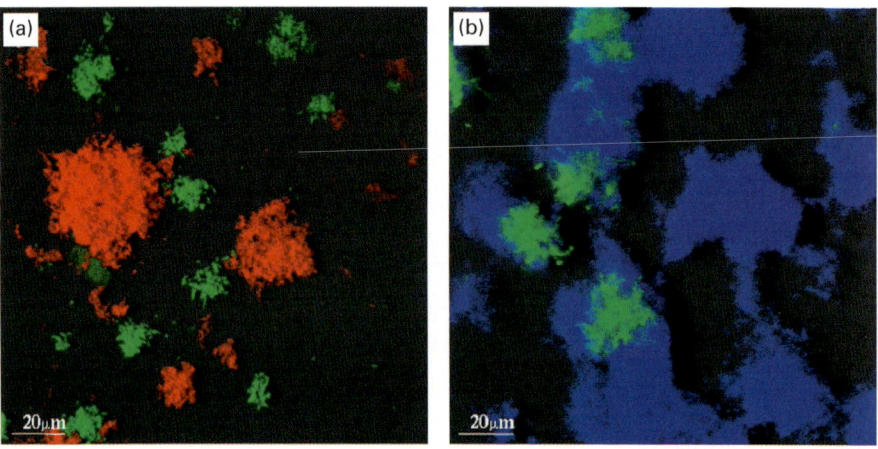

Fig. 1. Identification of organisms in mixed-species biofilm. Microcolonies of *Pseudomonas* sp. B13 and *Burkholderia* sp. LB400 growing in a flow-chamber biofilm and visualized in a confocal scanning laser microscope. (a) *In situ* rRNA hybridization of fixed cells in a polyacrylamide-embedded biofilm. Strain B13 is targeted with a specific rRNA probe labelled with the CY3 fluorophor (red), whilst strain LB400 is targeted with a specific probe labelled with fluorescein (green). (b) Strain LB400 was tagged with a transposon harbouring a fusion between a constitutive promoter and the *gfp* gene (green), whereas strain B13 was tagged with a *bfp* fusion (blue). No fixation or biofilm embedding was carried out.

both methods are limited with respect to how many different cells may be monitored simultaneously.

Structure

Microbial communities attached to surfaces grow in all three dimensions, and as shown in Fig. 2, the confocal scanning laser microscope offers excellent resolution power for the imaging of structures present in such communities (Caldwell *et al.*, 1992). It is important that the preparation steps needed for the fluorescence detection do not interfere with the overall structure pattern in the community, and, especially in connection with *in situ* hybridization, there is a need for ensuring that no drainage of water takes place in the course of the preparation.

Interactions

Bacteria are very good at relating to their environments (including other organisms) by processing chemical and physical signals, and in particular they often respond to very low concentrations of nutrients by metabolic reactions resulting in active growth (protein synthesis). Through design and application of molecular reporter systems which respond specifically or generally upon exposure to chemical signals, it is possible

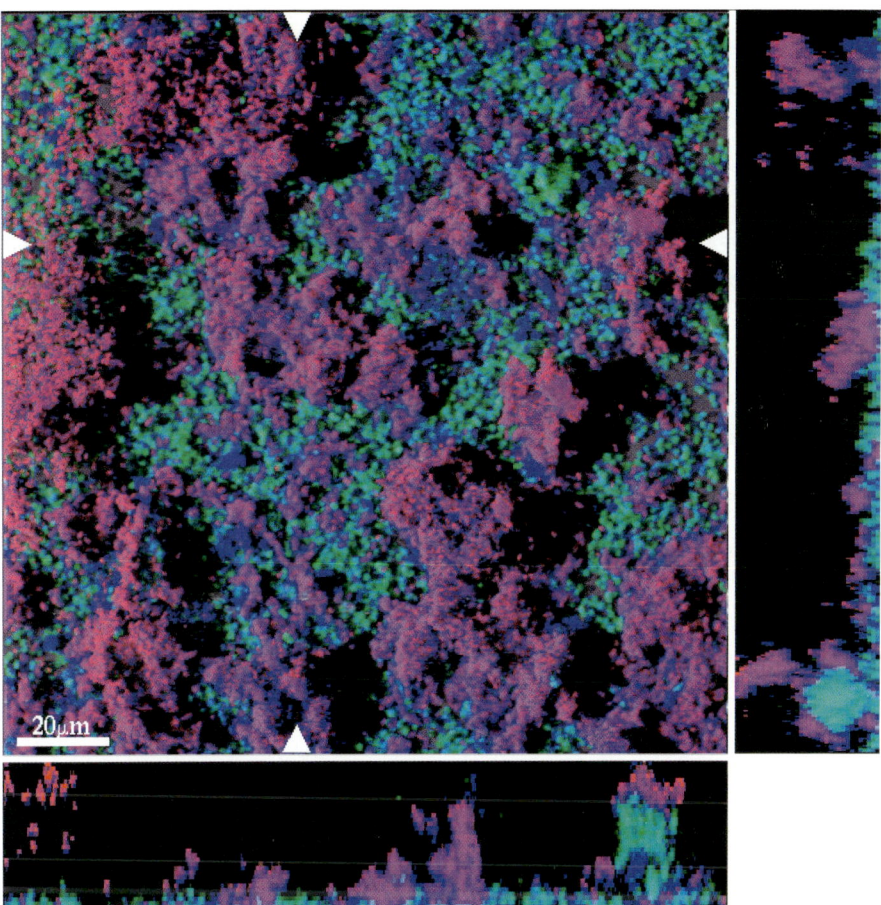

Fig. 2. Structure of multispecies biofilm. Confocal micrograph of a biofilm community with seven different species. The biofilm was fixed and embedded, and differentially labelled probes targeting *P. putida* R1 (red), *Acinetobacter* sp. C6 (green) and all other eubacteria (blue) were added. The frames below and on the right side of the main frame illustrate *zx* and *zy* scans, which indicate how the community is organized as microcolonies rising from the surface.

to construct cells which act as indicators for the presence of nutrients, antagonists, primary and secondary metabolites, and products of other bacteria. In Fig. 3, it is shown how bacteria in a biofilm may report on the presence of specific metabolites distributed unevenly through the community. These interactive reactions provide evidence of localized microenvironments in the relatively simple communities described here, and they support the notion of complexity and heterogeneity as common features of surface communities. Such interactions often cause couplings between organisms observed as mixed-species microcolonies in the communities.

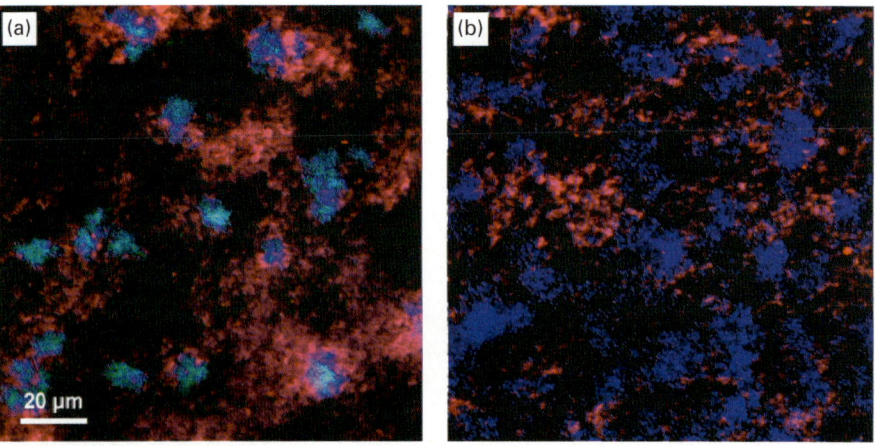

Fig. 3. Metabolic interactions in 3-chlorobiphenyl-degrading biofilm. The biofilm community comprises *Pseudomonas* sp. B13 (with a chromosomal insert of a gene fusion between the TOL Pm promoter and the *gfp* gene) and *Burkholderia* sp. LB400. The community was supplied with 3-chlorobiphenyl (a) (which is converted to 3-chlorobenzoate by LB400) or citrate (b) as carbon source. Strain B13 grows on 3-chlorobenzoate, which is also an inducer of the Pm promoter (cf. details described in Nielsen *et al.*, 2000). Red cells represent LB400 identified by specific rRNA hybridization, blue cells represent B13 targeted by a specific rRNA probe, and turquoise cells represent B13 expressing Gfp from the 3-chlorobenzoate-inducible Pm promoter. No Pm-promoted transcription was observed in the citrate-supplied community.

Communication

In our previous disclosure of the SCIO model (Molin & Molin, 1997), the letter C indicated Co-ordination. It now seems more correct to let it indicate Communication, since the former is not directly measurable. The consequence of this exchange is that co-ordination and control activities in microbial communities become indirect conclusions based on interpretation of the entire SCIO analysis rather than on direct empirical information. Fluorescent reporter systems compatible with confocal microscopy of biofilms have now been designed which directly monitor the presence of certain communication signals (alkylated homoserine lactones) in the local environment (unpublished). The application of such a reporter system in a biofilm community is presented in Fig. 4. Although these signals may be seen as specific examples of interaction signals already described in the previous paragraph, it is important to note that communication signals intervene in the cellular response repertoire at a higher regulatory level than metabolites and nutrients. Therefore, the presence and activity of these signals may have profound impacts on the entire community performance and development by exerting control over many community-related genetic functions.

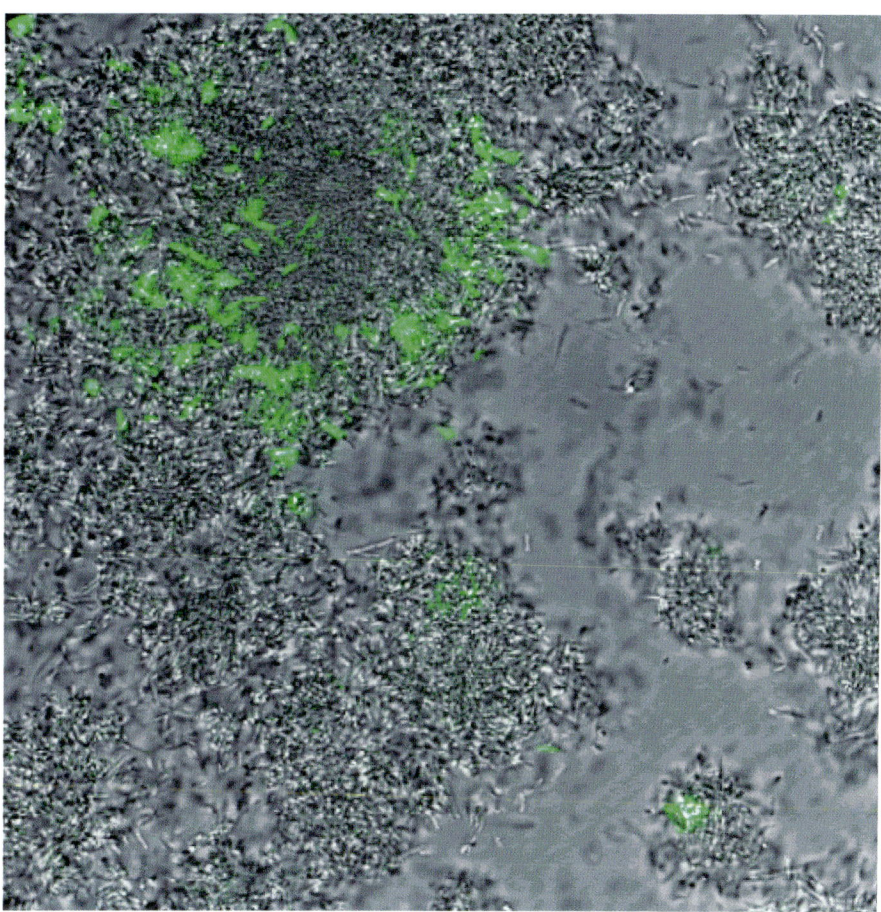

Fig. 4. Monitoring of cell-to-cell communication in a flow-chamber biofilm. *Pseudomonas aeruginosa* with an *N*-acyl homoserine lactone (AHL) reporter fusion consisting of a *gfp* gene located downstream of a *luxI* promoter (cf. Fuqua *et al.*, 1994 concerning the regulation of the promoter; the reporter system was constructed in our laboratory by Dr Jens Bo Andersen) was established in a flow-chamber biofilm. Scattered green-fluorescent cells located in the denser parts of the biomass indicate that in surface communities communication signals are produced, and other investigations show that the signals are also excreted by the producing cells (not shown).

In a first conclusion, it may be stated that significant traits and features of microbial communities, important for the performance and development of these, are in fact experimentally accessible *in situ*. Therefore, the following step in the analysis – the collection of evidence in favour of controlled activities at the community level – may proceed on the basis of experimental information.

FLOW-CHAMBER BIOFILM COMMUNITIES

The test tube (and the chemostat) has been a key environment for an infinite number of important investigations of the performance of bacteria, covering both physiological and genetic aspects of the organisms. The idealized scenario in the test tube is that of a strictly homogeneous and controlled environment in which single cells of a chosen clone will all be surrounded by identical environmental conditions. This means that signal/response activities in the entire population may be taken as evidence for activities in the single cells, i.e. the organism. In a surface-associated microbial community, the situation is quite different, no matter to what extent the external conditions are kept constant and defined. First, the bacteria growing in the surface community will develop heterogeneous conditions simply by forming structured biolayers. Second, nutrient availability will vary with biomass distribution and the local structures. Therefore, it is not possible to make conclusions concerning the individual cells on the basis of population data (and vice versa).

Flow-chambers with bacterial biofilms forming on surfaces under conditions of a constant flow of nutrients over the biofilms have been a preferred set-up for community analysis in laboratories all over the world, and although the actual designs may vary from group to group, it is probably fair to consider these to be the 'test tubes' of surface communities. In the major part of our studies, flow-chamber biofilms composed of a few different species capable of degrading pollutant organic compounds have been model cases of microbial communities, for which the SCIO analysis has been applied. The experimental characteristics of these studies will be described in connection with the data presentation below, and details of the flow-chamber biofilms have been described by Christensen *et al.* (1999).

COMPETITION, COMMENSALISM AND SYNTROPHY IN COMPOSITE MICROBIAL BIOFILMS

Community behaviour may be represented by the combined outcome of measurements of the Organism profile (O) and Interactions (I) between the organisms (and the exchanges with the environment, i.e. the added nutrients). In one mixed-species community comprising two toluene-degrading strains, *Pseudomonas putida* R1 and *Acinetobacter* sp. C6, the organisms formed a biofilm in flow-chambers supplied with benzylalcohol as the only available carbon and energy source (Møller *et al.*, 1996). The two strains grew together over long time periods (Molin *et al.*, 2000), and as such it could be concluded that if they competed for substrate, it did not result in exclusion of one or the other of the strains. It could further be shown that the *Acinetobacter* strain accumulated and excreted benzoate when metabolizing benzylalcohol (HPLC analysis, data not shown), indicating that the conversion of benzoate is a rate-limiting step in the total mineralization of toluene and benzylalcohol. In contrast, *P. putida* R1 converts benzoate very efficiently, and consequently there is the possibility that although the two organisms were expected to compete for the carbon source, they may in fact interact positively by improving the exploitation of the carbon source when growing together. This possibility was supported by the observation that biomass production in biofilms supplied with benzylalcohol was significantly increased if both strains were present compared to similar monospecies biofilms (data not shown).

In another mixed community, composed of *Pseudomonas* sp. B13 and *Burkholderia* sp. LB400, growth on and mineralization of chlorinated biphenyls require the combined metabolic activities of the two strains. Therefore, it was not surprising that both strains were present over long time periods in flow-chamber biofilms supplied with 3-chlorobiphenyl as the only carbon source. However, also under conditions where citrate (utilized by both organisms) was added as carbon source, they both remained in the community (Nielsen *et al.*, 2000). Similar observations have been made for other mixed communities, which indicate that in surface-associated communities there are bacterial properties other than nutrient utilization which are important for their continued presence, even under growth selection conditions.

The observed coexistence of two strains which could be expected to out-compete each other may reflect that under conditions where the cells are positioned as colonies in fixed locations on the substratum, one organism does not limit the presence of the other. This explanation is probably true for the community described above, where the continuous flow of citrate ensures that the small amount of biomass is receiving sufficient nutrient supply in all parts of the biofilm. If citrate is suddenly replaced by 3-chlorobiphenyl, the B13 strain will experience carbon source deficiency, since it is unable to utilize the biphenyl. However, after conversion of the biphenyl to 3-chlorobenzoate carried out by LB400, B13 may resume growth due to its capacity to metabolize this intermediate. In this situation, B13 therefore depends on the presence and activity of LB400 for its continued maintenance in the community. The overall composition and performance of this community are thus determined by the genotypes of the organisms (metabolic pathways) and the external conditions (nutrient composition).

It may be concluded that different strains of bacteria show sustained presence and activity in surface-associated communities under conditions where competition for nutrients might have favoured one or the other on the basis of cellular growth rates or yields. In addition, metabolic interactions between the organisms present may or may not take place; if interactions are favourable to one or more of the organisms there may be selective advantages to particular community members, or the entire community may benefit from the continued presence of several members.

COMMUNITY STRUCTURE AND ORGANIZATION

The flow-chambers used here are characterized by reproducibility of community performance if conditions are identical (inoculum size, nutrient composition, flow rate, chamber dimensions and design). This has been documented by repeated monitoring of a number of simple or complex communities (data not shown). Among the interesting findings from such monitoring schedules is the variation in community structure (three-dimensional) from organism to organism. Several interesting findings concerning the relationship between organism genotype and biofilm structure and development have been published over the last few years. Cell adhesion and motility seem to play important roles during initial phases of biofilm development (O'Toole & Kolter, 1998a, b; Pratt & Kolter, 1998), whilst cell-to-cell communication has been found to affect later stages of biofilm structure (Davies *et al.*, 1998).

In most of these investigations, the approach to understand structural development has been genetic, i.e. analysis of mutants affected in functions known to be connected to the particular phenotype of interest. This may be a very useful route if the mutation affects a gene whose function is specific and fully characterized. In contrast, changed biofilm properties concerning community organization and control, observed in cases of mutants affected in regulatory functions, may be more difficult to interpret due to the pleiotropic nature of such mutations. The question is whether development of specific structural features in microbial surface communities reflects a truly co-ordinated process serving the entire community, or whether the structural features represent the combined individual response patterns to nutrient gradients developing as biomass increases.

Monospecies biofilms

In the toluene-degrading community described above, the two community members were *P. putida* R1 and *Acinetobacter* sp. C6. As an alternative to the genetic approach, the structure-development process has been investigated by a physiological approach. In Fig. 5(a–c) is presented a series of micrographs showing how a monospecies biofilm of *Acinetobacter* C6 cells develops over time. Initially, the biofilm is fairly homogeneous with the biomass distributed evenly over the entire substratum, but after 2–3 days' growth a regular array of microcolonies develops, and this pattern is maintained for a long period thereafter. The carbon source in the presented example was benzylalcohol – a relatively poor nutrient. One interpretation of this community development may be derived from the cellular automaton model as discussed by Wimpenny & Colasanti (1997) and Picioreanu *et al.* (1998), in which the formation of microcolony and biofilm structure is mainly dependent on the concentration and quality of the nutrients. However, it has been shown that the *Acinetobacter* strain used in these experiments produces cell communication signals (N-acyl homoserine lactones, AHLs) at a time

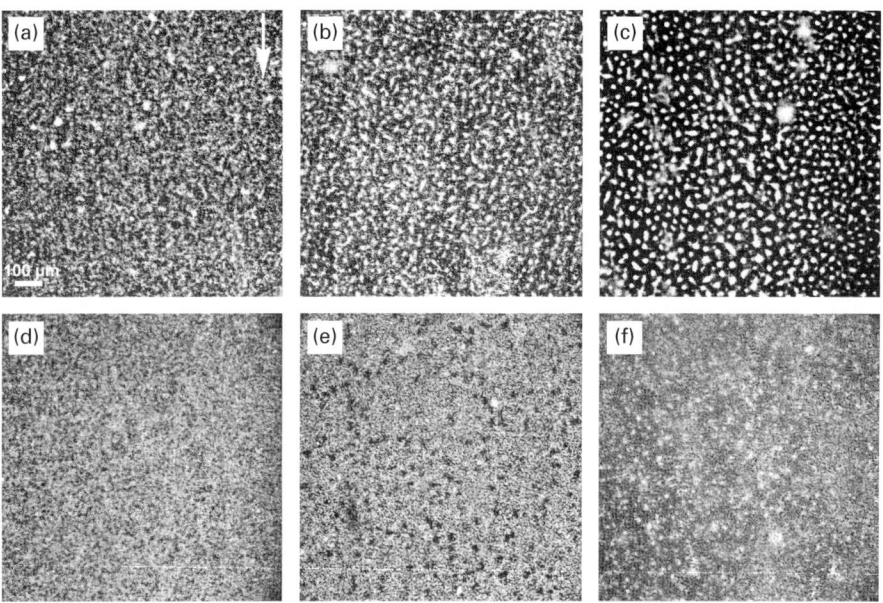

Fig. 5. Development of biofilm structure. *Acinetobacter* sp. C6 established in a flow-chamber with benzylalcohol (a–c) or glucose (d–f) as carbon source. Confocal scanning laser micrographs were recorded 24 h (a, d), 48 h (b, e) and 68 h (c, f) after inoculation of the flow-chambers.

close to that at which the colonies in the biofilm are formed (data not shown). It can, therefore, also be argued that the organization into the regular array of colonies may possibly be controlled by the accumulation of homoserine lactones in the community fluids.

It has so far not been possible to perform genetic analysis of the *Acinetobacter* strain due to a number of constraints of the organism. Instead, a number of tests based on environmental changes were carried out to determine the reproducibility of the colony formation pattern in the biofilms. It was found that patterns similar to those presented in Fig. 5(a–c) appeared every time the biofilms were supplied with either benzylalcohol or benzoate. If, however, glucose or citrate was the sole carbon source, the development of microcolonies was much less distinct, a large part of the biomass being found evenly distributed over the substratum (Fig. 5d–f). After a shift from benzylalcohol (at a time when the microcolonies had formed) to glucose, biomass grew up in the voids between the microcolonies.

One should be very careful when interpreting data such as these, and obviously there is a need to complement the data with a genetic analysis and with direct monitoring in the

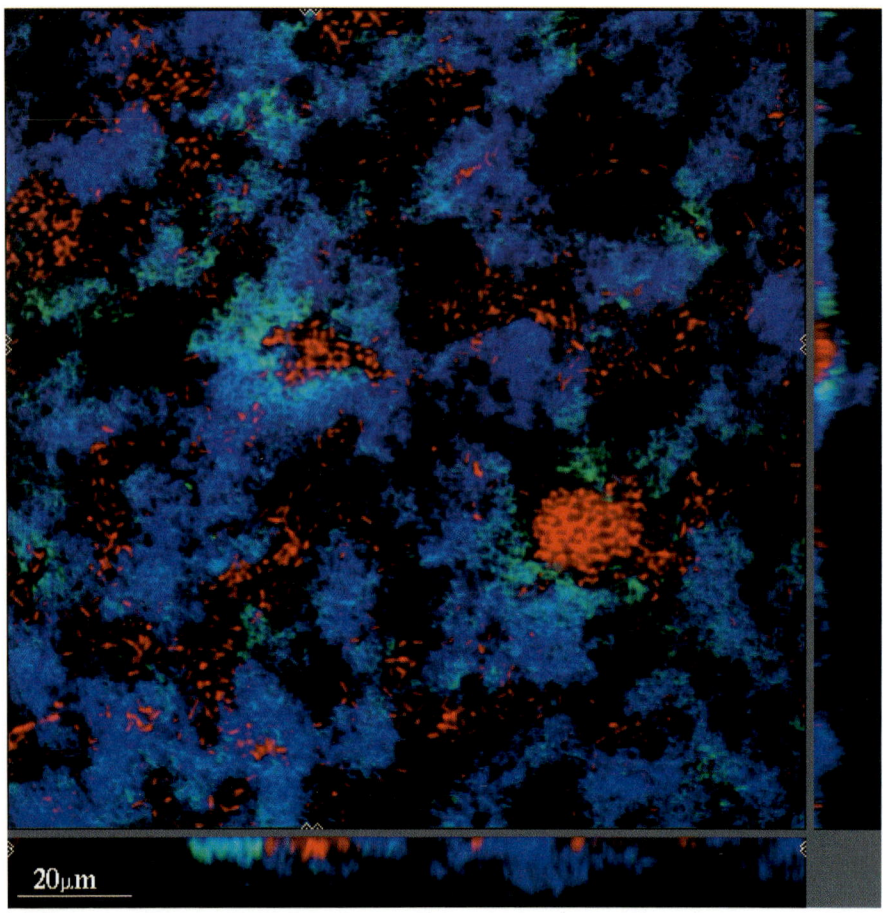

Fig. 6. Coupling between metabolically interacting cells. Flow-chamber community of *P. putida* R1 (blue cells) and *Acinetobacter* C6 (red cells) growing on benzylalcohol. The R1 strain carries a fusion between the rRNA promoter from the *Escherichia coli* rRNA B operon and a *gfp* gene expressing an unstable Gfp. This reporter fusion expresses green fluorescence in actively growing cells only (for a detailed description of this reporter system, see Sternberg *et al.*, 1999). Although R1 and C6 are located in a non-associated manner, it is clear that when the two organisms are coupled, R1 is growth-active (turquoise cells) due to excretion of benzoate from C6.

biofilm of cell communication signals. It is also important to extend the physiological analysis to cover more nutrient conditions, different carbon sources, different concentrations, different flow rates, etc. All this said, it is obvious from this preliminary investigation that biofilm structure and organization are not fixed predetermined features connected with a certain genotype, but rather contextual properties determined by the organism, the environmental conditions, and the available monitoring tools.

Multiple-species biofilms

When the *Acinetobacter* strain grew on a substrate with benzylalcohol as the carbon source, benzoate was excreted to the environment, and it was suggested that *P. putida* R1 could benefit from this extra carbon source. In an analysis of structure development in composite biofilms supplied with benzylalcohol, in which both organisms were present, two observations were made: (1) the microcolony pattern of *Acinetobacter* C6 changed from the regular array of small colonies to one where occasional very large colonies appeared among the more fluffy *P. putida* colonies; and (2) most of the large *Acinetobacter* colonies were covered by a layer of *Pseudomonas* cells (Fig. 6). Insertion of reporter cassettes into the chromosome of *P. putida* carrying either a fusion between the TOL Pm promoter and *gfp*, or between an rRNA promoter and *gfp* variants expressing unstable fluorescent proteins, made it possible to monitor the presence of benzoate (Pm) and active growth (rRNA promoter), respectively (Møller *et al.*, 1998; Sternberg *et al.*, 1999). Fig. 6 shows that the cells of *P. putida* associated with the large colonies of *Acinetobacter* are actively growing, in contrast to the *Pseudomonas* cells placed as separate colonies in the biofilm. Most likely this growth activity is induced by the excreted benzoate, which is able to induce the TOL Pm reporter fusion in the associated *Pseudomonas* cells, but not in the remotely placed cells (Møller et al., 1998).

In a parallel investigation of the 3-chlorobiphenyl-degrading community of *Pseudomonas* sp. B13 and LB400, it was demonstrated that B13 was fully associated with the LB400 colonies if 3-chlorobiphenyl was the sole carbon source, whereas in a citrate-supplied biofilm the two strains formed separate colonies (Fig. 7). The strong association between the two organisms under the former conditions obviously reflects the strong dependence of B13 on the metabolite 3-chlorobenzoate excreted by LB400 (Nielsen *et al.*, 2000). After a shift from citrate to 3-chlorobiphenyl, there was a rapid conversion of the colony pattern from totally separated microcolonies of the two strains to an increasing association, most likely brought about by movement of B13 cells in the direction of LB400. As described above for the TOL community, the insertion of a growth and a 3-chlorobenzoate reporter in B13 revealed that activity was only measurable in the close vicinity of the LB400 colonies (Nielsen *et al.*, 2000).

NATURE OF MICROBIAL COMMUNITIES

The introduction of molecular markers and reporters provides a first line of evidence that metabolic interactions are very important determinants of the way bacteria associate with each other in structured heterogeneous communities. The investigations of two-species biofilms have documented both expected interactions (the 3-chlorobiphenyl community) and surprising associations (the TOL community), but in both cases there are strong indications for metabolite gradients associated with colonies of cells. Chemotaxis and motility in response to such gradients (or more random

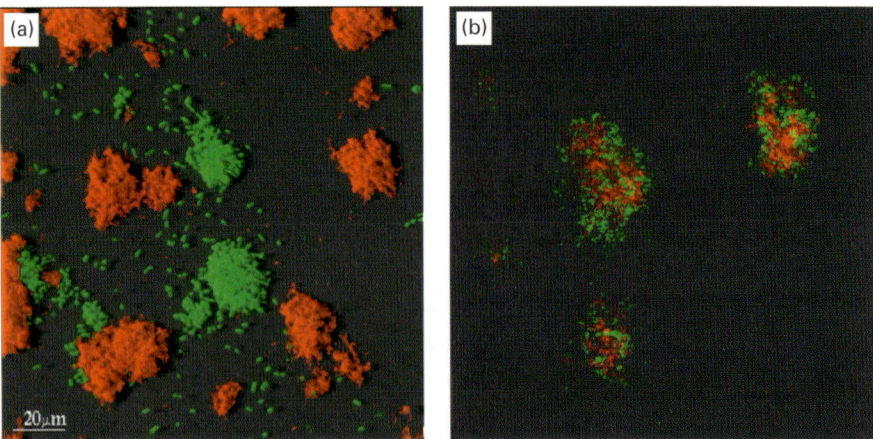

Fig. 7. Coupling between metabolically interacting cells. Flow-chamber community of *Burkholderia* sp. LB400 (red cells) and *Pseudomonas* sp. B13 (green cells) growing on citrate (a) or 3-chlorobiphenyl (b). There is tight, non-random coupling between the two organisms in (b) (cf. Nielsen *et al.*, 2000, from which the two pictures were taken and reproduced with permission from the publishers, Blackwell Science, Oxford, UK).

movements of cells) may explain the organization of species distribution in many cases in a satisfactory way, without assuming any community control or co-ordination based on cell-to-cell communication or other regulatory activities.

It is premature to conclude firmly that the colony arrays observed for *Acinetobacter* cells in flow-chamber biofilms (Fig. 5) are created as a consequence of nutrient gradients. The approaches used for two-species biofilms, in which the strains are distinguishable on the basis of rRNA sequences and molecular markers, are not directly transferable to a monospecies community analysis. When a two-species colony is formed due to the excretion of metabolites from one organism and a chemotactic motility of the other, the organizational activity of the community is traceable to activities of the individual cells. It is therefore possible that also in a monospecies community, there may be similar heterogeneous conditions, in which randomly formed aggregates of some cells create metabolite gradients, which are sensed by individual, but genetically identical, cells in the surroundings. In this way, a regular pattern may develop on the simple basis of nutrient limitation, small variations in biomass concentration and heterogeneous production of attractants/repellants from the cellular aggregates. Coupling of cells in surface-associated colonies may occur without any overall community regulation.

Through the use of molecular reporters inserted into surface-associated bacteria, it has been documented that communication signals (AHLs) are in fact produced in several biofilm settings by organisms known to carry out cell-to-cell communication (Fuqua & Greenberg, 1998). However, the significance of this production of signals remains to be determined, and it must be emphasized that the presence of signals is not sufficient for systemic co-ordination, although it may be required for effective co-ordination. The significant changes in community structure caused by simple shifts of nutrients (as discussed above for *Acinetobacter* C6) may indicate that expression of any cellular control system interfering directly or indirectly with nutrient sensing and uptake may lead to analogous community responses with no relation to overall community co-ordination. Moreover, it should be kept in mind that structural features alone are not direct evidence for multicellular activity and performance – only if functions are distributed in separate compartments within the community may we begin to consider it comparable to a composite higher organism.

The systematic analysis of bacterial surface communities has just begun. One important objective is to understand better the connection between the cellular repertoire (genotype), signal-transduction processes, community performance, and the structure/function relationships (organization). The interesting question is not necessarily *whether* microbial communities behave like multicellular organisms *or* like random aggregates of individuals. Instead, the question may be *when* a community behaves like one or the other, and *which parts* of a community do so. Perhaps a community is as many different things as we choose to focus on. Certainly, microbial communities are never the same when conditions change. We should therefore be cautious when we seek consensus explanations, and we should definitely avoid absolute interpretations and definitions.

ACKNOWLEDGEMENTS

The contributions from many of our colleagues in the Molecular Microbial Ecology Group are gratefully acknowledged. Grants from the Danish Biotechnology Programme supported the work.

REFERENCES

Amann, R. I., Ludwig, W. & Schleifer, K. H. (1995). Phylogenetic identification and *in situ* detection of individual microbial cells without cultivation. *Microbiol Rev* **59**, 143–169.

Caldwell, D. E., Korber, D. R. & Lawrence, J. R. (1992). Imaging of bacterial cells by fluorescence exclusion using scanning confocal laser microscopy. *J Microbiol Methods* **15**, 249–261.

Caldwell, D. E., Korber, D. R. & Lawrence, J. R. (1993). Analysis of biofilm formation using 2D vs 3D digital imaging. *J Appl Bacteriol Suppl* **74**, 52S–66S.

Christensen, B. B., Sternberg, C., Andersen, J. B., Palmer, R. J., Nielsen, A. T., Givskov, M. & Molin, S. (1999). Molecular tools for study of biofilm physiology. *Methods Enzymol* **310**, 20–42.

Costerton, J. W., Cheng, K. J., Geesey, G. G., Ladd, T. I., Nickel, J. C., Dasgupta, M. & Marrie, J. T. (1987). Bacterial biofilms in nature and disease. *Annu Rev Microbiol* **42**, 435–464.

Costerton, J. W., Lewandowski, Z., Caldwell, D. E., Korber, D. R. & Lappin-Scott, H. M. (1995). Microbial biofilms. *Annu Rev Microbiol* **49**, 711–745.

Davies, D. G., Parsek, M. R., Pearson, J. P., Iglewski, B. H., Costerton, J. W. & Greenberg, E. P. (1998). The involvement of cell-to-cell signals in the development of a bacterial biofilm. *Science* **280**, 295–298.

Fuqua, W. C. & Greenberg, E. P. (1998). Self-perception in bacteria: quorum sensing with acylated homoserine lactones. *Curr Opin Microbiol* **1**, 183–189.

Fuqua, W. C., Winans, S. C. & Greenberg, E. P. (1994). Quorum sensing in bacteria: the LuxR-LuxI family of cell density-responsive transcriptional regulators. *J Bacteriol* **176**, 269–275.

Massol-Deyá, A. A., Whallon, J., Hickey, R. F. & Tiedje, J. M. (1995). Channel structures in aerobic biofilms of fixed-film reactors treating contaminated groundwater. *Appl Environ Microbiol* **61**, 769–777.

Molin, J. & Molin, S. (1997). CASE: Complex Adaptive Systems Ecology. *Adv Microb Ecol* **15**, 27–80.

Molin, S., Nielsen, A. T., Christensen, B. B., Andersen, J. B., Licht, T. R., Tolker-Nielsen, T., Sternberg, C., Hansen, M. C., Ramos, C. & Givskov, M. (2000). Molecular ecology of biofilms. In *Biofilms*, pp. 89–120. Edited by James Bryers. New York: Wiley.

Møller, S., Pedersen, A. R., Poulsen, L. K., Arvin, E. & Molin, S. (1996). Activity and three-dimensional distribution of toluene degrading *Pseudomonas putida* in a multispecies biofilm assessed by quantitative *in situ* hybridization and scanning confocal laser microscopy. *Appl Environ Microbiol* **62**, 4632–4640.

Møller, S., Sternberg, C., Andersen, J. B., Christensen, B. B. & Molin, S. (1998). *In situ* gene expression in mixed-culture biofilms: evidence of metabolic interactions between community members. *Appl Environ Microbiol* **64**, 721–732.

Nielsen, A. T., Tolker-Nielsen, T., Barken, K. B. & Molin, S. (2000). Role of commensal relationships on the spatial structure of a surface-attached microbial consortium. *Environ Microbiol* **2**, 59–68.

O'Toole, G. A. & Kolter, R. (1998a). Initiation of biofilm formation in *Pseudomonas fluorescens* WCS365 proceeds via multiple, convergent signalling pathways: a genetic analysis. *Mol Microbiol* **28**, 449–461.

O'Toole, G. A. & Kolter, R. (1998b). Flagellar and twitching motility are necessary for *Pseudomonas aeruginosa* biofilm development. *Mol Microbiol* **30**, 295–304.

Picioreanu, C., van Loosdrecht, M. C. M. & Heijnen, J. J. (1998). Mathematical modeling of biofilm structure with a hybrid differential-discrete cellular automaton approach. *Biotechnol Bioeng* **58**, 101–116.

Pratt, L. A. & Kolter, R. (1998). Genetic analysis of *Escherichia coli* biofilm formation: roles of flagella, motility, chemotaxis and type I pili. *Mol Microbiol* **30**, 285–293.

Prosser, J. I. (1994). Molecular marker systems for the detection of genetically modified micro-organisms in the environment. *Microbiology* **140**, 5–17.

Sternberg, C., Christensen, B. B., Nielsen, A. T., Andersen, J. B., Givskov, M. & Molin, S. (1999). Distribution of bacterial growth activity in flow-chamber biofilms. *Appl Environ Microbiol* **65**, 4108–4117.

Wimpenny, J. W. T. & Colasanti, R. (1997). A unifying hypothesis for the structure of microbial biofilms based on cellular automaton models. *FEMS Microbiol Ecol* **22**, 1–16.

Gene transfer in biofilms

Laura J. Ehlers

National Research Council, Washington, DC, USA

INTRODUCTION

Horizontal gene transfer is a ubiquitous process that allows DNA sequences to be widely disseminated in natural microbial populations. During horizontal gene transfer, micro-organisms transfer genetic material to organisms other than their descendants, distinguishing the process from vertical gene transfer in which genetic material is passed to offspring via sexual or asexual reproduction. Horizontal gene transfer is most prevalent among bacteria but is not restricted to prokaryotes; highly unrelated organisms can participate in horizontal gene transfer including single- and multi-celled eukaryotes. Presently, little is known about the rates of horizontal gene transfer in both natural and engineered biofilm environments such as water distribution systems, water and wastewater treatment plants, wetlands and biologically active soils and sediments.

Evidence for horizontal gene transfer has been gathered steadily over the last 50 years. Its most prominent manifestation is the spread of antibiotic resistance among bacteria. The genes conferring antibiotic resistance can be quickly disseminated through bacterial populations via horizontal gene transfer, confounding the development of new drugs to treat bacterial infections. Another manifestation of horizontal gene transfer is tumour formation on the surfaces of plant roots after infection by *Agrobacterium tumefaciens*. This bacterium has been shown to transfer its DNA to plant cells, where the DNA directs the rapid growth of plant-cell tumours and the production of chemicals that nourish the bacteria. From these studies and others, several mechanisms of gene transfer have emerged, which appear to be highly conserved.

SGM symposium 59: Community structure and co-operation in biofilms. Editors D. Allison, P. Gilbert, H. Lappin-Scott, M. Wilson. Cambridge University Press. ISBN 0 521 79302 5 ©SGM 2000.

Greater understanding of horizontal gene transfer in the environment is needed for multiple purposes. In particular, there has been rapid progress in the creation of genetically engineered micro-organisms (GEMs) to be released into the environment. GEMs have been proposed for agricultural purposes, e.g. microbial pesticides, development of frost- and salt-resistant cultivars of crop plants, and enhancement of nitrogen fixation. Environmental applications of GEMs include mineralization of organic pollutants and detoxification of heavy metals. Current regulations require information on the fate and transport of GEMs and their recombinant DNA in order to assess risk to the native biological community. Thus accurate risk assessment is highly dependent on the availability of quantitative gene transfer rates in natural environments.

In the future, horizontal gene transfer could be a valuable tool in environmental remediation. The controlled dissemination of degradative or catabolic genes via horizontal gene transfer could be used to enrich bacterial populations capable of utilizing contaminants of concern. If the necessary degradative genes are not found in the indigenous pool of microbes at the site of contamination, GEMs could be used to supplement the gene pool, acting primarily as shuttles of new genetic material. Initial studies suggest that this strategy can be successful in accelerating degradation rates of pesticides (Top *et al.*, 1999). This strategy has also been suggested for toxic chemical degradation in suspended microbial systems like activated sludge units (Rittmann *et al.*, 1990). Further information on rates of horizontal gene transfer among bacteria should help determine the feasibility of promoting this process during remediation. In addition, identifying environmental parameters that influence transfer rates could enable the manipulation of chemical conditions to achieve the desired level of remediation.

This paper reviews literature on horizontal gene transfer among bacteria growing in a biofilm, the preferred mode of growth for bacteria in natural systems. It concentrates primarily on bacterial conjugation, the mechanism of horizontal gene transfer that has been the focus of most studies to date. Little research on conjugation has been conducted in natural biofilm systems, although over 99% of bacteria in nature are attached to surfaces. There is reason to believe that biofilm systems may be more favourable for conjugation than aqueous systems, perhaps as a consequence of increased cell contact time or stabilization of cells on the substratum. As discussed below, both qualitative and quantitative studies suggest that rates of conjugation on surfaces are often elevated compared to rates in liquid media. A wide range of factors, including nutrient conditions, shear stress, cell concentrations and temperature, control the extent of conjugation among biofilm bacteria.

MECHANISMS OF HORIZONTAL GENE TRANSFER

As shown in Fig. 1, three general mechanisms of horizontal gene transfer have been described – transformation, transduction and conjugation – that can facilitate the transfer of plasmid DNA, chromosomal DNA, or both. These are discussed in turn below, placing the greatest emphasis on conjugation. Transposition, also shown in Fig. 1, is an intracellular mechanism of gene transfer that can enhance the extent of transformation, transduction and conjugation, possibly by orders of magnitude. Transposons are genetic elements of variable size that can 'jump' from one random location to another within chromosomal and plasmid DNA. However, because it does not mediate intercellular gene transfer, transposition is not explored further in this paper.

Transformation

Transformation is the mechanism by which cells take up free DNA from their surroundings. Its occurrence is dependent on the amount of DNA available and bacterial 'competence' (defined as the physiological state in which a bacterium is able to receive DNA). For some bacteria, competence is a function of the growth cycle; competence in *Azotobacter*, *Acinetobacter* and *Pseudomonas* strains is highest during stationary phase (Carlson *et al.*, 1983; Veal *et al.*, 1992), while for *Haemophilus influenzae*, competence is induced by unbalanced growth (Stewart, 1989). Among the Gram-positive bacteria *Bacillus subtilis* and *Streptococcus* spp., a 'competence factor' is constitutively secreted which subsequently induces competence in neighbouring bacteria (Stewart, 1989). Very little information exists regarding the numbers of competent bacteria in the environment.

Competent cells quickly bind free DNA, which is internalized and recombined with the host genomic DNA. Most bacteria bind DNA nonspecifically, although a few bacterial types, such as *Haemophilus*, only bind DNA containing specific sequences. If the transforming DNA is chromosomal, it will only become integrated into the host genome if it contains homologous sequences. Transformation of plasmid DNA, on the other hand, imposes no homology requirements since plasmids are maintained outside the chromosome. The homology requirements imposed by the DNA binding and recombination processes may place substantial limitations on natural rates of transformation.

In addition to the requirement for competent bacteria, transformation is heavily dependent on the pool of free DNA in the environment. In aqueous environments, the ubiquitous bacterial DNA-degrading enzyme DNase limits the size of this free DNA pool. However, recent studies suggest that there is considerable DNA available for uptake. The work of Lorenz and coworkers has shown that DNA can be adsorbed to

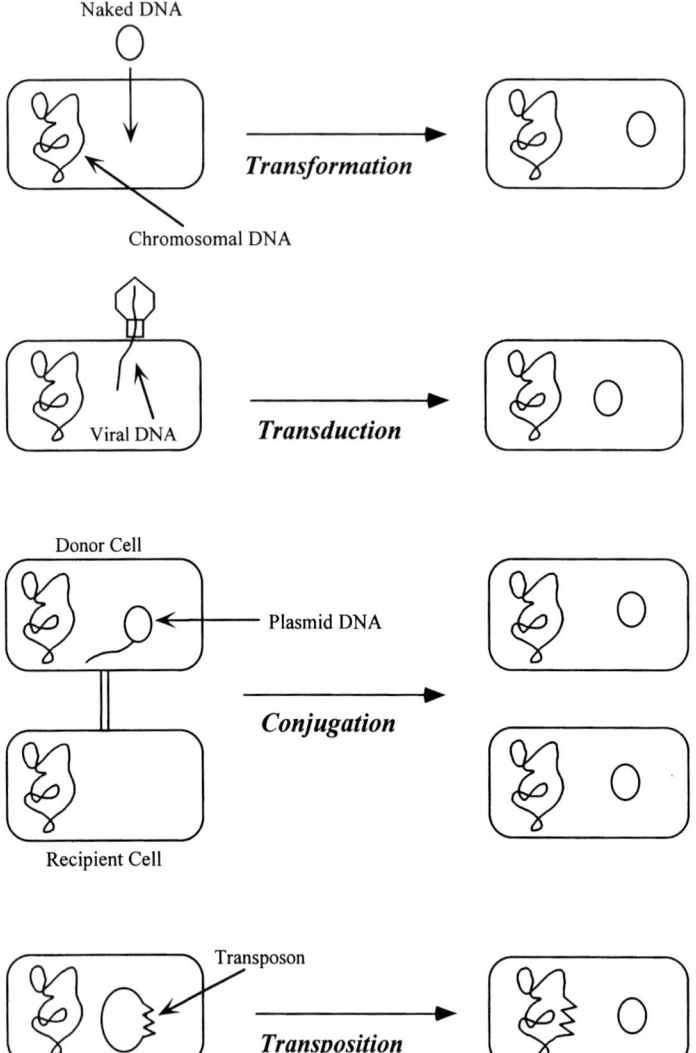

Fig. 1. Mechanisms of gene transfer. (1) During *transformation*, free DNA is taken up by the recipient bacteria. The DNA may become integrated into the host chromosome or, as is shown for plasmid DNA, reside independently of the chromosome. (2) Bacteriophages containing foreign DNA can mediate DNA transfer via *transduction*. The DNA is inserted into the recipient cell during viral infection and either integrated into the host chromosome or maintained extrachromosomally as shown. (3) *Conjugation* is the mechanism of DNA transfer involving cell–cell contact. Plasmid DNA from a donor cell is replicated and transferred to a recipient, after which both cells contain the plasmid. The bar connecting the cells represents the pilus of Gram-negative bacteria, a structure thought to bring the cells into close contact. (4) *Transposition*, as shown here, is actually a mechanism of intracellular gene transfer. The transposon initially residing on a plasmid relocates to the chromosome. Transposons dramatically affect the other three mechanisms of transfer and can, in some circumstances, mediate their own intercellular transfer.

sand grains, thereby reducing its susceptibility to DNase 100-fold (Aardema *et al.*, 1983; Lorenz & Wackernagel, 1987). This sand-adsorbed DNA was found to be capable of transformation into *Pseudomonas stutzeri* and *B. subtilis* (Lorenz *et al.*, 1988; Lorenz & Wackernagel, 1990). Other factors thought to influence transformation are the concentration of divalent cations (which have been found to stabilize the DNA–cell complex) and shearing forces (which may disrupt the complex).

There is a growing body of evidence that transformation occurs in many natural environments, including biofilm environments. Stewart & Sinigalliano (1990) detected transformation of bacteria in marine water and sediment columns and found transformation rates to be highest in sterile sediment with high organic matter. Paul *et al.* (1991) also investigated transformation in marine microcosms and found the highest rates to occur in sediments after nutrient addition. Lorenz *et al.* (1992) demonstrated that high-efficiency transformation in nonsterile soil and groundwater is dependent on Ca^{2+} and the presence of DNases. These studies and others suggest that transformation may be a significant mechanism of gene transfer in environments that afford DNA protection from DNase and contain nutrients sufficient for achieving bacterial competence.

Transduction

Transduction is the mechanism of gene transfer mediated by bacterial viruses (bacteriophages). Two types of transduction have been distinguished (Fig. 2). *Generalized transduction* is a consequence of the lytic growth phase of a bacteriophage. During lysis, phage-encoded nucleases break down the host bacterial DNA, some of which may be accidentally packaged into newly forming phage particles. This bacterial DNA may be transferred to other bacteria during subsequent viral infection. *Specialized transduction* is carried out by temperate phages that have become incorporated into the bacterial genome during lysogenic growth. (Most bacteria are polylysogenic for several distinct types of bacteriophage.) When the lytic cycle is entered, the phage DNA may be incorrectly excised from the genome, taking bacterial DNA with it. As before, the bacterial DNA may be packaged into new viral particles and transferred to other bacteria. Specialized transduction can only transfer genes near the excision site of the virus, while generalized transduction can transfer any genomic DNA sequences.

Transduction is limited by the *host range* of the bacteriophage, which ranges from very narrow (only one bacterial strain can be the host, as for λ virus) to relatively broad. Phages have been discovered in over 57 bacterial species from 24 genera (Veal *et al.*, 1992), suggesting that few, if any, bacterial species are completely immune to virus infection. The ability of phages to survive outside their bacterial hosts will also affect

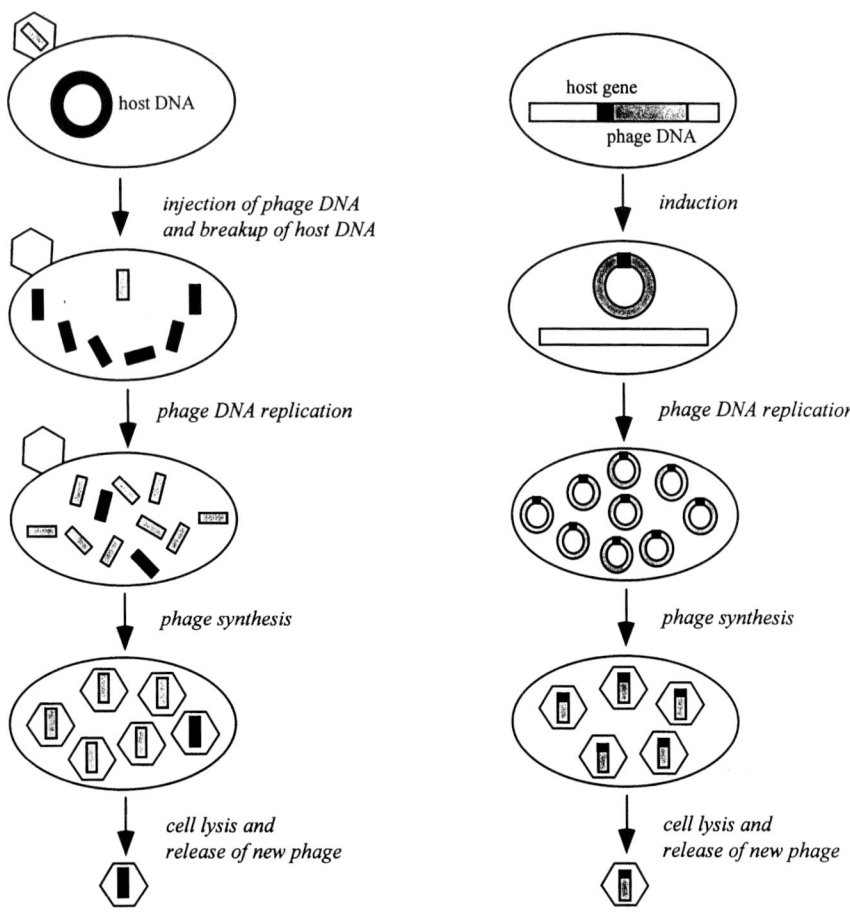

(a) Generalized Transduction　　　　　　　**(b) Specialized Transduction**

Fig. 2. Generalized versus specialized transduction. (a) Formation of generalized-transducing phage is illustrated. After contacting the bacterial host, the phage injects its DNA into the cytoplasm, which directs the subsequent breakdown of host chromosomal DNA. New phage DNA is replicated and packaged into phage particles. During assembly, bacterial DNA may accidentally be packaged if it is the correct size. Subsequent lysis of the cell releases new infectious phage (not shown) and transducing phage, which may inject its 'bacterial' DNA into a new host, completing transduction. (b) Specialized transduction is carried out by lysogenic phage. Lysogenized bacteria carry a copy of the phage DNA on their chromosome. Under certain environmental conditions, the lytic cycle of phage reproduction is induced and the viral DNA is excised from the bacterial chromosome. Improper excision may remove bacterial DNA sequences along with the phage DNA. The bacterial DNA is replicated and packaged with the phage DNA into new phage particles. Lysis of the cell releases transducing phage which may infect a new host.

the frequency of transduction. Although viruses require no nutrients or other growth substrates in their environment, certain ions (Mg^{2+}) may stabilize viral particles and delay inactivation.

Studies have shed some light on the environmental conditions conducive to transduction. Stotzky & Babich (1986) observed transduction of *Pseudomonas aeruginosa* in a freshwater reservoir over 10 days. Transduction was found to be more efficient when pure viruses were used instead of lysogenic donor bacteria. The same host/phage system was utilized by Saye *et al.* (1990), who demonstrated transduction between combinations of lysogenic and nonlysogenic strains of *P. aeruginosa* introduced into a freshwater habitat. The highest frequency of transduction was found to occur when both the donor and recipient strains were lysogenic, possibly because lysogeny has been shown to increase bacterial fitness.

The effect of surfaces on transduction has also been investigated, although there have been no studies of transduction in biofilm reactors. Kidambi *et al.* (1994) employed the *P. aeruginosa* phage/host system to study transduction on plant leaves. Introduced lysogenic bacteria transferred their DNA to indigenous bacteria on the same and nearby plants. Finally, Ripp & Miller (1995) found that both phage survival and transduction frequency in a freshwater environment were greatly enhanced in the presence of particulate matter. While these studies are a beginning, until more is known about the prevalence and host ranges of indigenous viruses, it will be difficult to assess the contribution of transduction to overall horizontal gene transfer in the environment.

Conjugation

Conjugation is a process that requires direct contact between bacteria. It is the most intensely studied mechanism of horizontal gene transfer because of its high frequencies under ideal laboratory conditions. During conjugation, plasmid DNA from a donor cell is unwound and one copy of single-stranded DNA is passed to a recipient cell, thereafter called a transconjugant. The single strands are then replicated in the donor and recipient organisms such that both cells possess a copy of the plasmid and are usually capable of further conjugation events. Conjugation between a transconjugant and a recipient is termed secondary transfer.

In order for a plasmid to be transferred from one cell to another via conjugation, it must contain specific transfer genes known as the *tra* genes. Because of the large number of required transfer genes (the F plasmid has 17), all self-transmissible (conjugative) plasmids are at least 30 kb in size (Veal *et al.*, 1992). Plasmids without the *tra* genes are termed nonconjugative and can only be transferred via conjugation in the presence of a conjugative plasmid (see discussion on mobilization below). The *tra* genes and transfer machinery of different conjugative plasmids are highly related and share considerable homology with the transfer genes of the Ti plasmids that are transferred from bacteria to plants (Lessl & Lanka, 1994).

Conjugation is dependent on the formation of a mating pair between donor and recipient cells. In Gram-negative bacteria, mating-pair formation is brought about by pili – hollow, proteinaceous strands extending from the donor cell that interact with the recipient cell surface. Pili are plasmid-encoded and come in a variety of shapes and sizes. Upon microscopic observation, they have been classified according to thickness and rigidity: (1) thin, flexible pili, (2) thick, flexible pili and (3) thick, rigid pili (Bradley *et al.*, 1980). Rigid pili have been shown to mediate plasmid transfer on solid surfaces much more efficiently than in liquid media, while plasmids encoding thin, flexible pili transfer equally well under both conditions (Bradley *et al.*, 1980). Although the exact mechanism of action is unknown, it is thought that after the initial contact between pili and the recipient cell surface, the pili are retracted via disassembly, pulling the donor and recipient cells together (Ippen-Ihler, 1989). Pili are the target of several bacteriophages such that their production can make a cell vulnerable to viral infection (Bradley & Williams, 1982; Haase *et al.*, 1995).

Among Gram-positive bacteria, mating-pair formation is entirely independent of pili. The recipient bacteria secrete sex pheromones that attract the donor bacteria (Frost, 1992). Pheromones are small peptides whose specific amino acid sequences induce the transfer of certain plasmids. Researchers speculate that pheromones accomplish mating-pair formation by binding a substance common to the surfaces of donor and recipient cells (Ippen-Ihler, 1989).

For both Gram-positive and Gram-negative bacteria, the specific interactions at the junction of the two cells have yet to be fully characterized but are known to involve membrane-bound proteins encoded by the *tra* genes (Haase *et al.*, 1995). (Surface exclusion, a phenomenon in which a conjugative plasmid can block entry of another plasmid into its host cell, is a consequence of faulty mating-pair formation and has been attributed to *tra* gene mutations.) With Gram-negative bacteria, it is not clear whether the DNA and accessory proteins to be transferred travel through the pilus, or whether the pili simply act to bring the cells together. Several studies indicate that pili are not an obligatory component at the moment of DNA transfer (Ippen-Ihler & Minkley, 1986; Haase *et al.*, 1995). In most cases, it is thought that an intermembrane channel formed during mating-pair formation facilitates transfer of the DNA.

Regulation of conjugation. The regulation of conjugation is a complex process that differs widely between plasmids and bacteria. The transfer genes of some plasmids are constitutively expressed, resulting in frequent conjugation. Such plasmids are often termed 'derepressed' to differentiate them from plasmids in which conjugation is repressed. Repression is thought to be the wild-type phenotype because conjugative plasmids isolated from nature usually transfer at very low frequencies. For Gram-negative bacteria, repression of conjugation is a plasmid-encoded process and usually operates by preventing pili synthesis (Guiney & Lanka, 1989). It is generally not known how environmental signals remove repression and induce plasmid transfer, as the chemical signal has only been identified in a handful of cases. The best-characterized signal is that which induces Ti-plasmid transfer from *Agrobacterium* to plant cells. Sugars, acids and phenolic compounds released from wounds in plant tissue have been found to directly regulate the transfer genes of the Ti plasmid, resulting in Ti transfer to plant cells and subsequent tumour formation (Winans, 1992). It has yet to be seen whether the pheromones that bring about mating-pair formation among Gram-positive bacteria also induce expression of plasmid transfer genes.

There are several possible reasons that plasmids are naturally repressed for transfer. Repression of conjugation may provide bacteria with a defence mechanism against pili-specific bacteriophages by preventing pili production. Another explanation may involve the need for energy conservation within the host cell, as plasmid replication and conjugation are energy-intensive processes. Indeed, derepressed mutants produce smaller colonies with longer doubling times than repressed cells (Meynell, 1973). Some authors have suggested that repression has evolved to provide feedback control within a population of plasmid-bearing cells (Lundquist & Levin, 1986), although this has yet to be proven experimentally. Because conjugation involves creating a passageway for DNA transfer, repression may serve to protect the cell from continual breaching of the cell membrane.

Plasmid host range. Aside from its state of repression, the host range of a plasmid is critical in determining the extent of its dissemination in a microbial population. Plasmids are classified as either broad or narrow host range, depending on the spectrum of organisms into which the plasmid can be transferred *and* maintained. Broad-host-range plasmids are able to transfer between many genera of Gram-negative bacteria. Versatile, nonspecific pili composition is thought to play a role in the broad-host-range phenotype, as it is unlikely that vastly different recipient cells would possess a common structure that is recognized during mating-pair formation (Guiney & Lanka, 1989).

Because the host range of a plasmid is important both in determining its ability to spread throughout a microbial community and in developing tools for use in molecular biology, it has been extensively investigated. Multiple-drug-resistance plasmids can transfer efficiently between diverse Gram-negative bacteria, including pathogens of human, animal and fish origin (Kruse & Sorum, 1994). The TOL plasmid, which encodes toluene degradation, was found to transfer via conjugation intragenerically between Gram-negative bacteria in soils and intergenerically in laboratory filter matings (Ramos-Gonzalez *et al.*, 1991). Conjugation of IncQ plasmids has been observed under anaerobic conditions from *Escherichia coli* to thermophilic green sulfur bacteria (Wahlund & Madigan, 1995). Efficient plasmid transfer from Gram-negative to Gram-positive bacteria (Schafer *et al.*, 1990) and vice versa (Trieu-Cuot *et al.*, 1988) has been demonstrated. DNA transfer via conjugation has even been accomplished between bacteria and yeast (Heinemann & Sprague, 1989; Nishikawa *et al.*, 1990; Inomata *et al.*, 1994). These studies are testimony to the versatility and potential of plasmid-mediated gene exchange via conjugation.

Mobilization. Gene transfer via conjugation is not restricted to self-transmissible plasmids. Nonconjugative plasmids and chromosomal DNA can also be transferred between organisms, provided that a conjugative plasmid resides in either the donor or recipient strain. This process, known as mobilization, significantly enhances the potential for horizontal gene transfer in a bacterial population.

Nonconjugative plasmids can be mobilized if they contain an origin of transfer (*oriT*) and mobilization (*mob*) genes, the products of which have been shown to catalyse cleavage at *oriT* (Haase *et al.*, 1995). The proteins needed to produce pili and catalyse the replication and transfer reactions are provided by a conjugative plasmid. Plasmids naturally containing the *mob* sequences are termed mobilizable. Nonmobilizable plasmids do not have these sequences and can be mobilized only if they acquire the genes via transposition or homologous recombination with another plasmid.

Mobilization can occur in one of three ways, depending on where the conjugative plasmid resides: (1) in the donor organism with the mobilizable plasmid (cotransfer), (2) in the recipient organism (retrotransfer) or (3) in a third strain known as the helper (three-way or triparental mating) (see Fig. 3). During cotransfer, both the conjugative and mobilizable plasmids are transferred from the donor to the recipient. The conjugative plasmid may or may not be expressed in the recipient. A three-way mating consists of two steps. Initially, the conjugative plasmid is transferred into the donor cell from a helper cell. It then mobilizes the nonconjugative plasmid into the recipient. Retrotransfer is a recently discovered phenomenon in which the recipient strain can provide the conjugative plasmid (i.e. it acts as the helper cell). During retrotransfer, the conjugative plasmid is first transferred to the donor, and then the conjugative plasmid mobilizes the nonconjugative plasmid from the donor to the recipient (Heinemann & Ankenbauer, 1993; Sia *et al.*, 1996).

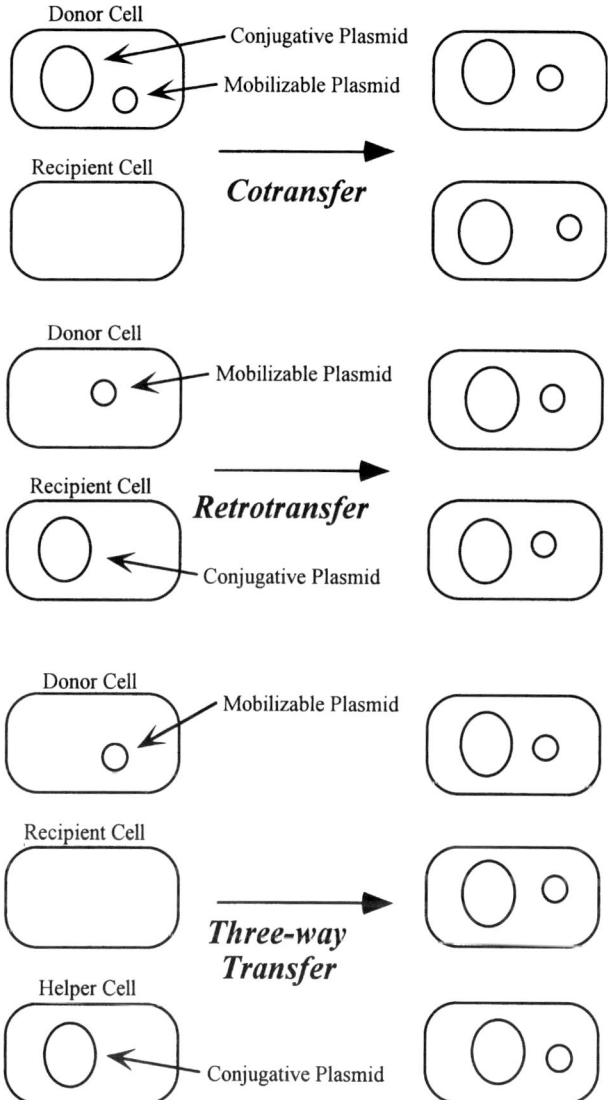

Fig. 3. Three mechanisms of mobilization. (1) During *cotransfer*, the mobilizable and conjugative plasmids reside in the donor. The conjugative plasmid transfers itself and the mobilizable plasmid from donor to recipient. (2) The recipient provides the conjugative plasmid during *retrotransfer*. Both the conjugative and mobilizable plasmids are transferred, resulting in both bacterial strains containing both plasmids. (3) In a *three-way or triparental mating*, the conjugative plasmid is supplied by a helper. The conjugative plasmid transfers itself into the donor cell and then mobilizes both itself and the mobilizable plasmid into the recipient. The helper may also receive a copy of the mobilizable plasmid.

The search for natural, conjugative plasmids capable of mobilizing other plasmids has uncovered a wealth of candidates. Bacterial strains indigenous to wastewater were found to contain plasmids that mobilized pBR325 between *E. coli* (McPherson & Gealt, 1986). Eight plasmids from bacteria living on river stones were found to mobilize the large recombinant plasmid pD10, which encodes 3-chlorobenzoate degradation (Hill *et al.*, 1992). Four of the eight were large, broad-host-range plasmids that carried mercury resistance. Polluted soils and sludges were found to contain plasmids that mobilized the transfer of the IncQ plasmid pMOL155 from *E. coli* to *Alcaligenes eutrophus* (Top *et al.*, 1994). The sludges from several wastewater treatment plants were especially rich in broad-host-range mobilizing plasmids of several incompatibility groups.

Several investigators have demonstrated mobilization in natural environments using both laboratory and indigenous strains of bacteria. Plasmids in *E. coli* were found to mobilize pBR322 into indigenous bacteria in a waste disposal system (Gealt *et al.*, 1985). The rates of transfer were significantly higher in broth matings than in sterilized wastewater. Soil was shown to be conducive for mobilization of the cloning vector pIJ702 between strains of *Streptomyces* (Rafi & Crawford, 1988). Many recipients were found to contain slightly altered versions of pIJ702, a phenomenon often observed during mobilization. Retrotransfer was demonstrated between strains of *E. coli* and *Alcaligenes eutrophus* in soil (Top *et al.*, 1990). Both mobilizable and nonmobilizable plasmids were retrotransferred to recipient strains containing the conjugative plasmid RP4. Smit *et al.* (1993) compared the relative frequencies of cotransfer, retrotransfer and triparental transfer of the mobilizable plasmid pSKTG. While retrotransfer was only found to occur on filters, cotransfer and triparental transfer were detectable in nonsterile soils with indigenous bacteria as recipients. As expected, plasmid mobilization via cotransfer had the highest frequency followed by triparental mating and retrotransfer. Finally, several studies have demonstrated the mobilization of chromosomal DNA by conjugative plasmids (Aronson & Beckman, 1987; Guiney & Lanka, 1989; Stout & Iandolo, 1990).

CONJUGATION IN THE ENVIRONMENT

One of the difficulties faced when conducting research on conjugation is the variety of ways in which its frequency is measured. Conjugation is typically tested in laboratory matings with donor (D), recipient (R) and transconjugant (T) concentrations measured by selective plating. Frequency of conjugation is generally defined as T/D, T/R or T/DR measured *after a predetermined period of time*, which can range from minutes to hours to days depending on the application. For example, the time to stationary phase growth might be chosen for conjugation experiments in liquid batch systems, while, for biofilm experiments, conjugation frequency is often measured when 'steady state' growth has been achieved. Because recipient and transconjugant physiologies are assumed to be similar, changes in T/R are thought to be the best indication of conjugation. During relatively short matings (minutes to hours), the values of D and R are initial values, and the contribution of secondary transfer can often be neglected. Longer matings, however, generally require measurements of D and R at the conclusion of the experiment and a consideration of secondary transfer.

Other methods of expressing conjugation efficiency include simply T or T/time. Mathematical models of conjugation generate transfer rate parameters which, when certain assumptions are made, can be compared to frequency ratios (Smets *et al.*, 1990). Conjugation has also been compared to enzyme kinetics, with determinations of the maximum conjugation rate (T/D/time) and the K_m value (R) (Andrup *et al.*, 1998; Andrup & Andersen, 1999). Because the frequency of conjugation is dependent on so many factors, it is rare for two independent studies to have been conducted in a similar manner. In addition, traditional detection methods (selective plating) cannot distinguish between transconjugants created via plasmid transfer and those created via transconjugant growth. It is generally difficult and sometimes futile to compare studies that do not utilize the same method. Saunders & Saunders (1992) have suggested that only order-of-magnitude differences should be considered significant when comparing data.

The following discussion mentions those environmental factors known to influence the extent of bacterial conjugation in laboratory and natural environments. Such studies have been carried out under a wide variety of conditions that represent suspended growth environments (e.g. liquid matings and freshwater microcosms), biofilm or attached-growth environments (e.g. agar plates and biofilm reactors), and environments in which both suspended and attached growth occur (e.g. soil, sediment slurries and packed columns).

Environmental factors affecting conjugation

Every bacterial strain grows optimally under a specific set of environmental conditions defined by parameters such as temperature, pH, salinity, nutrient concentration and pressure. For some parameters (e.g. pH), the range over which bacteria grow optimally can be quite broad, while for others (e.g. temperature) the range of optimal values is narrow. Presumably, bacteria are constrained to grow under conditions in which their enzymic machinery will function. Thus it is likely that the optimal conditions for conjugation are very similar to those for bacterial growth. The goal of many conjugation studies has been to determine what physical, chemical and biological parameters affect transfer frequency. In several instances, important trends have emerged.

Temperature. Conjugation requires the replication of plasmid DNA in both the donor and recipient cells, a process associated with cellular growth and reproduction. It is not surprising, then, that most bacteria have an optimal conjugation temperature identical to their optimal growth temperature. Frequencies of conjugation tend to decrease on both sides of the optimal temperature, a trend that has been observed in numerous plasmid transfer studies. Table 1 summarizes the extensive body of research on this topic.

The studies in Table 1 demonstrate that conjugation can occur over a very wide temperature range, which has significant ramifications for environmental gene transfer. Although most studies show a definite relationship, there are some instances in which temperature appears not to influence conjugation. For example, Fernandez-Astorga *et al.* (1992) observed the expected increase in T/D as temperature increased from 8 °C to 37 °C during mating in rich media. However, they also observed conjugation in the presence of low or no bacterial growth that was *not* temperature-dependent. It could be that a low-level constitutive expression of conjugation genes accounted for the constant conjugation rates observed; this constitutive expression may not be dependent on temperature-sensitive enzymes.

One of the direct effects of temperature on the conjugation process was discovered by Walmsley (1976), who found that suboptimal conjugation at extreme temperatures is a consequence of reduced mating-pair formation. Mating-pair formation in *E. coli*, measured visually under the microscope, was shown to increase gradually from 0% at 0 °C to 20% at 40 °C before declining steeply. Failure to form mating pairs may be the result of reduced pili synthesis or impaired pili attachment to recipient cells.

Nutrient concentration. Because conjugation requires substantial cellular energy, higher transfer frequencies are expected in more nutrient-rich environments. This trend is clearly demonstrated in Table 2, which summarizes studies that have shown a positive effect of nutrient addition on conjugation rates in several different systems. The source of nutrients in most conjugation experiments is the external environment. However, internal nutrient storage within bacteria can also supply the necessary energy for conjugation. Indeed, MacDonald *et al.* (1992) showed that the plasmid transfer rate increased dramatically with increased 'energy availability' in the form of nutrient amendments *or* internal storage materials accumulated during prior growth on rich media. Internal storage materials accumulated during previous culture conditions were also thought to account for the high rates of conjugation observed under nutrient-poor conditions in other studies (Fernandez-Astorga *et al.*, 1992). Clearly, a scarcity of external nutrients does not rule out the possibility for substantial transfer if the physiological state of the bacteria is favourable for conjugation.

Table 1. Temperature effects on frequencies of conjugation

Donor:recipient/plasmid	Mating conditions	Temp. range	Remarks*	Reference
E. coli strains/R1drd-19	Liquid matings Rich media	17–37 °C	10^5 increase in T as temp. increases	Singleton & Anson (1981)
E. coli strains from sewage/antibiotic-resistance plasmids	Liquid matings Rich media	15–35 °C	Optimum temp. = 22 °C; T/D 10^2–10^3× lower at 15 and 37 °C	Altherr & Keswick (1982)
Marine bacterium: *E. coli* K-12 C600/mercury-resistance plasmids	Liquid matings Rich media	16–37 °C	Optimum temp. = 30°C; frequencies are 10× lower at 25 and 37 °C	Gauthier *et al.* (1985)
Epilithic bacteria: *P. putida* KT2440/mercury-resistance plasmid pQM1	River water and laboratory microcosms	10–20 °C	Optimum temp. = 17 °C with detection as low as 10 °C	Bale *et al.* (1988)
Streptomyces lividans strains/pIJ303 (thiostrepton resistance)	Sterile soil	25–37 °C	Optimum temp. = 30 °C; T/R is 2× lower at 25 and 37 °C	Rafii & Crawford (1988)
E. coli: *R. fredii*/pBLK1-2(Tn5)	Soil	5–40 °C	Optimum temp. ranged from 20 to 28 °C; 5× change in T/R over the temp. range	Richaume *et al.* (1989)
Epilithic pseudomonads (five combinations)/mercury-resistance plasmids	Agar plate matings Rich media	4–41 °C	Optimum temp. ranged from 10 to 40 °C; T/R decreased on both sides of optimum temp.	Rochelle *et al.* (1989)
E. coli strains/antibiotic-resistance plasmids	Liquid matings Rich media	8–37 °C	10^4 increase in T/D over temp. range	Fernandez-Astorga *et al.* (1992)
	Liquid matings Distilled water	8–37 °C	Constant T/D over the temp. range	
Pseudomonas species/RP4	Steady state annular biofilm reactor in minimal media with 35 mg acetate l^{-1}	15–28 °C	T/R was 100× lower at 15 than at 28 °C, although donor and recipient concentrations were the same at the two temps	Ehlers & Bouwer (1999)

Notes: *T, Transconjugant concentration; R, recipient concentration; D, donor concentration.

Table 2. Studies demonstrating a positive effect of nutrients on frequency of conjugation

Donor::recipient/plasmid	Mating conditions	Nutrient concn*	Remarks*	Reference
E. coli strains/R plasmid	Sterile soil Stream water	With or without nutrient broth amendment	Conjugation observed only after nutrient amendment	Trevors & Oddie (1986)
Epilithic bacteria/mercury-resistance plasmid pQM1	River stones	Rich media or river water	T/R 10× higher in rich media	Bale et al. (1987)
P. cepacia strains/pR388::Tn1721	Soil slurries	Rich media or minimal media	Transfer frequency 100–1000× higher in rich media	Walter et al. (1989)
Streptomyces/pIJ303 (thiostrepton resistance)	Soil	With or without nutrient amendment	T/R 10³–10⁴× higher with nutrient amendment, but only when soil moisture content is 20 %	Bleakley & Crawford (1989)
P. fluorescens strains/RP4	Soil	With or without nutrient amendment	Conjugation observed only after nutrient amendment	Van Elsas et al. (1989)
E. coli::R. fredii/pBLK1–2(Tn5)	Sterile soil	Various amend-ments with organic matter	5 % organic matter supported 3× more conjugation than 0 %; greater than 5 % showed no improvement	Richaume et al. (1989)
E. coli::Alcaligenes eutrophus/metal-resistance plasmids	Sterile soil Nonsterile soil	With or without COD addition With or without COD addition	Addition of 1 mg COD l⁻¹ increased transfer frequency 1000× Transfer detected only after nutrient addition	Top et al. (1990)
E. coli strains/antibiotic-resistance plasmids	Liquid matings	1–11510 mg TOC l⁻¹	No change in T/D from 1 to 1139 mg TOC l⁻¹; at 11510 mg TOC l⁻¹, T/D is 10× higher	Fernandez-Astorga et al. (1992)

Notes: *T, Transconjugant concentration; R, recipient concentration; D, donor concentration; COD, carbonaceous oxygen demand; TOC, total organic carbon.

Having noted the importance of nutrients, however, it appears that conjugation is possible in the complete absence of either internal nutrient stores or external nutrients. Goodman *et al.* (1993) demonstrated the conjugal transfer of plasmid RP1 between strains of *Vibrio* and *E. coli* under oligotrophic marine conditions. The donor and recipient bacteria were starved of nutrients for at least 15 days prior to mating and no nutrients were provided during the 24 h mating. Similar experiments were performed by Dahlberg *et al.* (1998), in which cells were starved in nutrient-poor seawater for 24 h prior to conjugation experiments. Subsequent 3-day matings on filters floating in nutrient-free artificial seawater revealed conjugation frequencies between 2.5×10^{-2} and 3.4×10^{-1} T/D. Finally, some studies have not shown an association between increased nutrient levels and conjugation frequencies. For example, Hausner & Wuertz (1999) observed no difference in conjugation frequency (expressed as T/R/hour) for matings on a glass slide submerged in 1% Luria Broth (LB) versus 100% LB. Thus although nutrients generally enhance frequencies of conjugation, in some cases they are not an absolute requirement.

Selection pressure. It has been observed that bacteria from polluted environments contain more plasmids than those from nonpolluted areas (Grimes *et al.*, 1988). Presumably, these plasmids contain genes that enable the bacteria to survive in harsh environments, e.g. genes encoding heavy-metal resistance or xenobiotic degradation. The presence of contamination is a type of *selection pressure* – an environmental condition that promotes the appearance or 'selection' of bacteria with certain traits, often plasmid-encoded. There are two mechanisms by which selection pressure increases plasmid density in a bacterial population: (1) by increasing the relative growth rate of the plasmid-bearing bacteria or (2) by increasing plasmid transfer among the population. The latter mechanism is dependent on a chemically induced increase in the rate of conjugation.

In an attempt to identify possible chemical inducers of conjugation, many studies have consisted of measurements of transconjugant populations before and after addition of a selection pressure. Fulthorpe & Wyndham (1992) studied the effect of selection pressure on distribution of chlorobenzoate catabolic genes in freshwater microcosms. Microcosms dosed with 3-chlorobenzoate, 4-chloroaniline and 3-chlorobiphenyl developed larger populations of chlorobenzoate degraders than microcosms receiving no amendments. Nüßlein *et al.* (1992) observed similar effects of selection pressure during growth of *Pseudomonas* strains carrying a TOL plasmid derivative in activated sludge. The density of transconjugants was highest when xenobiotic compounds were present in the media. Transconjugants resulting from the transfer of the tetracycline-resistance plasmid pRAS1 from *Aeromonas salmonicida* to marine bacteria in a sediment microcosm were shown to increase in concentration by three to four orders of magnitude under selection pressure (Sandaa & Enger, 1994). However, in each of these studies, the authors were unable to differentiate between an increase in the rate of plasmid transfer and selective growth of transconjugants. To distinguish between these phenomena, short experiments must be performed during which transconjugant growth can be neglected. To date, few such experiments have been conducted and few chemical inducers of conjugation have been clearly identified.

Bacterial concentration. The relative concentrations of donor and recipient bacteria substantially affect the formation of transconjugants in both liquid and surface environments. Optimal ratios of donor to recipient generally range from 10:1 to 1:100. The optimum ratio is dependent on the mating time since secondary transfer can become important in longer experiments. Gauthier *et al.* (1985) found that 1:1 (D:R) ratios produced optimal mating frequencies after 2, 5 and 24 h liquid matings between marine bacteria and *E. coli*. O'Morchoe *et al.* (1988) found that an excess of recipients in mating mixtures of *P. aeruginosa* in freshwater chambers increased mating frequencies by 100. Filter matings between epilithic bacteria occurred maximally over a range of ratios (0.4 to 30 D:R) (Rochelle *et al.*, 1989).

Overall cell concentration is also important in determining the extent of plasmid transfer. In dense cultures, the likelihood of encountering a mate is high compared to dilute cultures. Rochelle *et al.* (1989) found that conjugation frequency on surfaces was optimal at initial and final cell concentrations of 10^5 and 10^8 c.f.u. cm^{-2}, respectively, and that no conjugation was detected when initial cell concentration was below 10^2 c.f.u. cm^{-2}. These trends are supported by Simonsen (1990), who also found that frequencies of conjugation on agar plates increased with increasing initial cell density. She showed that plates inoculated with dense cultures of donors and recipients yielded more transconjugants than plates inoculated with dilute cultures, although total cell concentration at the conclusion of mating was always 5×10^8 c.f.u. ml^{-1}. This was hypothesized to be because donor and recipient bacteria grow in relative isolation on dilute plates, forming microcolonies of each strain. When the bacterial concentration finally is large enough to allow contact between donor and recipient cells, only the edges of the microcolonies interact, leaving most of the bacteria uninvolved in the mating. This would not be the case in aqueous-phase matings because of the continual mixing possible in a liquid environment. In support of this, Simonsen (1990) demonstrated that liquid batch cultures inoculated with both low and high initial concentrations of donor and recipient bacteria gave identical numbers of donor, recipient and transconjugant bacteria after 24 h.

Bacterial growth phase. In liquid batch cultures, bacteria pass through several phases of growth including lag, exponential, stationary and decay phases. Cellular physiology and growth rate vary from phase to phase, with growth occurring predominantly during the exponential phase. Because both growth and conjugation require replication of DNA, the fastest rate of conjugation would be expected to occur during exponential phase. This was demonstrated by Levin *et al.* (1979) for three plasmids in batch cultures of *E. coli* and by MacDonald *et al.* (1992), which showed that plasmid transfer was highest when donor cells were harvested during the early exponential phase compared to stationary phase. Very different results were reported by Muela *et al.* (1994) for conjugation of *E. coli*. They found transfer frequencies to be *lowest* during early exponential phase and highest during late exponential and early stationary phases of the donor strain. They hypothesize that this is part of a strategy employed by plasmid-carrying bacteria to ensure the survival of genetic information under poor growth conditions. In addition, Levin *et al.* (1979) also observed conjugation during lag phase in the absence of parental strain growth, demonstrating that cell growth is not a prerequisite for plasmid transfer via conjugation. Extending these results to attached-growth systems is tenuous because it is difficult to define a growth phase for biofilm bacteria. However, it is clear that cellular physiology (of the donor strain especially) is important in determining the extent of conjugation.

Conjugation in aqueous phase systems

Liquid medium has been a traditional environment for conducting bacterial matings in the laboratory. Donor and recipient strains are cultivated independently and then combined for a defined period of time, either in batch or chemostat cultures. In general, the highest frequencies are obtained when high concentrations of bacteria are used at nutrient, temperature and pH values ideal for growth of the organisms. The advantages of liquid matings include (1) an optimal number of cell encounters, especially when cultures are well mixed, (2) ease of measurement of bacterial populations and (3) mathematical simplicity to aid in modelling efforts. When optimized, some liquid matings can produce 100% transfer, meaning that every available recipient receives a copy of the conjugative plasmid (e.g. Licht *et al.*, 1999). As described below, conjugation under laboratory conditions generally occurs at frequencies higher than conjugation in natural aquatic environments such as wastewater treatment plants, fresh waters and marine environments.

Wastewater. Wastewater treatment plants are nutrient-rich environments containing high numbers of bacteria representing a wide variety of genera (McClure *et al.*, 1990). In the future, they may receive GEMs with enhanced degradative capabilities, necessitating an investigation of their gene transfer potential. Horizontal gene transfer in wastewater treatment systems is highly probable. Altherr & Keswick (1982) observed antibiotic resistance transfer between strains of introduced *E. coli* in membrane diffusion chambers submerged in treatment tanks. Transfer frequencies were two orders of magnitude lower than under laboratory conditions. In similar experiments, Mach & Grimes (1982) demonstrated R-plasmid transfer between introduced donor and recipient strains in the primary and secondary settling tanks of a wastewater treatment plant. Mean transfer frequencies were approximately 4.9×10^{-5} (T/D), again about 100-fold lower than similarly timed experiments conducted under laboratory conditions. Wastewater and rich media were compared for their ability to support conjugation during three-way matings (Gealt et al., 1985). Introduced donor and helper strains were able to transfer the plasmid pBR322 to indigenous wastewater bacteria, with transfer frequencies being several orders of magnitude lower in sterilized wastewater than in rich media.

Laboratory-scale activated sludge tanks (continuous-flow microcosms) have been used by several researchers to understand gene transfer processes in wastewater. The transfer of a mobilizable 3-chlorobenzoate-degradative plasmid from *Pseudomonas putida* to indigenous bacteria in a wastewater microcosm was demonstrated by McClure et al. (1991). In similar experiments, Nüßlein et al. (1992) observed transfer of a TOL plasmid derivative between *Pseudomonas* strains introduced into an activated sludge microcosm. Conjugation frequencies in the microcosm were quite high (T/D = 10^{-1}) and were only 10-fold lower than in laboratory matings. Geisenberger et al. (1999) measured conjugal transfer of plasmid RP4 from introduced donors into indigenous bacteria in activated sludge microcosms. They were able to identify transconjugants *in situ* by using an RP4 derivative tagged with green fluorescent protein. Finally, Ashelford et al. (1995) conducted conjugation experiments in a pilot-scale percolating filter system, similar to a trickling filter used for wastewater treatment. *P. putida* bacteria carrying an IncP plasmid, pQKH6, were observed to conjugate in this flow-through system at frequencies around 6.43×10^{-5} (T/R). These studies suggest that wastewater treatment plants are active sites of conjugation and may be suitable for developing treatment strategies that rely on plasmid transfer.

Freshwater. Agricultural water is particularly important as a potential receiving body of GEMs, and it has been extensively used as a medium for plasmid transfer by Trevors and colleagues. In stream water and agricultural water, conjugation of RP4 between introduced *Pseudomonas fluorescens* was only observed after the addition of nutrients (Trevors & Oddie, 1986; Trevors et al., 1990). Failure to detect transconjugants was blamed on the poor survival of the introduced strains in the freshwater media. In other freshwater gene transfer studies, O'Morchoe et al. (1988) observed conjugation between introduced *P. aeruginosa* strains in a lake environment. Liquid matings in rich media and sterile lake water were comparable in frequency (T/D), while matings in nonsterile lake water were 10–100-fold lower. Fulthorpe & Wyndham (1991) studied the transfer of the catabolic plasmid pBRC60 from an introduced *Alcaligenes* strain into indigenous bacteria in a freshwater mesocosm containing both water and sediment. Transfer frequencies of 5×10^{-4} (T/D) were obtained and were comparable to the frequencies of laboratory filter matings. Finally, transfer of plasmids between introduced *E. coli* strains has been investigated in rich media and river water (Muela *et al.*, 1994; Arana *et al.*, 1997). Transfer frequencies in rich media were approximately 1×10^{-3} (T/D); those in river water were comparable for the first 12 h of incubation before plummeting below the detection limit. Thus conjugation in some natural aqueous systems can match levels of conjugation seen in optimal laboratory experiments, especially under conditions that allow for survival of the parent strains.

Seawater. Saline environments are capable of supporting detectable levels of conjugation (Angles *et al.*, 1993; Goodman *et al.*, 1993; Dahlberg *et al.*, 1998). Dahlberg *et al.* (1998) demonstrated conjugal plasmid transfer from *P. putida* to indigenous marine bacteria in artificial seawater, filtered sterile seawater and natural seawater. Frequencies of conjugation were lowest for natural seawater, with T/R ranging from 3.0×10^{-6} to 5.6×10^{-5}. High ionic strength, an important characteristic of seawater, has been shown to affect plasmid transfer in a handful of studies, although the findings demonstrate no consistent trends. Singleton & Anson (1983) showed that the transfer of a derepressed plasmid between *E. coli* was enhanced by NaCl concentrations common in estuaries. Gauthier *et al.* (1985) found that frequency of conjugation of a mercury-resistance plasmid from a marine pseudomonad to *E. coli* increased with salinity to an optimum value at 3% salinity. Venables *et al.* (1995) found that conjugation was completely inhibited above a salinity of 5%.

Conjugation in soils and sediments

Soils and sediment are natural environments in which both suspended and attached bacterial growth occurs. However, because of their heterogeneity, they are not typically thought of as biofilm systems in which the activities of attached bacteria can be clearly isolated and studied. Because of the potential impact of gene transfer from GEMs to indigenous soil microbes, soils have been intensely investigated and found to support substantial levels of conjugation. This paper briefly reviews these studies because they contribute to our understanding of bacterial conjugation in natural systems. It should be recognized that in most of the studies, the activities of biofilm bacteria have not been distinguished from those of planktonic bacteria, and many experiments have been performed in a soil or sediment microcosm that contains an overlying water column. Nevertheless, conjugation frequencies measured in soil are commonly compared to frequencies measured during mating on agar plates, the classic solid surface used to optimize conjugation in laboratory settings.

Rafii & Crawford (1988) compared the frequency of conjugation of plasmid pIJ303 on agar plates and in sterile soil. The ratio of transconjugants to recipients was approximately 30% on agar slants and only 1% in soil. Mobilization of a nonconjugative plasmid was also demonstrated, with frequencies in soil being two orders of magnitude lower than on agar. The TOL plasmid was shown to transfer among several genera of Gram-negative bacteria in agar and sterile soil (Ramos-Gonzalez *et al.*, 1991). Only intrageneric matings were successful in soil, possibly due to the poor survival of some strains. Plasmid mobilization between strains of *E. coli* was found to occur at lower frequencies (10–100-fold) in soil than in liquid matings (Selvaratnam & Gealt, 1992).

Moisture content. The influence of moisture content on plasmid transfer has been investigated by a number of researchers. Bleakley & Crawford (1989) observed antibiotic resistance transfer between strains of *Streptomyces* in sterile silt loam. Soil with 20% water-holding capacity supported the highest transfer frequencies. The authors hypothesized that dry soils possess frequent microsites where bacteria are in proximity, favouring conjugation. The identical moisture content was found by Lafuente *et al.* (1996) to be ideal for RP4 plasmid transfer between strains of *Rhizobium* in sterile soil microcosms. An even lower moisture content (8%) was found to be optimal for plasmid transfer between *E. coli* and *Rhizobium fredii* (Richaume *et al.*, 1989). Increasing the moisture content to 20% resulted in lower frequencies of conjugation even though the donor and recipient strains became more numerous.

Indigenous soil organisms. Both positive and negative effects of indigenous soil organisms on plasmid transfer have been observed. Several studies have demonstrated a decrease in conjugation frequencies as one moves from agar matings to sterile soil and finally to nonsterile soil (Top *et al.*, 1990; Smit *et al.*, 1993; Neilson *et al.*, 1994; Kozdrój, 1997). Neilson *et al.* (1994) postulated that low transfer frequencies observed in nonsterile soils were the result of dramatic decreases in the concentration of donor and recipient strains caused by predation from indigenous organisms. However, not all indigenous organisms are detrimental to conjugation. Walter *et al.* (1989) could find no statistical difference between conjugation frequencies obtained in sterile and nonsterile soil. Daane *et al.* (1996, 1997) observed *increased* levels of plasmid transfer in soil after the addition of earthworms. The increase was attributed to the mixing action of the burrowing earthworms, which increased cell contact.

Clay content. The presence of clay has been found to affect conjugation frequencies in soil. Because clay retains water better than sand or silt, soil coated with clay is frequently inhabited by bacteria. Stotzky & Babich (1986) observed an enhancement of conjugation between *E. coli* in soil after the addition of montmorillonite. Van Elsas & Trevors (1990) were able to detect transfer of RP4 between *P. fluorescens* strains in nonsterile soil only after the addition of 10% bentonite clay. A range of clay content (5–50%) was evaluated for its effect on plasmid transfer by Richaume *et al.* (1989), with optimal frequencies obtained at 15% montmorillonite. Below 15%, the high specific surface area of clay was thought to contribute to its beneficial effects by promoting cell sorption and cell–cell contact. At clay contents above 15%, however, excessive cell immobilization was hypothesized to decrease conjugation frequency. In contrast, Singleton (1983) found that small clay particles (0.22 μm diameter) greatly reduced conjugation in liquid matings by effectively coating the bacteria and hampering their contact. Thus the effect of clay on conjugation depends on clay concentration, mean particle size, and whether clay is adsorbed or in suspension.

GENE TRANSFER IN BIOFILM SYSTEMS

It has been known for some time that optimal frequencies of conjugation often occur on solid surfaces (for poignant example, see Genther *et al.*, 1988). This has led to the extensive use of agar plates with or without filters for performing laboratory matings. Generally, however, the goal of such matings is to produce a transconjugant with unique properties, not to quantify or optimize the mating procedure. Why agar plates and other solid surfaces are favourable for conjugation has only recently been investigated.

There appear to be several advantages of solid media over liquid in promoting gene transfer via conjugation. Attachment to a surface may stabilize mating-pair formation. Bradley *et al.* (1980) demonstrated this for conjugation of plasmids encoding thick, rigid pili [incompatibility group (Inc) M, N, P and W plasmids]. Such plasmids were found to transfer more than 2000 times more efficiently on agar plates than in liquid media, while plasmids encoding thin, flexible pili (IncI and IncK plasmids) transferred equally well in both media. Cell–cell contact time may be increased on solid media. Although well-mixed liquid medium supports more encounters, it may not provide enough time for mating to occur. Kröckel & Focht (1987) attributed their ability to form rare transconjugants in a glass bead column, but not in a chemostat, to the increased cell residence time in the column. Attached bacteria may experience physiological changes that increase their natural rate of conjugation. Cells attached to beads were found to have better plasmid maintenance than planktonic cells (de Taxis du Poet *et al.*, 1986).

One of the difficulties with characterizing conjugation in surface environments is their complex mathematical description. Modelling transfer on agar plates is formidable and rarely attempted (see Simonsen *et al.*, 1990). The use of other surface growth systems, such as biofilm reactors, for studying conjugation has only recently been developed. In general, though, the frequencies of conjugation achieved in those systems are less than those achieved on agar plates. The handful of gene transfer studies done in well-defined biofilm reactors is discussed shortly.

Biofilm structure

Bacteria in natural aquatic environments have a marked tendency to interact with surfaces, leading to the formation of a biofilm. A biofilm is a group of cells immobilized at a substratum and frequently embedded in an organic polymer matrix of microbial origin (Characklis & Marshall, 1990). The term can also apply to aggregates of cells in suspension. Bacteria in almost every natural environment have developed ways to colonize surfaces. Depending on the environment, attached bacteria can outnumber suspended (planktonic) bacteria by four orders of magnitude (Costerton *et al.*, 1987). Not only are sessile cells more predominant, they also constitute the bulk of the bacterial metabolic activity observed in natural systems. Biofilms form on both natural and industrial surfaces in aquatic systems, including the human body, river stones, ship hulls and the pipes of water distribution systems. Deleterious effects of microbial growth on industrial surfaces include corrosion of metal pipes, prevention of efficient heat exchange and clogging of filter beds. Pathogenic biofilms that grow on tissue and medical materials intended for implantation are particularly problematic because of their resistance to conventional antibiotic treatment. Many biofilms, however, are beneficial to society. Biofilm bacteria have been used for years in wastewater treatment plants in both trickling filters and rotating biological contactors. Subsurface biofilms have been shown to biodegrade trace contaminants and are exploited during *in situ* bioremediation.

The development of *in situ* confocal scanning laser microscopy (CSLM) has significantly enhanced our understanding of biofilm structure (Caldwell *et al.*, 1992). As shown in the cross-section in Fig. 4, a biofilm can be conceptualized by several layers, including a substratum, base biofilm, surface biofilm, bulk liquid phase characterized by flow, and gas phase. Viewed areally, biofilms have historically been thought to be a homogeneous layer of bacteria of constant density evenly distributed over a surface. However, new microscopic techniques have revealed changes in biofilm density with depth (Lawrence *et al.*, 1991), extensive surface roughness, and heterogeneous fractal-like patterns in all dimensions (Zahid & Ganczarczyk, 1994). Biofilms are now known to have deep channel structures that permit convective flow of water throughout the biofilm. The term 'biofilm architecture' has been coined to describe this structural complexity, which is thought to facilitate cell–cell communication, exchange of nutrients and metabolic specialization (Costerton *et al.*, 1995). In fact, knowledge of biofilm structure has led some to postulate that the evolutionary success of bacteria may be a consequence of the physiological flexibility afforded by growth in a biofilm (Costerton *et al.*, 1995).

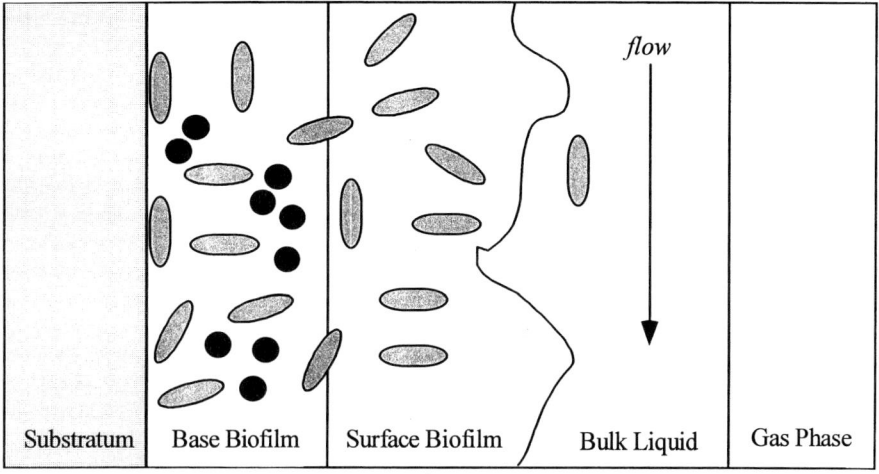

Fig. 4. Conceptual biofilm cross-section. A vertical cross-section through two-species biofilm illustrates the phases of biofilm structure. The substratum provides the solid support for biofilm formation. The homogeneous layer of bacteria adjacent to the substratum is known as the base biofilm. For modelling purposes, it is often assumed to be of constant volumetric and areal density. The surface biofilm can exhibit considerable heterogeneity including channels and pores. Detachment from the biofilm is assumed to occur only in the surface biofilm. The bulk liquid phase abuts the surface biofilm and provides the biofilm with nutrients while taking away waste products. In some cases, a gas phase exists from which oxygen, carbon dioxide and other compounds important for biofilm growth are exchanged with the liquid phase. As illustrated in this figure, distinct species within biofilms are sometimes spatially separated from one another.

Biofilm versus planktonic cells

There are major physical and biological differences between cells grown in suspension and those in a biofilm. Some of the unique physiological characteristics of biofilm cells have been correlated with differential gene expression of the bacterial genome. The best example is the production and secretion of polymers by biofilm cells. The exopolysaccharide matrix surrounding biofilm bacteria provides a sticky, protective layer through which chemicals must diffuse. Bacteria attached to glass surfaces have been shown to express 19 times more alginate biosynthesis genes, which are responsible for exopolysaccharide production, than suspended bacteria (Davies & Geesey, 1995). Other examples of differential gene expression in biofilm cells are sure to follow, as 30% of cellular proteins have been found to be differentially expressed in biofilm and planktonic bacteria of the same species (Costerton *et al.*, 1995). Differential gene expression may be controlled by a recently discovered σ factor similar to that which regulates sporulation in Gram-positive bacteria (Deretic *et al.*, 1994). These studies suggest that bacteria possess a sophisticated mechanism for adapting to changing environments by switching between planktonic and biofilm growth. It should be noted that for most modelling purposes, biological reactions of biofilm bacteria are assumed to be identical to those of planktonic bacteria, and differences between biofilm and suspended systems are assumed to result from the substantial mass transfer and diffusional resistances encountered by chemicals within the biofilm.

An example of such metabolic flexibility is the bacterial response to oligotrophic growth conditions. Biofilms only form in nutrient-rich environments because polymer production and secretion, which initiate and maintain biofilm formation, are energy-intensive processes. This does not mean, however, that biofilms are not found in oligotrophic environments. If nutrient conditions improve temporarily, planktonic starved cells will produce exopolysaccharides and attach to surfaces, forming food reserves in the biofilm. When oligotrophic conditions return, cells in the biofilm will persist on the entrapped nutrients and exopolysaccharides.

Another important difference between suspended and biofilm cells is the relative hydrophobicity of bacteria growing in biofilms. Hydrophobic forces are thought to aid in the initial attachment of bacteria to surfaces by overcoming the electrostatic repulsion present between negatively charged cells and surfaces. Although conclusive work remains to be done, recent studies suggest that bacteria may alter their surface hydrophobicity in response to environmental changes, thereby promoting or discouraging biofilm development. Cells that detach from biofilms were shown to be significantly more hydrophilic than bacteria remaining in the biofilm (Allison *et al.*, 1990). Ascon-Cabrera *et al.* (1995) observed an increase in biofilm cell hydrophobicity as the dilution rate in a complete-mix reactor increased, suggesting that biofilm bacteria 'respond' to increased fluid shear forces by strengthening their hold on surfaces (thought to be mediated by hydrophobic forces).

A major consequence of the physiological differences between planktonic and biofilm bacteria is the increased resistance of biofilms to disinfectants and antibiotics. For example, biofilms of *P. aeruginosa* were found to be entirely resistant over 7 h to concentrations of tobramycin that produced complete killing of suspended cells in 2 h (Anwar *et al.*, 1989). Because of the slow diffusion of compounds through the biofilm matrix, one would expect that the time necessary for killing biofilm bacteria would be longer than for suspended cells. Another factor contributing to the resistance of biofilms to chemical treatment is the entrapping and neutralization of chemicals by exopolysaccharides surrounding the bacteria. For both chlorine and povidone-iodine, reactions with extracellular material within biofilms have been shown to reduce their penetration and effective concentrations (DeBeer *et al.*, 1994; Brown *et al.*, 1995).

Conjugation in biofilms

Gene transfer among biofilm bacteria has only recently caught the attention of scientists conducting studies of biofilm structure and function. After a biofilm conference in 1989, it was proposed that 'frequencies of gene transfer in biofilms should be investigated in pure cultures, mixed cultures, and the natural environment' (Stal, 1989). Researchers have speculated that gene transfer between biofilm bacteria, particularly via conjugation, takes place at enhanced levels (Bauda *et al.*, 1990), and evidence existed from the high frequencies of conjugation observed in agar plate matings. However, until 1987 there were no attempts to quantify gene transfer in reactors that were realistic representations of natural biofilms. In particular, there were no studies of bacterial conjugation in a system of attached growth that includes an overlying, flowing water column. The section below discusses recent studies on conjugation in biofilm systems ranging from river stones to complex biofilm reactors.

Plants. Plant surfaces were some of the first to be investigated for their ability to support conjugation. This research has been motivated by a need to better understand plant-associated plasmid transfer that might accompany the use of microbial pesticides and other GEMs intended for agricultural applications. Strains of *Xanthomonas* and *Erwinia herbicola* were able to conjugally transfer plasmid RP4 after inoculation in hazelnut tissues (Manceau *et al.*, 1986), and Knudsen *et al.* (1988) observed the appearance of transconjugants after strains of *Pseudomonas cepacia* were applied to plant leaves and root systems. Wheat roots were found to enhance RP4 conjugation in soil (Van Elsas & Trevors, 1990), although the mechanism of enhancement was not known. Extremely high frequencies have been measured for conjugation among strains of *Pseudomonas* on the leaves of bush bean plants. Bjorklof *et al.* (1995) recorded a conjugation frequency of T/D = 0.1 between introduced *Pseudomonas syringae*, while Normander *et al.* (1998) measured conjugation frequencies after 96 h as high as T/R = 0.3 among *P. putida*. In the latter study, conjugation frequencies on bush bean leaves were as much as 30 times higher than on control filters. Plasmid transfer was most favourable under high humidity, apparently due to leaching of nutrients on the leaf surface, which stimulated conjugation, and in leaf microniches. As in many of the biofilm experiments discussed below, transconjugant populations levelled off at a concentration less than the recipient concentration. Although plant leaves and roots clearly support the attached growth of bacteria and conjugation, these systems do not include a free-flowing liquid phase and thus are not directly comparable to the biofilm systems discussed below.

River stones. River stones have been shown to support detectable levels of conjugation. To simulate a biofilm system, Fry and colleagues filtered donor and recipient bacteria onto membranes that were placed face down on sterile or nonsterile stones and submerged in river water. Initial experiments demonstrated the dependence of conjugation on the presence of indigenous epilithic bacteria: transfer rates were significantly higher on scrubbed, sterile stones than on unscrubbed, nonsterile stones (Bale *et al.*, 1987). Frequencies of conjugation were generally three orders of magnitude lower on river stones than on agar plates, and there were no attempts to compare transfer rates on the stones to rates in suspension.

These studies were extended to evaluate the effect of fluid flow on conjugation by comparing quiescent batch and continuous-flow reactors containing membrane-coated stones (Hill *et al.*, 1994). Transfer of the catabolic plasmid pD10 was observed in both types of reactors under oligotrophic conditions and no selective pressure. Although no information was provided on the fluid shear stress, reactor flow rate or growth rates of the bacteria, these initial experiments demonstrated that conjugation was possible among bacteria attached to surfaces in the presence of a flowing aqueous phase.

Submerged filters, slides and beads. Several other investigators have also employed filters, slides and beads submerged in liquid media to study bacterial conjugation. Angles *et al.* (1993) observed the transfer of the broad-host-range plasmid RP1 in reactors filled with either hydrophilic or hydrophobic glass beads. Donor-strain cells were inoculated into reactors containing a recipient-strain biofilm and allowed to interact under no-flow conditions for 24 h. The frequency of conjugation (T/RD) in the biofilm (comprised of loosely and tightly attached cells) was one to two orders of magnitude higher than the frequency in the surrounding aqueous phase. In addition, controls excluding glass beads, but with comparable donor and recipient numbers, contained no transconjugants, implying that those found in the aqueous phase surrounding the biofilm probably originated in the biofilm and subsequently detached. If the tightly bound transconjugants alone are considered, then the frequency of conjugation (T/RD = 2.9×10^{-10}) was as much as three orders of magnitude higher than in suspension. Interestingly, matings conducted on traditional plates produced no detectable transconjugants. The results of this study implied that transfer rates in biofilm systems may be significantly elevated over those in suspended systems. However, this study was conducted in relatively rich media under no-flow conditions.

Dahlberg *et al.* (1998) observed conjugation between a *P. putida* strain and different seawater-indigenous recipient strains. Donor and recipient bacteria were simultaneously loaded on filters that were then submerged in media for 1–3 days. Although the liquid media surrounding the slides was nutrient-poor (artificial seawater, sterile filtered seawater or seawater), rates of conjugation as high as 3.4×10^{-1} T/D were observed. As in the Angles *et al.* (1993) study, the rates of conjugation measured on the filters were higher than rates measured for bacteria in the bulk liquid, suggesting that attached-growth systems favour conjugation. These experiments took advantage of noninvasive fluorescent detection techniques to monitor the transconjugant population *in situ* over time. As discussed below, other researchers have subsequently used these techniques to distinguish between transconjugant growth and plasmid transfer.

Hausner & Wuertz (1999) investigated plasmid transfer on slides under somewhat different conditions. In these experiments, recipient bacteria were grown on a slide to form a biofilm, which was then submerged in either 100% or 1% LB. Donor and helper cells were added and allowed to contact the biofilm for either 2 or 24 h. Conjugation frequencies ranged from 2.5×10^{-4} to 3.4×10^{-3} (T/R/hour). Interestingly, there was no significant difference between conjugation rates in 100% versus 1% LB, although the biofilm morphology under the two conditions was clearly different. The most intriguing aspect of this work was the use of CSLM and fluorescent techniques to observe bacterial populations *in situ* over time. Under 100% LB conditions, transconjugants were detected mainly as single cells, while at 1% LB, transconjugants were detected as clusters located in distinct regions of the slide. These clusters were speculated to have favoured plasmid transfer under lower nutrient conditions because of the close proximity of cells.

None of the experiments described above utilized flow-through biofilm systems in which planktonic bacteria can be washed out. For this reason, it is impossible to conclude that the observed results were due entirely to conjugation in the biofilm rather than conjugation in the aqueous phase, although evidence points to this conclusion in some cases. In the experiments described below, conjugation is observed in biofilm systems with a flowing liquid phase.

Flow channels. Licht *et al.* (1999) observed conjugation between *E. coli* strains harbouring the derepressed plasmid R1drd19 in three systems: a chemostat, a simple, continuous-flow biofilm reactor and mouse intestine. For both the chemostat and biofilm reactor experiments, a steady state population of the recipient strain was established in the reactor before the donor strain was introduced. All populations of bacteria were then measured over 15 days as the reactors were run on continuous-flow mode, using minimal media with glycerol as the sole carbon source. In the biofilm reactor, the flow rate was sufficient to wash out any planktonic cells, such that all cells detected in the effluent were derived from the biofilm.

In the chemostat, plasmid transfer proceeded as predicted by a simple mass balance model, i.e. the transconjugant population increased until all of the recipients were converted into transconjugants (Fig. 5). In the biofilm reactor, the transconjugant population increased extremely rapidly but never reached or exceeded the concentration

(a)

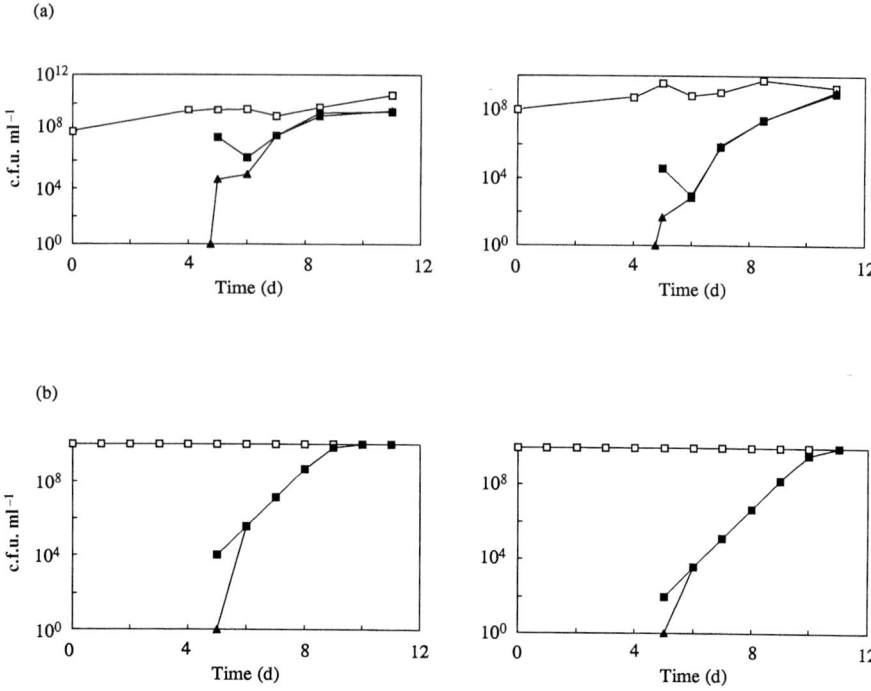

Fig. 5. Conjugation in a chemostat. (a) Chemostat mating experiments in minimal medium supplemented with glycerol (reproduced from Licht *et al.*, 1999, with permission). Recipients were introduced at day 0. At day 5, either 10^9 (left) or 10^6 (right) donor cells were introduced 6 h before sampling of the effluent. Recipients + transconjugants (□), donors + transconjugants (■) and transconjugants (▲) were quantified by plating of effluent samples on selective plates. (b) Mathematical prediction of transfer kinetics in a chemostat, where the bacteria are subjected to perfect mixing. Numbers of recipients + transconjugants (□), donors + transconjugants (■) and transconjugants (▲) were calculated according to Licht *et al.* (1999).

of the recipients, and mean conjugation frequencies were 10^{-4} (T/R) (Fig. 6). Results from the mouse intestine were similar to those observed in the biofilm rather than the chemostat. The authors hypothesized that conjugation in the biofilm reactor and mouse intestine was limited by the spatial separation between donors and recipients; that is, after donor cells attach to the biofilm and encounter the first layer of recipients, their further penetration is limited. In addition, their results showed that newly formed transconjugants do not immediately transfer the plasmid to adjacent biofilm cells. Thus this study suggests that plasmid transfer in biofilms is initially more rapid than in aqueous-phase systems, perhaps because mating-pair formation is stabilized, but eventually limited in extent due to spatial separation of plasmid-bearing and plasmid-free cells.

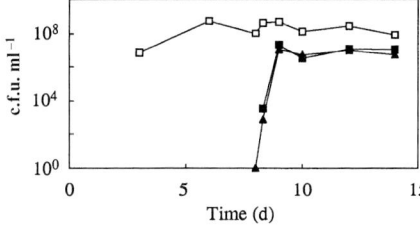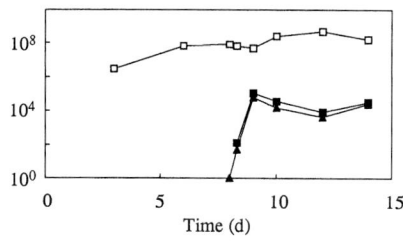

Fig. 6. Conjugation in a biofilm. Biofilm mating experiments in minimal medium supplemented with glycerol (reproduced from Licht *et al.*, 1999, with permission). Recipients were introduced at day 0. At day 8, either 2×10^8 (left) or 2×10^4 (right) donor cells were introduced 5 h before sampling of the effluent. Recipients + transconjugants (□), donors + transconjugants (■) and transconjugants (▲) were quantified by plating of effluent samples on selective plates.

Experiments conducted by Christensen *et al.* (1998) using the identical biofilm system shed further light on these conclusions. Conjugation of the TOL plasmid was observed between different strains of *P. putida* in a biofilm reactor. Like the Licht *et al.* (1999) experiments, a steady state population of the recipient strain was established in the reactor before the donor strain was introduced. All populations of bacteria were then measured over time (15 days) as the reactors were run on continuous-flow mode, using minimal media with benzyl alcohol as the sole carbon source. The conjugation frequencies observed over time varied considerably, depending on the ability of the donor strain to survive in the biofilm. In particular, when the donor became well-established in the biofilm, transconjugant numbers were low. When the donors did not survive well in the biofilm, the transconjugant population increased quickly, with conjugation frequencies between 10^{-3} and 10^{-1} (T/R). In another set of experiments, the recipient population was varied by several orders of magnitude. Transconjugants only became well-established at the highest recipient concentrations.

Fluorescence microscopy was used to visualize the bacterial populations and help interpret the biofilm data. First, the recipient strain biofilm was not uniformly spread over the substratum but consisted of distinct clusters of bacteria protruding from the substratum. After an initial plasmid transfer event, a small microcolony of transconjugants was observed to form. Most of these transconjugant microcolonies were found on top of recipient microcolonies rather than away from recipient clusters. There was no widespread plasmid transfer throughout an entire microcolony of recipients. These observations led the authors to speculate that most transconjugants arise from growth of microcolonies rather than from conjugation. As in other experiments, conjugation in the biofilm reactor was shown to be limited in extent and influenced by the locations of donor, recipient and transconjugant populations.

Annular biofilm reactor. The most complex studies of conjugation in biofilm reactors to date have been conducted in an annular biofilm reactor (see Chen *et al.*, 1993, for a description). This system affords experimental flexibility because flow rate, shear stress, nutrient levels and substratum materials can be controlled independently. Ehlers & Bouwer (1999) used the annular biofilm reactor to compare rates of conjugation between species of *Pseudomonas* harbouring the broad-host-range plasmid RP4 in liquid, agar plate and biofilm environments. Temperature, acetate concentration and fluid shear stress were varied to determine their effect on conjugation frequency.

In liquid media, transconjugants were only detected under nutrient-rich conditions with frequencies of 5.23×10^{-7} (T/R). As expected, frequencies of conjugation (T/R and T/D) were 1000–50 000 times higher on agar plates in comparison to liquid media. [RP4 encodes rigid, thick pili, which have been demonstrated to mediate transfer much more efficiently on solid surfaces (Bradley *et al.*, 1980).] A major difference between liquid and agar plate matings is the large fluid shear present in shaking liquid media, suggesting that fluid shear at the biofilm–liquid interface may be a factor in controlling conjugation in the biofilm reactor. To test this hypothesis, the annular biofilm reactor was operated at multiple shear stresses by varying the annular drum speed to observe their influence on the formation of transconjugants. Transconjugants were not detected at an annular drum speed of 50 r.p.m. (0.284 N m^{-2}) but were found at 15 and 6.5 r.p.m. (0.0851 and 0.0369 N m^{-2}). Conjugation frequencies at the lower shear stresses were 1.74×10^{-2} and 1.64×10^{-2} (T/R), respectively, similar to their values on agar plates.

There are several explanations for the observed results. First, strong shear forces may prevent conjugation by disrupting cell–cell contact, perhaps through the breakage of pili. Second, at annular drum speeds of 6.5 and 15 r.p.m., flow within the annular reactor is laminar, while at 50 r.p.m. flow is turbulent. Third, low shear (0.07 N m^{-2}) biofilms have been shown to be thicker, less volumetrically dense, more areally dense, and structurally more heterogeneous than high shear (0.99 N m^{-2}) biofilms, resulting in a significant increase in biofilm surface area (Cao & Alaerts, 1995). Along with the laminar flow regime surrounding the biofilm, these heterogeneous structures may have increased the number of cell–cell contacts within the annular reactor. This study was the first to directly implicate fluid shear stress as a factor controlling conjugation in biofilm systems.

The annular biofilm reactor was also used to study mobilization via cotransfer and retrotransfer (Beaudoin *et al.*, 1998a). Unlike the previous study, the main parameter that was varied was succinate concentration rather than fluid shear stress (no values for fluid shear or annular drum rotation are given). In addition, experiments were conducted over both short timescales to determine primary conjugation rates and long time periods to determine secondary transfer rates. For cotransfer, primary transfer rates were consistently higher than secondary transfer rates. The kinetics of retrotransfer in the biofilm reactor were considerably more complex, with transfer rates initially increasing before dropping to zero.

Because the annular reactor is amenable to mathematical modelling, the above experimental results were used to validate a model of bacterial conjugation in a biofilm reactor (Beaudoin *et al.*, 1998b). Experimental data were first used to fit unknown parameters for bacterial attachment, bacterial detachment, primary transfer rate and secondary transfer rate. These parameters, and other measured or estimated parameters, were then used as input for the AQUASIM program, which accurately simulated bacterial growth, primary transfer and secondary transfer in the annular reactor. In addition, the model predicted spatial distributions of donors, recipients and transconjugants, which may account for the limitations on conjugation previously observed in biofilm systems. In particular, the model predicted a steady state biofilm in which plasmid-bearing and plasmid-free cells are spatially separated and have limited interaction. At steady state, the transconjugant population was predicted to be less than the recipient population.

Alginate beads. Finally, conjugation has been studied in cells immobilized within alginate beads. Although alginate beads are structurally different from slides, filters or the annular biofilm reactor, conjugation within alginate beads has been shown to follow many of the trends observed in other biofilm systems. Mater *et al.* (1999) investigated the transfer of the broad-host-range plasmid pLV1017 between strains of *P. putida*. The beads were formed by mixing suspensions of both donor and recipient cells with alginate and other chemicals. The beads were then placed in either batch reactors or continuous-flow reactors containing LB. The structure of the beads was varied such that some beads were relatively smooth and surface-colonized, while others contained multiple fractures and pores, which were the main sites of colonization. Both batch and continuous-flow cultures supported conjugation, with frequencies around 2×10^{-2} T/R for batch cultures, about 5×10^{-2} T/R for 2% alginate continuous cultures, and about 5×10^{-3} for 1% alginate continuous cultures. As in all of the previous biofilm studies described above, conjugation in alginate beads did not lead to the conversion of all recipient cells into transconjugants.

CONCLUSIONS

Anecdotal evidence from the last 20 years has suggested that rates of horizontal gene transfer via conjugation may be elevated among bacteria attached to surfaces compared to planktonic bacteria. In particular, those plasmids encoding thick, rigid pili are known to transfer thousands of times more efficiently between bacteria grown on agar plates and filters compared to bacteria grown in liquid media. The beneficial effects of surfaces on conjugation are thought to include increased cell contact time and the stabilization of mating pairs.

Biofilm environments contain elements of both liquid media and solid surfaces; that is, bacteria grow attached to a substratum, but are exposed to a flowing liquid phase that may impose considerable fluid shear stress at the biofilm–liquid interface. It is reasonable to suppose that, for certain plasmids, rates of conjugation in biofilm systems may fall between those observed in liquid and solid-surface environments.

This review has shown that conjugation occurs in multiple attached-growth environments at detectable frequencies, including soil, sediment, plant surfaces, river stones, filters, glass slides and beads, and biofilm reactors. In all systems, environmental conditions such as nutrient levels, temperature, pH, salinity and bacterial concentration influence the extent of conjugation.

There are several environmental parameters that are particularly important in controlling conjugation in biofilm systems. First, available studies suggest that conjugation in biofilm systems is limited by the spatial separation of donor, recipient and transconjugant cells. In almost all studies examined, including those conducted on plant leaves (Normander *et al.*, 1998), transconjugant populations never matched recipient concentrations at steady state, nor did all recipient bacteria receive copies of the plasmid. This is in direct contrast to conjugation experiments conducted in chemostats, in which all of the recipients eventually receive copies of the plasmid (Licht *et al.*, 1999). Immobilization of cells on the substratum, growth patterns that result in microcolonies and clusters, and limited numbers of encounters between plasmid-bearing and plasmid-free cells account for the trends observed in biofilm systems. In addition, it is possible that the exopolysaccharide matrix plays some role in limiting conjugation, perhaps by preventing the movement of cells and chemical inducers of conjugation within the biofilm.

Fluid shear is a second environmental factor of particular importance in biofilm systems. Although most studies of conjugation in flowing biofilm systems do not consider the effect of fluid shear, Ehlers & Bouwer (1999) demonstrated that conjugation is favoured in low shear environments characterized by laminar flow. It is not known whether the dependence of conjugation on fluid shear extends to plasmids encoding thin, flexible pili, which have previously been shown to conjugate equally well in liquid and solid-surface environments.

There are undoubtedly factors yet to be identified which strongly influence conjugation in biofilm environments. As discussed above, the physiology of biofilm cells is markedly different from that of planktonic cells. Differential gene expression among biofilm bacteria is likely to affect conjugation in biofilm systems, perhaps by altering expression of *tra* genes or other plasmid-encoded proteins needed to facilitate conjugation. The effects of variable cell physiology need to be better incorporated in models of biofilm growth. Although there is no definitive body of evidence to support the practice, for simplicity current models of biofilm growth usually assume that growth and decay rates among biofilm and planktonic bacteria of the same strain are identical. The studies reviewed here suggest that making similar assumptions for bacterial conjugation in biofilm versus suspended systems is not possible. Models of bacterial conjugation in biofilms will require the development of more accurate estimates of primary and secondary plasmid transfer rates that take such factors as fluid shear, nutrient concentration, the concentration of chemical inducers of conjugation, and temperature into account.

This review considered primarily conjugation, thought to be the predominant mode of horizontal gene transfer in the environment. Results from the relatively small body of work focused on transformation and transduction suggest that these mechanisms are not occurring at frequencies comparable to conjugation, although such mechanisms have yet to be rigorously studied in biofilm systems. However, it is too early to discount their contribution to overall horizontal gene transfer frequencies, especially given the difficulties inherent in measuring these processes in environmental samples. It is likely that combinations of conjugation, transduction, transformation and transposition operate in concert, both in suspended and attached-growth systems, accomplishing significantly higher levels of gene transfer than any single mechanism.

ACKNOWLEDGEMENTS

Drafting of this paper was supported by funds from an Office of Naval Research graduate fellowship and an Environmental Protection Agency STAR Fellowship.

REFERENCES

Aardema, B. W., Lorenz, M. G. & Krumbein, W. E. (1983). Protection of sediment-adsorbed transforming DNA against enzymatic inactivation. *Appl Environ Microbiol* **46**, 417–420.

Allison, D. G., Brown, M. R. W., Evans, D. E. & Gilbert, P. (1990). Surface hydrophobicity and dispersal of *Pseudomonas aeruginosa* from biofilms. *FEMS Microbiol Lett* **71**, 101–104.

Altherr, M. R. & Keswick, K. L. (1982). In situ studies with membrane diffusion chambers of antibiotic resistance transfer in *Escherichia coli. Appl Environ Microbiol* **44**, 838–843.

Andrup, L. & Andersen, K. (1999). A comparison of the kinetics of plasmid transfer in the conjugation systems encoded by the F plasmid from *Escherichia coli* and plasmid pCF10 from *Enterococcus faecalis*. *Microbiology* **145**, 2001–2009.

Andrup, L., Schmidt, L., Andersen, K. & Boe, L. (1998). Kinetics of conjugative transfer: a study of the plasmid pXO16 from *Bacillus thuringiensis* subsp. *israelensis*. *Plasmid* **40**, 30–43.

Angles, M. L., Marshall, K. C. & Goodman, A. E. (1993). Plasmid transfer between marine bacteria in the aqueous phase and biofilms in reactor microcosms. *Appl Environ Microbiol* **59**, 843–850.

Anwar, H., Van Biesen, T., Dasgupta, M., Lam, V. & Costerton, J. W. (1989). Interaction of biofilm bacteria with antibiotics in a novel in vitro chemostat system. *Antimicrob Agents Chemother* **33**, 1824–1826.

Arana, I., Justo, J. I., Muela, A., Pocino, M., Iriberri, J. & Barcina, I. (1997). Influence of a survival process in a freshwater system upon plasmid transfer between *Escherichia coli* strains. *Microb Ecol* **33**, 41–49.

Aronson, A. I. & Beckman, W. (1987). Transfer of chromosomal genes and plasmids in *Bacillus thuringiensis*. *Appl Environ Microbiol* **53**, 1525–1530.

Ascon-Cabrera, M. A., Thomas, D. & Lebeault, J. M. (1995). Activity of synchronized cells of a steady state biofilm recirculated reactor during xenobiotic biodegradation. *Appl Environ Microbiol* **61**, 920–925.

Ashelford, K. E., Fry, J. C. & Learner, M. A. (1995). Plasmid transfer between strains of *Pseudomonas putida*, and their survival, within a pilot scale percolating-filter sewage treatment system. *FEMS Microbiol Ecol* **18**, 15–26.

Bale, M. J., Fry, J. C. & Day, M. J. (1987). Plasmid transfer between strains of *Pseudomonas aeruginosa* on membrane filters attached to river stones. *J Gen Microbiol* **133**, 3099–3107.

Bale, M. J., Fry, J. C. & Day, M. J. (1988). Transfer and occurrence of large mercury resistance plasmids in river epilithon. *Appl Environ Microbiol* **54**, 972–978.

Bauda, P., Menon, P., Block, J. C., Lett, M. C., Roux, B. & Hubert, J. C. (1990). Laboratory standardized biofilms as a tool to investigate genetic transfers in water systems. In *Bacterial Genetics in Natural Environments*, pp. 81–88. Edited by J. C. Fry & M. J. Day. London: Chapman & Hall.

Beaudoin, D. L., Bryers, J. D., Cunningham, A. B. & Peretti, S. W. (1998a). Mobilization of broad host range plasmid from *Pseudomonas putida* to established biofilm of *Bacillus azotoformans*. I. Experiments. *Biotechnol Bioeng* **57**, 272–279.

Beaudoin, D. L., Bryers, J. D., Cunningham, A. B. & Peretti, S. W. (1998b). Mobilization of broad host range plasmid from *Pseudomonas putida* to established biofilm of *Bacillus azotoformans*. II. Modeling. *Biotechnol Bioeng* **57**, 280–286.

Bjorklof, K., Suoniemi, A., Haahtela, K. & Romantschuk, M. (1995). High frequency of conjugation versus plasmid segregation of RP1 in epiphytic *Pseudomonas syringae* populations. *Microbiology* **141**, 2719–2727.

Bleakley, B. H. & Crawford, D. L. (1989). The effects of varying moisture and nutrient levels on the transfer of a conjugative plasmid between *Streptomyces* species in soil. *Can J Microbiol* **35**, 544–549.

Bradley, D. E. & Williams, P. A. (1982). The TOL plasmid is naturally derepressed for transfer. *J Gen Microbiol* **128**, 3019–3024.

Bradley, D. E., Taylor, D. E. & Cohen, D. R. (1980). Specification of surface mating systems among conjugative drug resistance plasmids in *Escherichia coli* K-12. *J Bacteriol* **143**, 1466–1470.

Brown, M. L., Aldrich, H. C. & Gauthier, J. J. (1995). Relationship between glycocalyx and povidone-iodine resistance in *Pseudomonas aeruginosa* (ATCC 27853) biofilms. *Appl Environ Microbiol* **61**, 187–193.

Caldwell, D. E., Korber, D. R. & Lawrence, J. R. (1992). Imaging of bacterial cells by fluorescence exclusion using scanning confocal laser microscopy. *J Microbiol Methods* **15**, 249–261.

Cao, Y. S. & Alaerts, G. J. (1995). Influence of reactor type and shear stress on aerobic biofilm morphology, population and kinetics. *Water Res* **29**, 107–118.

Carlson, C. A., Pierson, L. S., Rosen, J. J. & Ingraham, J. L. (1983). *Pseudomonas stutzeri* and related species undergo natural transformation. *J Bacteriol* **153**, 93–99.

Characklis, W. G. & Marshall, K. C. (1990). *Biofilms*. New York: Wiley.

Chen, C. I., Griebe, T. & Characklis, W. G. (1993). Biocide action of monochloramine on biofilm systems of *Pseudomonas aeruginosa*. *Biofouling* **7**, 1–17.

Christensen, B. B., Strenberg, C., Andersen, J. B., Eberl, L., Møller, S., Givskov, M. & Molin, S. (1998). Establishment of new genetic traits in a microbial biofilm community. *Appl Environ Microbiol* **64**, 2247–2255.

Costerton, J. W., Cheng, K. J., Geesey, G. G., Ladd, T. I., Nickel, J. C., Dasgupta, M. & Marrie, T. J. (1987). Bacterial biofilms in nature and disease. *Annu Rev Microbiol* **41**, 435–464.

Costerton, J. W., Lewandowski, Z., Caldwell, D. E., Korber, D. R. & Lappin-Scott, H. M. (1995). Microbial biofilms. *Annu Rev Microbiol* **49**, 711–745.

Daane, L. L., Molina, J. A. E., Berry, E. C. & Sadowsky, M. J. (1996). Influence of earthworm activity on gene transfer from *Pseudomonas fluorescens* to indigenous soil bacteria. *Appl Environ Microbiol* **62**, 515–521.

Daane, L. L., Molina, J. A. E. & Sadowsky, M. J. (1997). Plasmid transfer between spatially separated donor and recipient bacteria in earthworm-containing soil microcosms. *Appl Environ Microbiol* **63**, 679–686.

Dahlberg, C., Bergström, M. & Hermansson, M. (1998). In situ detection of high levels of horizontal plasmid transfer in marine bacterial communities. *Appl Environ Microbiol* **64**, 2670–2675.

Davies, D. G. & Geesey, G. G. (1995). Regulation of the alginate biosynthesis gene *algC* in *Pseudomonas aeruginosa* during biofilm development in continuous culture. *Appl Environ Microbiol* **61**, 860–867.

DeBeer, D., Srinivasan, R. & Stewart, P. S. (1994). Direct measurement of chlorine penetration into biofilms during disinfection. *Appl Environ Microbiol* **60**, 4339–4344.

Deretic, V., Schurr, M. J. & Boucher, J. C. (1994). Conversion of *Pseudomonas aeruginosa* to mucoidy in cystic fibrosis: environmental stress and regulation of bacterial virulence by alternative sigma factors. *J Bacteriol* **176**, 2773–2780.

De Taxis du Poet, P., Dhulster, P., Barbotin, J. N. & Thomas, D. (1986). Plasmid inheritability and biomass production: comparison between free and immobilized cell cultures of *Escherichia coli* BZ18(pTG201) without selection pressure. *J Bacteriol* **165**, 871–877.

Ehlers, L. J. & Bouwer, E. J. (1999). RP4 plasmid transfer among species of *Pseudomonas* in a biofilm reactor. *Water Sci Technol* **39**, 163–171.

Fernandez-Astorga, A., Muela, A., Cisterna, R., Iriberri, J. & Barcina, I. (1992). Biotic and abiotic factors affecting plasmid transfer in *Escherichia coli* strains. *Appl Environ Microbiol* **58**, 392–398.

Frost, L. S. (1992). Bacterial conjugation: everybody's doin' it. *Can J Microbiol* **38**, 1091–1096.

Fulthorpe, R. R. & Wyndham, R. C. (1991). Transfer and expression of the catabolic plasmid pBRC60 in wild bacterial recipients in a freshwater ecosystem. *Appl Environ Microbiol* **57**, 1546–1553.

Fulthorpe, R. R. & Wyndham, R. C. (1992). Involvement of a chlorobenzoate-catabolic transposon, Tn*5271*, in community adaptation to chlorobiphenyl, chloroaniline, and 2,4-dichlorophenoxyacetic acid in a freshwater ecosystem. *Appl Environ Microbiol* **58**, 314–325.

Gauthier, M. J., Cauvin, F. & Breittmayer, J. P. (1985). Influence of salts and temperature on the transfer of mercury resistance from a marine pseudomonad to *Escherichia coli*. *Appl Environ Microbiol* **50**, 38–40.

Gealt, M. A., Chai, M. D., Alpert, K. B. & Boyer, J. C. (1985). Transfer of plasmids pBR322 and pBR325 in wastewater from laboratory strains of *Escherichia coli* to bacteria indigenous to the waste disposal system. *Appl Environ Microbiol* **49**, 836–841.

Geisenberger, O., Ammendola, A., Christensen, B. B., Molin, S., Schleifer, K. H. & Eberl, L. (1999). Monitoring the conjugal transfer of plasmid RP4 in activated sludge and in situ identification of the transconjugants. *FEMS Microbiol Lett* **174**, 9–17.

Genther, F. J., Chatterjee, P., Barkay, T. & Bourquin, A. W. (1988). Capacity of aquatic bacteria to act as recipients of plasmid DNA. *Appl Environ Microbiol* **54**, 115–117.

Goodman, A. E., Hild, E., Marshall, K. C. & Hermansson, M. (1993). Conjugative plasmid transfer between bacteria under simulated marine oligotrophic conditions. *Appl Environ Microbiol* **59**, 1035–1040.

Grimes, D. J., Somerville, C. C., Straube, W., Roszak, D. B., Ortiz-Conde, B. A., MacDonell, M. T. & Colwell, R. R. (1988). Plasmid mobility in the ocean environment. *Aquat Toxic Hazard Assess* **10**, 37–42.

Guiney, D. G. & Lanka, E. (1989). Conjugative transfer of IncP plasmids. In *Promiscuous Plasmids of the Gram Negative Bacteria*, pp. 27–56. Edited by C. Thomas. London & San Diego: Academic Press.

Haase, J., Lurz, R., Grahn, A. M., Bamford, D. H. & Lanka, E. (1995). Bacterial conjugation mediated by plasmid RP4: RSF1010 mobilization, donor-specific phage propagation, and pilus production require the same Tra2 core components of a proposed DNA transport complex. *J Bacteriol* **177**, 4779–4791.

Hausner, M. & Wuertz, S. (1999). High rates of conjugation in bacterial biofilms as determined by quantitative in situ analysis. *Appl Environ Microbiol* **65**, 3710–3713.

Heinemann, J. A. & Ankenbauer, R. G. (1993). Retrotransfer in *Escherichia coli* conjugation: bi-directional exchange or de novo mating? *J Bacteriol* **175**, 583–588.

Heinemann, J. A. & Sprague, G. F. (1989). Bacterial conjugative plasmids mobilize DNA transfer between bacteria and yeast. *Nature* **340**, 205–209.

Hill, K. E., Weightman, A. J. & Fry, J. C. (1992). Isolation and screening of plasmids from the epilithon which mobilize recombinant plasmid pD10. *Appl Environ Microbiol* **58**, 1292–1300.

Hill, K. E., Fry, J. C. & Weightman, A. J. (1994). Gene transfer in the aquatic environment: persistence and mobilization of the catabolic recombinant plasmid pD10 in the epilithon. *Microbiology* **140**, 1555–1563.

Inomata, K., Nishikawa, M. & Yoshida, K. (1994). The yeast *Saccharomyces kluyveri* as a recipient eukaryote in transkingdom conjugation: behavior of transmitted plasmids in transconjugants. *J Bacteriol* **176**, 4770–4773.

Ippen-Ihler, K. A. (1989). Bacterial conjugation. In *Gene Transfer in the Environment*, pp. 33–72. Edited by S. B. Levy & R. V. Miller. New York: McGraw-Hill.

Ippen-Ihler, K. A. & Minkley, E. G. (1986). The conjugation system of F, the fertility factor of *Escherichia coli. Annu Rev Genet* **20**, 593–624.

Kidambi, S. P., Ripp, S. & Miller, R. V. (1994). Evidence for phage-mediated gene transfer among *Pseudomonas aeruginosa* strains on the phylloplane. *Appl Environ Microbiol* **60**, 496–500.

Knudsen, G. R., Walter, M. V., Porteous, L. A., Prince, V. J., Armstrong, J. L. & Seidler, R. J. (1988). Predictive model of conjugative plasmid transfer in the rhizosphere and phyllosphere. *Appl Environ Microbiol* **54**, 343–347.

Kozdrój, J. (1997). Survival, plasmid transfer, and impact of *Pseudomonas fluorescens* introduced into soil. *J Environ Sci Health* **A32**, 1139–1157.

Kröckel, L. & Focht, D. D. (1987). Construction of chlorobenzene-utilizing recombinants by progenitive manifestation of a rare event. *Appl Environ Microbiol* **53**, 2470–2475.

Kruse, H. & Sorum, H. (1994). Transfer of multiple drug resistance plasmids between bacteria of diverse origins in natural microenvironments. *Appl Environ Microbiol* **60**, 4015–4021.

Lafuente, R., Maymó-Gatell, X., Mas-Castellà, J. & Guerrero, R. (1996). Influence of environmental factors on plasmid transfer in soil microcosms. *Curr Microbiol* **32**, 213–220.

Lawrence, J. R., Korber, D. R., Hoyle, B. D., Costerton, J. W. & Caldwell, D. E. (1991). Optical sectioning of microbial biofilms. *J Bacteriol* **173**, 6558–6567.

Lessl, M. & Lanka, E. (1994). Common mechanisms in bacterial conjugation and Ti-mediated T-DNA transfer to plant cells. *Cell* **77**, 321–324.

Levin, B. R., Stewart, F. M. & Rice, V. A. (1979). The kinetics of conjugative plasmid transmission: fit of a simple mass action model. *Plasmid* **2**, 247–260.

Licht, T. R., Christensen, B. B., Krogfelt, K. A. & Molin, S. (1999). Plasmid transfer in the animal intestine and other dynamic bacterial populations: the role of community structure and environment. *Microbiology* **145**, 2615–2622.

Lorenz, M. G. & Wackernagel, W. (1987). Adsorption of DNA to sand and variable degradation rates of adsorbed DNA. *Appl Environ Microbiol* **53**, 2948–2952.

Lorenz, M. G. & Wackernagel, W. (1990). Natural genetic transformation of *Pseudomonas stutzeri* by sand-adsorbed DNA. *Arch Microbiol* **154**, 380–385.

Lorenz, M. G., Aardema, B. W. & Wackernagel, W. (1988). Highly efficient genetic transformation of *Bacillus subtilis* attached to sand grains. *J Gen Microbiol* **134**, 107–112.

Lorenz, M. G., Reipschläger, K. & Wackernagel, W. (1992). Plasmid transformation of naturally competent *Acinetobacter calcoaceticus* in non-sterile soil extract and groundwater. *Arch Microbiol* **157**, 355–360.

Lundquist, P. D. & Levin, B. R. (1986). Transitory derepression and the maintenance of conjugative plasmids. *Genetics* **113**, 483–497.

McClure, N. C., Fry, J. C. & Weightman, A. J. (1990). Gene transfer in activated sludge. In *Bacterial Genetics in Natural Environments*, pp. 111–129. Edited by J. C. Fry & M. J. Day. London: Chapman & Hall.

McClure, N. C., Fry, J. C. & Weightman, A. J. (1991). Survival and catabolic activity of natural and genetically engineered bacteria in a laboratory-scale activated-sludge unit. *Appl Environ Microbiol* **57**, 366–373.

MacDonald, J. A., Smets, B. F. & Rittmann, B. E. (1992). The effects of energy availability on the conjugative-transfer kinetics of plasmid RP4. *Water Res* **26**, 461–468.

Mach, P. A. & Grimes, D. J. (1982). R-plasmid transfer in a wastewater treatment plant. *Appl Environ Microbiol* **44**, 1395–1403.

McPherson, P. & Gealt, M. A. (1986). Isolation of indigenous wastewater bacterial strains capable of mobilizing plasmid pBR325. *Appl Environ Microbiol* **51**, 904–909.

Manceau, C., Gardan, L. & Devaux, M. (1986). Dynamics of RP4 plasmid transfer between *Xanthomonas campestris* pv. *corylina* and *Erwinia herbicola* in hazelnut tissues, in planta. *Can J Microbiol* **32**, 835–841.

Mater, D. D. G., Nava Saucedo, J. E., Truffaut, N., Barbotin, J. N. & Thomas, D. (1999). Conjugative plasmid transfer between *Pseudomonas* strains within alginate bead microcosms: effect of the internal gel structure. *Biotechnol Bioeng* **65**, 34–43.

Meynell, G. G. (1973). *Bacterial Plasmids: Conjugation, Colicinogeny and Transmissible Drug-Resistance*, pp. 50–51. Cambridge, MA: MIT Press.

Muela, A., Pocino, M., Arana, I., Justo, J. I., Iriberri, J. & Barcina, I. (1994). Effect of growth phase and parental cell survival in river water on plasmid transfer between *Escherichia coli* strains. *Appl Environ Microbiol* **60**, 4273–4278.

Neilson, J. W., Josephson, K. L., Pepper, I. L., Arnold, R. B., DiGiovanni, G. D. & Sinclair, N. A. (1994). Frequency of horizontal gene transfer of a large catabolic plasmid (pJP4) in soil. *Appl Environ Microbiol* **60**, 4053–4058.

Nishikawa, M., Suzuki, K. & Yoshida, K. (1990). Structural and functional stability of IncP plasmids during stepwise transmission by trans-kingdom mating: promiscuous conjugation of *Escherichia coli* and *Saccharomyces cerevisiae*. *Jpn J Genet* **65**, 323–334.

Normander, B., Christensen, B. B., Molin, S. & Kroer, N. (1998). Effect of bacterial distribution and activity on conjugal gene transfer on the phylloplane of the bush bean (*Phaseolus vulgaris*). *Appl Environ Microbiol* **64**, 1902–1909.

Nüßlein, K., Maris, D., Timmis, K. & Dwyer, D. F. (1992). Expression and transfer of engineered catabolic pathways harbored by *Pseudomonas* spp. introduced into activated sludge microcosms. *Appl Environ Microbiol* **58**, 3380–3386.

O'Morchoe, S. B., Ogunseitan, P., Sayler, G. S. & Miller, R. V. (1988). Conjugal transfer of R68.45 and FP5 between *Pseudomonas aeruginosa* strains in a freshwater environment. *Appl Environ Microbiol* **54**, 1923–1929.

Paul, J. H., Frischer, M. E. & Thurmond, J. M. (1991). Gene transfer in marine water column and sediment microcosms by natural plasmid transformation. *Appl Environ Microbiol* **57**, 1509–1515.

Rafii, F. & Crawford, D. L. (1988). Transfer of conjugative plasmids and mobilization of a nonconjugative plasmid between *Streptomyces* strains on agar and in soil. *Appl Environ Microbiol* **54**, 1334–1340.

Ramos-Gonzalez, M. I., Duque, E. & Ramos, J. L. (1991). Conjugational transfer of recombinant DNA in cultures and in soils: host range of *Pseudomonas putida* TOL plasmids. *Appl Environ Microbiol* **57**, 3020–3027.

Richaume, A., Angle, J. S. & Sadowsky, M. J. (1989). Influence of soil variables on in situ plasmid transfer from *Escherichia coli* to *Rhizobium fredii*. *Appl Environ Microbiol* **55**, 1730–1734.

Ripp, S. & Miller, R. V. (1995). Effects of suspended particulates on the frequency of transduction among *Pseudomonas aeruginosa* in a freshwater environment. *Appl Environ Microbiol* **61**, 1214–1219.

Rittmann, B. E., Smets, B. F. & Stahl, D. A. (1990). The role of genes in biological processes. *ES & T* **24**, 23–29.

Rochelle, P. A., Fry, J. C. & Day, M. J. (1989). Factors affecting conjugal transfer of plasmids encoding mercury resistance from pure cultures and mixed natural suspensions of epilithic bacteria. *J Gen Microbiol* **135**, 409–424.

Sandaa, R. A. & Enger, O. (1994). Transfer in marine sediments of the naturally occurring plasmid pRAS1 encoding multiple antibiotic resistance. *Appl Environ Microbiol* **60**, 4234–4238.

Saunders, J. R. & Saunders, V. A. (1992). The estimation of gene transfer in natural environments. In *Genetic Interactions Among Microorganisms in the Natural Environment*, pp. 258–263. Edited by E. M. H. Wellington & J. D. van Elsas. Oxford & New York: Pergamon Press.

Saye, D. J., Ogunseitan, O. A., Sayler, G. S. & Miller, R. V. (1990). Transduction of linked chromosomal genes between *Pseudomonas aeruginosa* strains during incubation in situ in a freshwater habitat. *Appl Environ Microbiol* **56**, 140–145.

Schafer, A., Kalinowski, J., Simon, R., Seep-Feldhaus, A. H. & Puhler, A. (1990). High-frequency conjugal plasmid transfer from gram-negative *Escherichia coli* to various gram-positive coryneform bacteria. *J Bacteriol* **172**, 1663–1666.

Selvaratnam, S. & Gealt, M. A. (1992). Recombinant plasmid gene transfer in amended soil. *Water Res* **26**, 39–43.

Sia, E. A., Kuehner, D. M. & Figurski, D. H. (1996). Mechanism of retrotransfer in conjugation: prior transfer of the conjugative plasmid is required. *J Bacteriol* **178**, 1457–1464.

Simonsen, L. (1990). Dynamics of plasmid transfer on surfaces. *J Gen Microbiol* **136**, 1001–1007.

Simonsen, L., Gordon, D. M., Stewart, F. M. & Levin, B. R. (1990). Estimating the rate of plasmid transfer: an end-point method. *J Gen Microbiol* **136**, 2319–2325.

Singleton, P. (1983). Colloidal clay inhibits conjugal transfer of R-plasmid R1*drd-19* in *Escherichia coli*. *Appl Environ Microbiol* **46**, 756–757.

Singleton, P. & Anson, A. E. (1981). Conjugal transfer of R-plasmid R1*drd-19* in *Escherichia coli* below 22 °C. *Appl Environ Microbiol* **42**, 789–791.

Singleton, P. & Anson, A. E. (1983). Effect of pH on conjugal transfer at low temperatures. *Appl Environ Microbiol* **46**, 291–292.

Smets, B. F., Rittmann, B. E. & Stahl, D. A. (1990). The role of genes in biological processes. *ES & T* **24**, 162–169.

Smit, E., Venne, D. & van Elsas, J. D. (1993). Mobilization of a recombinant IncQ plasmid between bacteria on agar and in soil via cotransfer or retrotransfer. *Appl Environ Microbiol* **59**, 2257–2263.

Stal, L. J. (1989). Group report: cellular physiology and interactions of biofilm organisms. In *Structure and Function of Biofilms: Report of the Dahlem Workshop on Structure and Function of Biofilms*, Berlin 1988, November 27–December 2, pp. 269–286. Edited by W. G. Characklis & P. A. Wilderer. Chichester & New York: Wiley.

Stewart, G. J. (1989). The mechanism of natural transformation. In *Gene Transfer in the Environment*, pp. 139–164. Edited by S. B. Levy & R. V. Miller. New York: McGraw-Hill.

Stewart, G. J. & Sinigalliano, C. D. (1990). Detection of horizontal gene transfer by natural transformation in native and introduced species of bacteria in marine and synthetic sediments. *Appl Environ Microbiol* **56**, 1818–1824.

Stotzky, G. & Babich, H. (1986). Survival of, and genetic transfer by, genetically engineered bacteria in natural environments. *Adv Appl Microbiol* **31**, 93–138.

Stout, V. G. & Iandolo, J. J. (1990). Chromosomal gene transfer during conjugation by *Staphylococcus aureus* is mediated by transposon-facilitated mobilization. *J Bacteriol* **172**, 6148–6150.

Top, E., Mergeay, M., Springael, D. & Verstraete, W. (1990). Gene escape model:

transfer of heavy metal resistance genes from *Escherichia coli* to *Alcaligenes eutrophus* on agar plates and in soil samples. *Appl Environ Microbiol* **56**, 2471–2479.

Top, E., De Smet, I., Verstraete, W., Dijkmans, R. & Mergeay, M. (1994). Exogenous isolation of mobilizing plasmids from polluted soils and sludges. *Appl Environ Microbiol* **60**, 831–839.

Top, E. M., Maila, M. P., Clerinx, M., Goris, J., De Vos, P. & Verstraete, W. (1999). Methane oxidation as a method to evaluate the removal of 2,4-dichlorophenoxyacetic acid (2,4-D) from soil by plasmid-mediated bioaugmentation. *FEMS Microbiol Ecol* **28**, 203–213.

Trevors, J. T. & Oddie, K. M. (1986). R-plasmid transfer in soil and water. *Can J Microbiol* **32**, 610–613.

Trevors, J. T., van Elsas, J. D., Starodub, M. E. & van Overbeek, L. S. (1990). *Pseudomonas fluorescens* survival and plasmid RP4 transfer in agricultural water. *Water Res* **24**, 751–755.

Trieu-Cuot, P., Carlier, C. & Courvalin, P. (1988). Conjugative plasmid transfer from *Enterococcus faecalis* to *Escherichia coli*. *J Bacteriol* **170**, 4388–4391.

Van Elsas, J. D. & Trevors, J. T. (1990). Plasmid transfer to indigenous bacteria in soil and rhizosphere: problems and perspectives. In *Bacterial Genetics in Natural Environments*, pp. 188–199. Edited by J. C. Fry & M. J. Day. London: Chapman & Hall.

Van Elsas, J. D., Nikkel, M. & van Overbeek, L. S. (1989). Detection of plasmid RP4 transfer in soil and rhizosphere, and the occurrence of homology to RP4 in soil bacteria. *Curr Microbiol* **19**, 375–381.

Veal, D. A., Stokes, H. W. & Daggard, G. (1992). Genetic exchange in natural microbial communities. *Adv Microb Ecol* **12**, 383–430.

Venables, W. A., Wimpenny, J. W. T., Ayres, A., Cook, S. M. & Thomas, L. V. (1995). The use of two-dimensional gradient plates to investigate the range of conditions under which conjugal plasmid transfer occurs. *Microbiology* **141**, 2713–2718.

Wahlund, T. M. & Madigan, M. T. (1995). Genetic transfer by conjugation in the thermophilic green sulfur bacterium *Chlorobium tepidum*. *J Bacteriol* **177**, 2583–2588.

Walmsley, R. H. (1976). Temperature dependence of mating-pair formation in *Escherichia coli*. *J Bacteriol* **126**, 222–224.

Walter, M. V., Porteous, L. A. & Seidler, R. J. (1989). Evaluation of a method to measure conjugal transfer of recombinant DNA in soil slurries. *Curr Microbiol* **19**, 365–370.

Winans, S. C. (1992). Two-way chemical signaling in *Agrobacterium*-plant interactions. *Microbiol Rev* **56**, 12–31.

Zahid, W. & Ganczarczyk, J. (1994). A technique for a characterization of RBC biofilm surface. *Water Res* **28**, 2229–2231.

Population dynamics in microbial biofilms

Andrew J. McBain, David G. Allison and Peter Gilbert

School of Pharmacy and Pharmaceutical Sciences, University of Manchester, Oxford Road, Manchester M13 9PL, UK

INTRODUCTION

A perusal of the published literature might lead the naïve to believe that microbial biofilms are definable entities each of which, whilst possessing characteristics that are markedly dissimilar from their planktonic equivalents, has its own defined physiology and architecture. Thus it is easy to imagine, rather like for our own bodies, that the visible, physical manifestation has a degree of permanency. The reality is that the structures known as biofilms, as well as the majority of tissues that make up our bodies, are in a state of dynamic flux/turnover. Modern imaging techniques generally give us snapshots in time and only rarely give an indication of the turmoil within (Costerton *et al.*, 1995).

It is the intention of this article to dispel any view the reader might have entertained of biofilms being static entities. Dynamics within microbial biofilm communities will be considered in terms of their spatial stability (movement/drift), and changes in biomass, genetic diversity and community function. In the latter respects, population dynamics will be considered for both short-term events, such as the formation and establishment of biofilm communities on 'virgin' surfaces (Geesey *et al.*, 1992), and long-term events, with biofilms being considered as units of proliferation/evolution (Caldwell *et al.*, 1997).

Dynamics in microbial communities can be considered on a number of distinct levels, many of which have been considered separately in this symposium volume. The first of these relates to the spatial stability of the biofilm community. Matrix polymers not only

SGM symposium 59: Community structure and co-operation in biofilms. Editors D. Allison, P. Gilbert, H. Lappin-Scott, M. Wilson. Cambridge University Press. ISBN 0 521 79302 5 ©SGM 2000.

glue the biofilm to the surface but also enable spatial organization to be imposed on the community. The polymers are plastic and may be physically deformed through the imposition of shear forces from the fluid phase. In turbulent flow, this causes dynamic motion to be exhibited by parts of the biofilm (Stoodley *et al.*, 1998, 1999a, b) and in some instances spatial drift of the entire community (Moore and others, this volume). If the degree of deformation exceeds the elastic limit of the matrix then catastrophic sloughing will occur leading to the dispersal of the community through the liquid phase to virgin surfaces (Moore and others, this volume). The material and mechanical properties of the glycocalyx (rheology, viscoelasticity; Flemming and others, this volume) vary with the chemical composition of the matrix, and with the nature and species complexity of the community. In this fashion, community composition will have profound influences upon the displayed architecture and the physical attributes of the biofilm (Picioreanu and others, this volume).

The second aspect of population dynamics in microbial communities relates to temporal changes in the magnitude of the attached population (colonization density). Clearly, the numbers of cells associated with a surface increase once a virgin surface is introduced to the natural world. During this accumulation phase, the number of attaching organisms, and the growth and retention of the attached cells, must exceed the rate of detachment and loss from the surface. Whilst those factors that influence the initial attraction of bacterial cells to surfaces and their attachment have been well studied and reviewed (Marshall, 1985, 1992), the processes associated with detachment are less well understood. What is apparent is that as biofilms grow, then not only do the physiological and physico-chemical heterogeneities within them increase, but their physical stability decreases. Increases in biofilm thickness/mass beyond critical limits, imposed by the fluid dynamic environment and the nature of the substratum, lead to catastrophic events and dramatic losses of biomass. Eventually, pseudo-steady states, with respect to colonization density, are achieved which are dependent upon the physical, chemical and topological nature of the substratum, the prevailing fluid dynamic forces and the availability of nutrients (Boyle *et al.*, 1999). In a mature biofilm community, the sequestration of planktonic cells and the growth and division of the attached cells are counterbalanced by losses through dispersal and cell death (Willcock *et al.*, 1997).

At maturity, biofilms are often subjected to challenge with planktonic cells. These might constitute individual cells that have grown in suspension or ones that have been derived from biofilms that are upstream of the community. Even though the numbers of individual cells associated with a particular biofilm community might be constant, sequestered organisms will enhance the species and genetic diversity within them. Whether or not an immigrant organism can survive and grow within the community will depend upon its ability to displace, compete or co-operate effectively with the resident population (Marsh & Bowden, this volume). This ability will change as a function of time as the biofilm becomes physiologically and nutritionally more heterogeneous. Thus with the progression of time, the complexity of species within naturally occurring biofilm communities will increase, with competition leading to a selection of successful partnership, or consortium, phenotypes. Such selection pressures will be exerted disproportionately throughout the biofilm, leading to the establishment of mosaics of sub-communities within the global biofilm architecture.

COMMUNITY DYNAMICS DURING *DE NOVO* BIOFILM FORMATION

Earlier contributions to this volume have considered the physico-chemical forces of attraction between microbial cells and conditioned surfaces that lead to attachment and microcolony formation (Busscher & van der Mei, this volume). Similarly, Davies (this volume) has described the sensing of surface proximity and of attachment, through the localized concentration of various chemical signals that lead to the induction of a biofilm rather than a planktonic phenotype.

In most instances in nature, micro-organisms that encounter a surface will encounter one that is already conditioned and which may already be colonized by micro-organisms, rather than one that is 'sterile'. During the early stages of biofilm formation upon fresh/cleaned substrate, virtually all planktonic cells that contact the surface will attach and have the potential to form microcolonies (primary colonizers). Secondary colonizers will interact with either vacant sites on the surface or the primary colonizers. There are three possible outcomes to the encounter between a potential immigrant and a newly colonized surface. (i) The surface may be refractory to the potential colonizer due to lack of available/unoccupied binding sites and the immigrant will therefore fail to bind. (ii) The immigrant cells may displace physically, from the surface, one of the early colonizers by virtue of a possibly higher binding affinity for a common binding site. This is most likely to occur during the initial attachment phase of film formation and before the deposition of polymer cements. The duration of this phase will therefore be indirectly related to the metabolic potential at each colonized site. (iii) Both the immigrant species and the primary colonizer are retained on the surface, either at separate sites or attached to each other (for coaggregation, see the chapter by Kolenbrander and others, this volume) or to the matrix polymers. Where a surface is co-colonized, then the degree of interaction between the colonizers will be minimal in the first instance but will increase as the community grows and adjacent microcolonies come into closer proximity. Such interactions might be mediated through the production of cell–cell signalling compounds (Davies *et al.*, 1998), specific, and non-specific inhibitors (Gibson & Wang, 1994; Bernet-Camard *et al.*, 1997) or they might involve a competition for available nutrients. Alternatively, new substrate might become mobilized through co-operative degradation (Macfarlane *et al.*, 1986) and cross-feeding partnerships (Degnan & Macfarlane, 1995).

Complementary microbial activity is responsible for enhanced catabolism of nutrients and can provide protection from adverse conditions by maintaining a favourable local environment (Caldwell *et al.*, 1997; Marsh & Bradshaw, 1998). The outcome of such interactions might be the elimination of one or other of the initial colonizers, a neutral coexistence or the formation of synergistic partnerships (Marsh & Bowden, this volume). As the biofilm thickens and population density increases further, then the overall heterogeneity of the community will become more marked through the imposition of nutrient, gaseous and pH gradients generated by the metabolic activity of the enveloped cells (Wimpenny, 1995; Vroom *et al.*, 1999). This will provide for a variety of selection pressures, operational at different points in the developing biofilm, that influence the outcome of interactions occurring between adjacent species. The biofilm is therefore in a dynamic equilibrium with respect to its community membership and ethnicity. Throughout this process of maturation, the possibility of microbial succession within the community remains but it becomes increasingly unlikely and is dependent upon the successful immigration of autochthonous species.

As a consequence of their development in association with intense selection pressures, the climax communities are remarkably resistant to further colonization (colonization resistance/microbial homeostasis) (Alexander, 1971; Marsh & Bradshaw, 1999). Colonization resistance of climax communities has been elegantly demonstrated utilizing gnotobiotic animals. Such animals are devoid of a natural, commensal microflora and are therefore considerably more susceptible to infection by autochthonous bacteria (Boureau *et al.*, 1994). Similarly, the increased susceptibility towards infection of animals which possess a microflora that has become degraded by antibiotics or radiotherapy further supports such hypotheses (Larson & Welch, 1993; Brook *et al.*, 1993). Colonization events associated with the human alimentary tract therefore provide a suitable context in which to examine dynamics during biofilm formation.

Population dynamics during biofilm formation in the alimentary tract

The microbial community that inhabits the adult gastrointestinal tract is characterized by a high population density, a wide variety of metabolic activity, and considerable diversity in both microbial species composition and ecological niches. Distinct and separate microbial flora are associated with the nasopharynx, mouth, tongue, stomach, appendix and large intestine. In adults, these flora are relatively stable, provide a degree of colonization resistance to pathogens, and are comprised of complex, interacting communities of micro-organisms. The reviews by Marsh & Bowden and Kolenbrander and others (this volume) together give a comprehensive account of such interactions involved in the formation of biofilms associated with the oral cavity and in their associated pathologies. In considering dynamic changes during the establishment of commensal biofilms, the present contribution will therefore draw on examples provided by the intestinal tract and the progression from a neonatal to adult gut flora. Although the intestinal ecosystem is usually viewed as a single entity, individual assemblages of bacteria exist within the lumen in a multiplicity of different microhabitats and metabolic niches. Biofilm communities include not only those attached to and growing in close association with the mucosa, but also those growing in an intimate association with the surface of food particles (Englyst *et al.*, 1987; Macfarlane *et al.*, 1997). In this respect, the colon may be viewed as a form of fluidized-bed reactor.

The human colon contains as many as 10^{11} bacterial cells per g contents and comprises several hundred different species. As many as 99% of these bacteria are obligate anaerobes (Savage, 1977) and are incapable of colonizing the microbe-free, neonatal gut. From the moment of birth and throughout life, however, the digestive tract is exposed to vast numbers of micro-organisms, which vary in their colonization potential according to the current status of the individual. Colonization commences with a direct exposure to the mother's vaginal and faecal microflora during parturition, providing a simple and effective mechanism for the vertical transmission of microbes (Tannock, 1994; Mackie *et al.*, 1999).

Four distinct phases of microbial succession, relating to the progression to adult status, can be described (Cooperstock & Zedd, 1983). Phase one involves the initial bacterial colonization and lasts approximately 2 weeks. Although there is considerable inter-individual variation, the early colonizers normally comprise facultative anaerobes, such as *Escherichia coli* and streptococci, which attain population densities of 10^8–10^{10} per g faeces. Importantly, although the colon of the newly born infant is aerobic, these bacteria consume the available oxygen and create a sufficiently electronegative luminal redox potential for anaerobic genera to establish. Phase two occurs only in breast-fed infants prior to weaning or bottle-feeding, and is related to a preponderance of bifidobacteria over bacteroides, streptococci and *E. coli*. Once off the breast (phase three), the microflora quickly assumes the profile of a formula-fed infant with bacteroides replacing bifidobacteria as the dominant genus. Phase four commences when the infant is weaned onto solids. In phase four, there is a marked increase in complexity with respect to community profile (Conway, 1997), and the gut microflora starts to resembles that of the adult. The transition is usually complete by the second year of life.

It is apparent that until these communities mature, there is a distinct succession of microbes. The modulation of the growth environment, for example lowering of redox potential by aerobes and facultative species, illustrates the importance of community dynamics during *de novo* community formation. The importance of feeding regime on the developing microflora demonstrates the influence of environmental factors on community development. Furthermore, community complexity increases over time and this complexity is associated with overall population stability (hence colonization stability) (Alexander, 1971; Marsh & Bradshaw, 1999).

Population dynamics within pseudo-steady-state biofilms

A steady state is one that remains unchanged so long as all extrinsic factors are maintained constant. Such conditions are rare, if not impossible, in biology since growth and division will inevitably lead to increased genetic diversity, and even a constant selection pressure will always favour the most competitive genotype. In microbiology, chemostats are often regarded as being in steady state, yet even within such controlled cultures, inoculated with a single clone, there is a natural genetic drift towards the most competitive phenotype/genotype (Dykhuizen & Hartl, 1983). This is especially true of microbial biofilms.

In well-mixed planktonic populations, each individual cell has equal access to oxygen and substrate. In nutrient-rich environments, organization of bacteria as biofilms will reduce the overall rate of growth of the population through the creation of nutrient and gaseous gradients (Wimpenny, 1995). Since biofilms are ubiquitous (Costerton *et al.*, 1987), even to nutritionally rich situations, then close proximity of cells to one another or to a surface must bring with it significant advantage (Wimpenny, this volume). Increased cell density within biofilms will therefore enhance exchange of genetic information and facilitate cross-feeding partnerships, in addition to providing a limited protection against predation and dehydration. In oligotrophic settings, attached bacteria will be able to exploit nutrients that have become adsorbed and concentrated at the surface, enabling the organisms to grow faster than in planktonic mode. Continued growth leads to biofilm formation with an inevitable competition for nutrients. As this competition increases, then many of the advantages of life within the biofilm will be lost. There is, therefore, a requirement for cells to detach from the biofilm (Gilbert *et al.*, 1993) in order to colonize new, virgin sites and to avoid starvation (emigration). Similarly, if biofilms subsisted solely through growth and detachment, then the community would become highly 'inbred', with opportunities for community evolution and expansion of the genetic pool being severely limited. An essential part of community dynamics within biofilm therefore relates to the immigration of planktonic cells. It is inconceivable that all of the species represented within a climax community had been present within the primary colonizing group.

New species must be sequestered into the biofilms to enable community structures to evolve. Conversely, there is a need for individual cells/cellular-consortia within each biofilm to become dispersed throughout the environment in order to maximize the overall genetic diversity of the community and to promote the competitive demands of the gene (Dawkins, 1976). As with any community, therefore, there is a need for both emigration and immigration, without which inbreeding will lead to a lack of competitiveness.

A holistic view of a biofilm associated with a surface generally reveals it to be relatively constant with respect to cell number, thickness and community composition. At discrete locations on such surfaces, however, dramatic fluctuations in colonization density may be apparent with time (Davies *et al.*, 1998). Such temporal fluctuations reflect localized sloughing and dispersal events on the one hand, and a selection pressure towards removal resistance and competitiveness during the recovery phase. Whilst such selection pressures will apply equally to the residual biofilm community and to the immigrant cells, it is likely that a successful immigration would be more probable immediately after a sloughing event. Biofilms, at any particular time, should therefore be regarded as being in a quasi- or pseudo-steady state, rather than in a true one, for which variable selection pressures will apply. The following sections will consider separately (i) the biological imperatives of detachment/dispersal of individual cells and clusters of cells from biofilm communities, (ii) their sequestration/immigration into mature biofilm and onto virgin surfaces and (iii) the selection pressures that are exerted upon the residual quasi-steady-state biofilm community.

DISPERSAL OF BIOFILM

Attachment of cells to surfaces, and formation of a stable biofilm structure, involve forces of adhesion and cohesion mediated through structures such as pili and fimbriae, and, in the latter stages of biofilm development, through extracellular polymeric cements (Busscher & van der Mei, this volume; Flemming and others, this volume). Provided that the forces of adhesion and cohesion outweigh the forces of dispersion presented by the bulk phase, then the attached population will increase. As these developments take place, the biofilm will become thicker, less uniform in structure, and thereby heterogeneous in its susceptibility to the forces of dispersion. At some positions in the developing structure, critical points will be reached where the dispersion forces have a greater influence on the biofilm than do those of adhesion and cohesion. This will cause a localized detachment of cells and, as a consequence, a temporary physical stability. An alteration in the physico-chemical environment at the locale of detachment will result in a temporary loss of colonization resistance. Biofilm population dynamics at quasi-steady state therefore involve a fine balance between the forces of attachment and those associated with detachment and colonization resistance of the community.

Detachment of cells from a biofilm can be active as well as passive. For active dispersal of cells from a biofilm to occur, the cohesive forces holding it together and on the surface must be weakened. Such weakening may be as a result of physical stress (i.e. cell motility), predation or involve the active synthesis and release of enzymes such as proteases and polysaccharidases which might selectively weaken the glycocalyx and enhance sloughing (Boyd & Chakrabarty, 1994). Detachment may also be enhanced for individual cells at particular times in their cellular division cycle or at times of stress through starvation (emigration).

Regardless of the underlying cause, the process of cellular detachment from biofilms may be described as either erosive (a continuous loss of single cells, cell clusters and matrix) or fragmentative (a periodic, catastrophic loss of clusters of cells and matrix). Either of these processes may reflect combinations of physiological, enzymic and/or mechanical processes.

Erosion

Erosion is a process by which a surface is gradually reduced through the separate removal of small component parts. With respect to biofilms, these parts are either individual cells or small groups of cells. For this to occur, then the forces of dispersion must be greater than the forces of cohesion and adhesion that bond the parts together or to the substratum. Erosion may be brought about mechanically either by increasing the dispersion force or by the inclusion of abrasives (Chang *et al.*, 1991) within the fluid phase or by a programmed physiological change occurring within the biofilm itself.

(i) Mechanical erosion and abrasion. Fluid flow within a bulk phase is an essential element of many industrial processes. In natural, oligotrophic environments, movement of the liquid phase increases the flux of nutrients to a sessile biofilm. Particulate matter suspended within the bulk liquid will possess momentum. Under turbulent flow, or when directional changes are forced upon the movement of the fluid phase by either pipe-work or obstruction, then this momentum is transferred to the surface on collision as an abrasive force. The mass and density of the particles, properties such as hardness and elasticity, and collision speed will affect abrasion force. Particulate material will vary dramatically in its mechanical property and size and will include, for example, bacterial cells (1–5 μm), fine grit sands (100–250 μm), peas (5–10 mm), stones and pebbles (big).

For liquids and solutions containing particles of colloidal dimension, then the nature of the individual components is less relevant to mechanical abrasion/erosion than are bulk fluid properties such as viscosity and specific gravity. The tortuosity of the surface together with the rate of flow of the liquid and the Reynolds number determine whether the flow is laminar or turbulent. In both instances, the frictional drag of the mobile phase (shear) against the surface of the biofilm will cause erosion to take place. The degree of erosion will be affected, in turn, by the rate of movement of the liquid phase and tortuosity in the surface.

(ii) Physiologically mediated erosion. For erosion to be physiologically driven, the adhesion of individual cells to the surface, to each other or to the biofilm matrix must be weakened. High-level expression and localization of extracellular polysaccharidases, and readoption of a motile phenotype will enable individual cells to emerge from the glycocalyx (Boyd & Chakrabarty, 1994). Detachment of bacteria from surfaces and from biofilms has also been strongly associated with events occurring during cell division. In this respect, Allison *et al.* (1990a, b) reported that single cells spontaneously shed from *E. coli* and *Pseudomonas aeruginosa* biofilms divide synchronously when transferred to fresh medium, are uniformly hydrophilic, and bear a high negative surface charge relative to cells of the parent biofilm. Similar observations have been made with Gram-positive bacteria such as *Staphylococcus epidermidis* (Gilbert *et al.*, 1991). Such dispersed cells are relatively 'unsticky' with respect to both biofilm-coated and virgin surfaces. Collections of these cells are unable to adhere to mature biofilms unless they have been allowed to grow through at least two generations. In natural environments, this will ensure that the dispersed daughter cells are relocated some distance away from the parent biofilm before they can reattach (Gilbert *et al.*, 1993). In this respect, it is notable that only laterally flagellated cells of *Vibrio parahaemolyticus* are able to recolonize mature biofilms of this organism (Lawrence *et al.*, 1992). Previously, Lawrence *et al.* (1987) had described a 'recolonization phase' for biofilms which occurred some hours after the initial attachment to a substratum and which involved an increase in the proportion of motile cells.

Sloughing

Sloughing can be defined as the detachment of large aggregates of cells together with their intercellular matrix. The physiological properties of the sloughed material are therefore likely to mimic those of the biofilm rather than those of denuded cells eroded by a physiological mechanism. Large sloughs will contain representative members of the parent community and will continue to provide protection for obligate anaerobes against oxidative damage. Sloughed material, being larger than eroded particulates, is more likely to sediment from the bulk phase and will therefore tend to spread the community downstream rather than to establish new sites. As for erosive processes, sloughing might be brought about by growth and a mechanical instability of the biofilm in relation to shear, or it might be physiologically mediated through enzyme activation/induction.

(i) **Mechanical sloughing.** There is no single unifying structure for microbial biofilm. Many observations, made in diverse habitats, suggest that a wealth of structures are possible, each optimized and reflecting the particular characteristics of the growth environment. In nutritionally rich environments such as the oral cavity, biofilms are thick mats of cells spatially organized with respect to the component species (Marsh & Bradshaw, 1998). In environments that are moderately rich in nutrients, then mushroom-like structures predominate, with the passages formed beneath the canopies allowing for the circulation and passage of fluid. The latter has been likened to a primitive circulatory system (Costerton *et al.*, 1994). In oligotrophic environments, stack-like pinnacles emerge from a diffuse layer of cells colonizing the surface and maximize the adsorption and utilization of dissolved nutrients (Keevil, 1993). Such structures can be replicated in simple computer simulations (Wimpenny & Colasanti, 1997) which relate the diffusion and consumption of carbon substrate to growth kinetics (Monod, 1949).

Even within the biofilm matrix, cell density will vary as a function of distance from the liquid interface. The glycocalyx is elastic and will be deformed by mechanical forces exerted upon it. Elastic deformation will lead to fracturing of the matrix when critical forces are exceeded (Stewart, 1993). Heterogeneity within individual biofilms gives rise to positions in the structure that are particularly weak and will fracture when subjected to stress (fault lines). These might relate to regions with reduced elasticity (i.e. with particularly high cell densities), or inherent weaknesses in the physical structure. Where cell–cell adhesion is greater than the adhesion of the primary colonizers to the substratum, fracturing will occur due to cohesion failure at the substratum or conditioning film.

Adsorption of cations within the glycocalyx has a marked effect upon the physico-chemical characteristics of the polymers. Thus cohesive strength is increased (Applegate & Bryers, 1991) and porosity is decreased (Hoyle *et al.*, 1992) with divalent cation concentration. Accordingly, if biofilms are formed on surfaces which leach cations (i.e. lime-scale associated with hard-water distribution systems), then the properties of the glycocalyx will vary with distance from the surface. This will affect not only the susceptibility of the enveloped cells towards antimicrobial agents, but also the likely location of shear planes.

Let us now consider the nature of the forces exerted upon the biofilm from the fluid phase. Provided that the interactive surface between the biofilm and mobile fluid phase is perfectly flat at a molecular level and that fluid flow is laminar, then erosion alone will mechanically reduce the extent of biofilm development. Any irregularities in the surface of the biofilm due to its innate structure or through the topography of the substratum will have the potential to disturb the flow.

Small irregularities, whilst not disturbing the laminosity of flow, will be subject to compression on the upstream side and decompression downstream. For an elastic surface, the irregularity will become deformed and further increase the stresses imposed. This is demonstrated by the formation of 'streamers' that emanate from biofilm matrices that are subjected to laminar flows of liquid (Stoodley *et al.*, 1998). Fluid dynamic forces would be greater at the extremities of the streamer and might lead to fracturing at the tips where the limits of cohesion and elasticity for the biofilm matrix may be exceeded. Similarly, for less elastic structures, catastrophic sloughing will occur wherever the elastic limit (Hookes point) is exceeded. This is more likely to occur at 'fault lines' within the matrix (above) than at positions of strength (Moore and others, this volume).

Major irregularities in the topography of the surface and in the biofilm itself are likely to affect turbulence in the flow of liquid past the biofilm. Such turbulence will create rapid fluctuations in the direction of flow in the immediate vicinity of the biofilm surface. This will then be subject to cycles of compressive and decompressive deformation with associated decreases in the mechanical strength of the matrix. Since solutions of polysaccharide are often thixotrophic, then an additional effect of turbulent flow might be to reduce the elasticity of the biofilm matrix and promote sloughing.

(ii) Physiological sloughing. In order for sloughing to occur, the biofilm matrix must be cleaved. This will most probably occur along a line of weakness within the glycocalyx (above). Such lines of weakness might be inherent within the structure or brought about by local modifications within the glycocalyx. The latter might be through a repeated mechanical stress imposed by the turbulent flow of liquid past the biofilm or might result from chemical modification.

Hydrolytic enzymes will hydrolyse polysaccharide, leading to decreases in viscosity, cohesiveness and elasticity (Boyd & Chakrabarty, 1994; Xun *et al.*, 1990). Release of such enzymes by bacteria growing within a biofilm might lead to localized changes in the mechanical properties of the glycocalyx. Overexpression of polysaccharidases by cells positioned at or near to an interactive surface within the biofilm will lead to physiologically mediated erosion. The co-ordinated production of such enzymes by groups of cells within the biofilm matrix will weaken the structure. Whilst several candidate enzymes have been identified and associated with microbial biofilms (Boyd & Chakrabarty, 1994; Xun *et al.*, 1990), their regulation is unclear. If expression of such enzymes is regulated through an environmental trigger, possibly associated with growth rate or nutrient availability or a general stress response (sigma), then the weaknesses in structure are likely to conform to 'physiological fault lines', across or through the matrix. These will correspond to definable positions within the established nutrient/oxygen gradient. In such respects, the disaggregatase of *Methanosarcina mazei* has been reported to be overexpressed under conditions that do not favour growth (Xun *et al.*, 1990). Similarly, Allison *et al.* (1998) have shown a polysaccharidase generated during stationary phase, and under a form of quorum sensing, to be responsible for the dispersal of biofilms of *Pseudomonas fluorescens*.

An alternative strategy by which large aggregates of biofilm-derived materials might be shed from a surface involves surfactants. Whilst surfactants may be deliberately added to a system in order to cleanse a surface, they are also produced by bacteria in order to solubilize hydrophobic nutrients (Pines & Gutnick, 1986), and have been implicated in cellular detachment processes associated with microbial biofilm (Gutnick *et al.*, 1993). In this respect, a variety of amphipathic molecules have been identified which are either produced as extracellular virulence factors or are an intrinsic component of microbial capsules (Gutnick *et al.*, 1993). Many of such molecules have been investigated for their use in industrial processes (Gutnick *et al.*, 1993) and in prevention of dental plaque (Eigen & Simone, 1996). Amphipathic molecules have been identified which are ester-linked to the capsular polysaccharides but which can be released through a cell-surface esterase (Shabtai & Gutnick, 1985). In this respect, esterase activity and 'emulsan' release were enhanced in the presence of growth inhibitory concentrations of chloramphenicol (Rubinovitz *et al.*, 1982). This suggests a possible regulation through nutrient starvation and stringency, which might promote liberation of 'release agents' deep within the biofilm at the interface with the substratum.

Many factors can be seen to impinge upon the stability of biofilms. These clearly involve not only fluid dynamic effects of the surrounding medium but also the physiological status of the cells. In the latter respect, many processes can be identified which might moderate the programmed shedding of biofilm cells to the fluid phase. Detachment from biofilms is advantageous to the organisms not only in terms of promoting genetic diversity and escaping famine, but also in terms of colonizing new niches. In this respect, since it is unlikely that virgin surfaces will be encountered in nature, these released cells must be able to adhere to and immigrate into mature biofilm communities.

IMMIGRATION/SEQUESTRATION OF CELLS TO BIOFILM

Without the opportunity for planktonic cells, or those derived from other biofilm communities, to become integrated into matured populations, biofilms would become highly 'inbred', and severely limited in their scope for community evolution and expansion of the genetic pool. It is inconceivable that all of the species found within a climax community have been present since the initial colonization events took place.

New species and genotypes must be sequestered into the biofilms in order that community structures might evolve other than by mutation/selection. Thus pathogens to the gastro-intestinal tract must contend with the protective commensal biofilm that lines the large intestine. Similarly, *Burkholderia cepacia* infections of cystic fibrosis patients rarely, if ever, precede a *P. aeruginosa* infection; rather *B. cepacia* colonizes an existing *P. aeruginosa* biofilm (Govan & Deretic, 1996). Thus planktonic bacteria are often equally as capable of colonizing a mature biofilm as they are capable of colonizing a virgin surface. In this respect, the ability to produce extracellular products such as antibiotics and bacteriocins, and the polygamous nature of the conjugative exchange of plasmids, take on new significance. Similarly, the ability to express virulence factors such as proteases, lipases and polysaccharidases might equally well enable an organism to establish within a mature biofilm as they promote tissue invasion. Al-Bakri *et al.* (1999) established biofilm populations of laboratory and clinical isolates of *P. aeruginosa* and *B. cepacia* and challenged each with suspension cultures of the other. Stable mixed communities could only be established under antibiotic selection or when surfaces were co-inoculated. Matured *P. aeruginosa* biofilms completely resisted colonization by *B. cepacia*, yet *B. cepacia* communities were rapidly succeeded by *P. aeruginosa* when challenged. In order to promote genetic diversity other than by mutation/selection, a biofilm must be as capable of sequestering planktonic cells of already represented species as it is of unrepresented ones. Gilbert *et al.* (1997) challenged matured biofilms of *Enterococcus faecalis* with genetically marked but otherwise isogenic strains that had either been grown planktonically, grown as biofilms and resuspended, or had been shed spontaneously from such biofilms during their culture in a perfusion model (Hodgson *et al.*, 1995). Each of these culture types possessed SDS-PAGE envelope profiles that were substantially different and which required at least three divisions in the alternate growth environment in order to convert. Challenge of matured biofilm with resuspended biofilm (analogous to encountering a slough) resulted in an almost complete integration of the challenge material into the matured biofilm to give an overall increase in biomass. Daughter cells spontaneously dispersed from biofilms, on the other hand, were unable to adhere to the challenged community and passed through it unhindered. When late exponential phase, planktonic cells were used as the challenge population, however, an overall reduction in the size and vitality of the biofilm community was observed. The extent of this reduction in viability was dependent upon the ratio of planktonic to biofilm cells at the time of challenge. Confocal microscopy using BacLight vitality stain showed that binding had occurred between the planktonic and biofilm phenotypes and that 2–3 h later led to the death and lysis of the original biofilm-associated cells. This interaction was shown to require direct cell–cell contact and the presence of a metabolizable carbon source. It could be demonstrated not to involve the production of extracellular virulence factors such as bacteriocins and did not appear to involve activation of lysogenized phage. In a natural environment, the numbers of planktonic bacteria immigrating to the biofilm will be small. Following integration within the biofilm community, the incoming cells will adopt a biofilm phenotype and the lethal interaction will cease. In such instances, lysis of the adjacent biofilm cells will be limited in extent but will serve to liberate nutrients to the immigrating cells. Such fostering of the immigrant community will greatly enhance the gene pool within the biofilm. We therefore believe that the phenomenon represents a mechanism by which planktonic cells are actively sequestered into a biofilm rather than representing an aggressive takeover. Cells spontaneously dispersed from the biofilm during its normal growth and development were unable to reattach or to bring about the lethal interaction. Such cells required at least two passages in broth culture in order for the aggressive planktonic phenotype to be restored. Such a mechanism would ensure that cells dispersed from a natural biofilm would relocate to a new site rather than disrupt the parent community. Whether or not the phenomenon reported here represents a general property of biofilms or whether it is peculiar to the enterococci remains to be determined. It is noteworthy in these respects that starvation and induction of the general stress response in *E. coli* (Zambrano & Kolter, 1995) can lead to a phenotype that is not only more competitive in its growth than non-stressed cells but which can also directly bring about the death of non-stressed cells (Zambrano *et al.*, 1993). Whilst such phenomena have been reported in variously grown batch cultures, adoption of a killer phenotype by either the biofilm or émigrés from a biofilm would facilitate colonization resistance or invasiveness, respectively. In this respect, the general stress response can be associated with the separate regulation of at least 30 distinct proteins (Zambrano & Kolter, 1995), some of which might be assigned as binding receptors/cell-bound bacteriocins.

CHAOTIC STABILITY

Preceding sections of this article have considered dynamic aspects of the *de novo* formation of biofilm and quasi-steady-state biofilms. Dynamics reflect relative rates of binding, dissociation and growth of individual bacteria at surfaces. From a holistic standpoint, biofilms undergo a formation stage during which numbers and colonization density increase, followed by a maturation phase which can be described mathematically as a pseudo-steady state but which reflects a gradual increase in species complexity. Many formulae have been developed, by those inclined to engineering concepts, which are capable of ascribing the thickness, viscoelasticity, cell density, intercellular spacing, nutrient consumption and chemical gradients associated with biofilms to the prevailing nutrient and fluid dynamic environment (Geesey *et al.*, 1994). Only rarely have such concepts been used to predict the development of structural heterogeneities within biofilm (Kreft *et al.*, 1999; Wimpenny & Colasanti, 1997). It is only when such heterogeneities are fully recognized that the dynamic nature of the quasi-steady state can be fully appreciated. Within a defined segment of biofilm, some fragments will be at maturity and coincidentally colonization resistant, whereas others will have been subject to recent catastrophic collapse. Catastrophic collapse may be regarded as analogous to the physical cleansing of surfaces by either mechanical cleaning or chemical intervention. Residual cells associated with such fragments will have undergone marked changes in nutrient and gaseous environment and in their localized cell density. Such circumstances will render them highly susceptible to successful challenge either with newly sequestered planktonic cells or by sloughs derived from upstream populations. Individual fragments of the biofilm will therefore undergo cycles of 'boom and bust' whilst the net community will appear to be relatively stable.

DYNAMICS IN BIOFILM EVOLUTION

In his monograph '*The Origin of Species by Means of Natural Selection, or the Preservation of Favoured Races in the Struggle for Life*', Darwin applied the concept of selection to individual species (Darwin, 1859), and considered evolutionary outcome in terms of the survival of the 'victorious' races. Thus a gradual, yet spontaneous, variation in the genotype, brought about by mutation, genetic recombination or gene transfer, makes capable the expression of different phenotypes. Some of these new phenotypes affect survival and, thereby, the likelihood and success of propagation of the associated genotype. Such concepts of natural selection are often referred to as 'the survival of the fittest' and imply a destruction of the less fit. From a microbiologist's perspective, this concept is elegantly demonstrated by the emergence of antibiotic-resistant bacteria, which evolve as a direct result of the selection pressure of antibiotic use/abuse (Ferber, 1998; Stuart, 1998). It is, however, difficult to apply such principles to the conjoint success of genes, communities or species (Mayr, 1993). A contemporary hypothesis (Lamarkism), to that of Darwin, was the belief in continual adaptation of individuals to their environment, and passage of the adapted characteristics to the next generation. Realization that changes brought about to a phenotype during its 'lifetime', either through adaptation or mutation in the somatic cells, did not affect the germ line make adaptive hypotheses such as this untenable. Since prokaryotes do not possess a separate germ line and divide by binary fission, microbiology possibly represents the last refuge for adaptive hypotheses. This is particularly the case when considering microbial communities, which might proliferate and disperse as conjoint entities (sloughs), and where the gene pool is substantially larger than that contained within a single community member. In order to accommodate the failure of Darwinism to embrace the competitive success of consortia, a proliferation hypothesis has been proposed (Caldwell *et al.*, 1997). This sets out to recognize propagative and reproductive success at levels of biological organization (community level) other than at the level of the individual organism (species level). In common with the selection and adaptation hypotheses, the proliferation hypothesis recognizes that organisms may proliferate more effectively by adaptation through genetic mutation and recombination, but an important additional tenet is that self-replicating units sometimes proliferate more effectively in partnerships. The rate of evolutionary change in microbes is, however, a million- to a billion-fold more rapid than that of 'higher' organisms (Caldwell *et al.*, 1997) so evolution of microbial communities can take place within the span of a laboratory experiment/human lifetime.

Some prokaryotes might proliferate more effectively if they organized, through endocytosis and symbiosis, as eukaryotic units (Margulis & Guerrero, 1991; Schwemmler, 1989), and eukaryotic cells sometimes proliferate more effectively if they come together as communities. Further differentiation and specialization lead to the development of metazoans, and, most importantly, communities themselves sometimes proliferate most effectively if associated into ecosystems. Whilst Koch's postulates (Koch, 1881, 1884) have been important tenets in our understanding of infectious disease, it is now widely accepted that reductionism in the use of pure cultures is often inappropriate, and in many cases highly misleading. In this fashion, whilst the primary aetiological agents of many infectious diseases are single species, occasional pathologies may be related to the presence of inimical combinations of cells and the aetiological agent is recognized as the climax community. Thus pathologies such as dental caries (Sissons *et al.*, 1996) or ulcerative colitis (Levine *et al.*, 1998) and colon cancer (Roberton, 1993) have been attributed, at least in part, to dynamic changes in the composition and metabolic activity of microbial communities. Proliferation through adaptation and association is therefore postulated as being a primary source of biological diversification and complexity, as opposed to natural selection and competition between individual lineages. Proliferation hypotheses and Darwinian argument are brought into starker contrast when genes rather than individual species, lineages or races are considered as the base unit (Dawkins, 1976). Darwinian theory has, however, been appended in order to explain certain aspects of community behaviour. For example, Maynard Smith (1982) introduced the concept of evolutionary stable strategies. These are adopted by most members of a population or community when they cannot be bettered by alternative individual strategies. Since in this approach each member of the population is an individual and trying to maximize its own chances of survival, then only those community strategies which cannot be bettered by deviant individual strategies will persist (when in Rome . . .).

Biofilms as units of proliferation

Mixed-species biofilms associated with metal corrosion (Hamilton, 1997), dental caries, periodontitis (Bradshaw *et al.*, 1989; Sissons *et al.*, 1996) and biodegradation would fail any test of causality based on Koch's postulates of pathogenicity. In this respect, it is highly fortuitous that the concept of microbial biofilms was introduced (Costerton *et al.*, 1978) and that it has been so embraced (Costerton *et al.*, 1987, 1995). Biofilms are functional consortia of many different individual species, often from very different lineages (moulds, algae, protozoa, nematodes, bacteria, etc.). These often have collective properties that are greater than the sum of the individuals. Structural organization is driven, in such communities, not only by differentiation as it is in metazoan tissues, but also by the spatial arrangement of individual community members and is re-enforced through coaggregation and chemoreceptors (Lawrence *et al.*, 1996), cross-feeding (Allard *et al.*, 1992; Madsen & Aamand, 1992) and cell–cell signalling at an inter- (Davies *et al.*, 1998) and intra-species (Heys *et al.*, 1997; McKenney *et al.*, 1995) level. The provision of numerous, customized microniches within which individual consortia members may flourish (Costerton *et al.*, 1994) forces us to consider natural biofilms as 'units of proliferation' (Caldwell *et al.*, 1997) and catastrophic sloughing rather than the erosion of single cells as reproductive success. If we accept the microbial biofilm as a unit of proliferation, then it would also be pertinent to re-examine the validity of selection and adaptation hypotheses with respect to biofilm evolution and development.

CONCLUSIONS

Biofilm communities must be regarded as complex units of proliferation, which are subject to a continual dynamic change in community structure, composition and expression. Isolation of individual community members, and their separate study, whilst enhancing our understanding at a species level, does little to enhance our understanding of community physiology. As we begin a new millennium, it is to be hoped that recognition of communities, rather than single agents, as the causal agents of disease, fouling, biodegradation and corrosion will increase. Understanding of the inter-relationships within such communities, particularly as they adapt to natural and man-made change, is likely to open up new avenues for their control, and give new insight into their resistance towards antibiotics and biocides.

REFERENCES

Al-Bakri, G., Gilbert, P. & Allison, D. G. (1999). Mixed species biofilms of *Burkholderia cepacia* and *Pseudomonas aeruginosa*. In *Biofilms, the Good, the Bad and the Ugly*, pp. 327–337. Edited by J. Wimpenny, P. Gilbert, J. Walker, M. Brading & R. Bayston. Cardiff: BioLine.

Alexander, M. (1971). *Microbial Ecology*. New York: Wiley.

Allard, A.-S., Hynning, P. A., Remberger, M. & Neilson, A. H. (1992). Role of sulphate concentration in dechlorination of 3,4,5-trichlorocatechol by stable enrichment cultures grown with coumarin and flavanone glycones and aglycones. *Appl Environ Microbiol* **58**, 961–968.

Allison, D. G., Brown, M. R. W., Evans, D. J. & Gilbert, P. (1990a). Surface hydrophobicity and dispersal of *Pseudomonas aeruginosa* from biofilms. *FEMS Microbiol Lett* **71**, 101–104.

Allison, D. G., Brown, M. R. W., Evans, D. J. & Gilbert, P. (1990b). Possible involvement of the division cycle in dispersal of *Escherichia coli* from biofilms. *J Bacteriol* **172**, 1667–1669.

Allison, D. G., Ruiz, B., SanJose, C., Jaspe, A. & Gilbert, P. (1998). Extracellular products as mediators of the formation and detachment of *Pseudomonas fluorescens* biofilms. *FEMS Microbiol Lett* **167**, 179–184.

Applegate, D. H. & Bryers, I. D. (1991). Effects of carbon and oxygen limitations and calcium concentrations on biofilm removal processes. *Biotechnol Bioeng* **37**, 17–25.

Bernet-Camard, M. F., Lievin, V., Brassart, D., Neeser, J. R., Servin, A. L. & Hudault, S. (1997). The human *Lactobacillus acidophilus* strain LA1 secretes a nonbacteriocin antibacterial substance(s) active *in vitro* and *in vivo*. *Appl Environ Microbiol* **63**, 2747–2753.

Boureau, H., Salanon, C., Decaens, C. & Bourlioux, P. (1994). Caecal localization of the specific microbiota resistant to *Clostridium difficile* colonization in gnotobiotic mice. *Microb Ecol Health Dis* **7**, 1111–1117.

Boyd, A. & Chakrabarty, A. M. (1994). Role of alginate lyase in cell detachment of *Pseudomonas aeruginosa*. *Appl Environ Microbiol* **60**, 2355–2359.

Boyle, J. D., Dodds, I., Lappin-Scott, H. & Stoodley, P. (1999). Limits to growth and what keeps a biofilm finite. In *Biofilms, the Good, the Bad and the Ugly*, pp. 303–315. Edited by J. Wimpenny, P. Gilbert, J. Walker, M. Brading & R. Bayston. Cardiff: BioLine.

Bradshaw, D. J., McKee, A. S. & Marsh, P. D. (1989). Effects of carbohydrate pulses and pH on population shifts within oral communities *in-vitro*. *J Dent Res* **68**, 1298–1302.

Brook, I., Tom, S. P. & Ledney, G. D. (1993). Quinolone and glycopeptide therapy for infection in mice following exposure to mixed-field neutron-γ-irradiation. *Int J Radiat Biol* **64**, 771–777.

Caldwell, D. E., Wolfaardt, G. M., Korber, D. R. & Lawrence, J. R. (1997). Do bacterial communities transcend Darwinism? *Adv Microb Ecol* **15**, 105–191.

Chang, H. T., Rittmann, B. E., Amar, D., Heim, R., Ehlinger, O. & Lesty, Y. (1991). Biofilm detachment mechanisms in a liquid-fluidized bed reactor. *Biotechnol Bioeng* **38**, 499–506.

Conway, P. (1997). Development of the intestinal microbiota. In *Gastrointestinal Microbiology*, pp. 3–38. Edited by R. I. Mackie, B. A. White & R. E. Isaacson. New York: Chapman & Hall.

Cooperstock, M. S. & Zedd, A. J. (1983). Intestinal flora of infants. In *Human Intestinal Microflora in Health and Disease*, p. 79. Edited by D. J. Hentges. London: Academic Press.

Costerton, J. W., Geesey, G. G. & Cheng, K. J. (1978). How bacteria stick. *Sci Am* **238**, 86–95.

Costerton, J. W., Cheng, K. J., Geesey, G. G., Ladd, T. I., Nickel, J. C. & Dasgupta, M. (1987). Bacterial biofilms in nature and disease. *Annu Rev Microbiol* **41**, 435–464.

Costerton, J. W., Lewandowski, Z., Caldwell, D. E., Korber, D. R. & Lappin-Scott, H. M. (1994). Biofilms: the customised microniche. *J Bacteriol* **176**, 2137–2142.

Costerton, J. W., Lewandowski, Z., Caldwell, D. E., Korber, D. R. & Lappin-Scott, H. M. (1995). Microbial biofilms. *Annu Rev Microbiol* **49**, 711–745.

Darwin, C. (1859). *The Origin of the Species by Means of Natural Selection, or the*

Preservation of Favoured Races in the Struggle for Life. New York: New American Library.

Davies, D. G., Parsek, M. R., Pearson, J. P., Iglewski, B. H., Costerton, J. & Greenberg, E. P. (1998). The involvement of cell-to-cell signals in the development of a bacterial biofilm. *Science* **280**, 295–298.

Dawkins, R. (1976). *The Selfish Gene*. Oxford: Oxford University.

Degnan, B. A. & Macfarlane, G. T. (1995). Arabinogalactan utilization in continuous cultures of *Bifidobacterium longum* – effect of co culture with *Bacteroides thetaiotaomicron*. *Anaerobe* **2**, 103–112.

Dykhuizen, D. E. & Hartl, D. L. (1983). Selection in chemostats. *Microbiol Rev* **47**, 150–168.

Eigen, E. & Simone, A. J. (1996). Control of Dental Plaque and Caries. US Patent No. 4 619825.

Englyst, H. N., Hay, S. & Macfarlane, G. T. (1987). Polysaccharide breakdown by mixed populations of human faecal bacteria. *FEMS Microbiol Ecol* **95**, 163–171.

Ferber, D. (1998). New hunt for the roots of resistance. *Science* **280**, 280–289.

Geesey, G. G., Stupy, M. W. & Bremer, P. J. (1992). The dynamics of biofilms. *Int Biodeterior Biodegrad* **30**, 135–154.

Geesey, G. G., Lewandowski, Z. & Flemming, H.-C. (1994). *Biofouling and Biocorrosion in Industrial Water Systems*. Boca Raton, FL: Lewis.

Gibson, G. R. & Wang, X. (1994). Regulatory effects of bifidobacteria on the growth of other colonic bacteria. *J Appl Bacteriol* **77**, 412–420.

Gilbert, P., Evans, D. J., Evans, E., Duguid, J. G. & Brown, M. R. W. (1991). Surface characteristics and adhesion of *Escherichia coli* and *Staphylococcus epidermidis*. *J Appl Bacteriol* **71**, 72–77.

Gilbert, P., Evans, D. J. & Brown, M. R. W. (1993). Formation and dispersal of bacterial biofilms in vivo and in situ. *J Appl Bacteriol* **74** (suppl.), 67S–68S.

Gilbert, P., Allison, D. G., Jacob, A., Korber, D., Wolfaardt, G. & Foley, I. (1997). Immigration of planktonic *Enterococcus faecalis* cells into mature *E. faecalis* biofilms. In *Biofilms: Community Interactions and Control*, pp. 133–142. Edited by J. W. T. Wimpenny, P. Handley, P. Gilbert, H. M. Lappin-Scott & M. Jones. Cardiff: BioLine.

Govan, J. R. W. & Deretic, V. (1996). Microbial pathogenesis in cystic fibrosis: mucoid *Pseudomonas aeruginosa* and *Burkholderia cepacia*. *Microbiol Rev* **60**, 539–574.

Gutnick, D. L., Avigad, R., Blatt, Y., Minas, W. & Allon, R. (1993). Amphipathic microbial capsules as industrial products. *J Appl Bacteriol* **74** (suppl.), 125S–135S.

Hamilton, W. A. (1997). Microbially influenced corrosion as a suitable model for understanding the true nature of biofilm community interactions. In *Biofilms: Community Interactions and Control*, pp. 143–148. Edited by J. Wimpenny, P. Handley, P. Gilbert, H. Lappin-Scott & M. Jones. Cardiff: BioLine.

Heys, S. J. D., Gilbert, P. & Allison, D. G. (1997). Homoserine lactones and bacterial biofilms. In *Biofilms: Community Interactions and Control*, pp. 103–112. Edited by J. W. T. Wimpenny, P. Handley, P. Gilbert, H. M. Lappin-Scott & M. Jones. Cardiff: BioLine.

Hodgson, A. E., Nelson, S. J., Brown, M. R. W. & Gilbert, P. (1995). A simple *in vitro* method for growth control of bacterial biofilms. *J Bacteriol* **79**, 87–93.

Hoyle, B. D., Wong, C. K. W. & Costerton, J. W. (1992). Disparate efficacy of tobramycin on Ca^{2+}-treated, Mg^{2+}-treated, and HEPES-treated *Pseudomonas aeruginosa* biofilms. *Can J Microbiol* **38**, 1214–1218.

Keevil, C. W. (1993). Methods for assessing the activity of biofilm *in situ*. In *Bacterial Biofilms and their Control in Medicine and Industry*, pp. 45–47. Edited by J. Wimpenny, W. Nichols, D. Stickler & H. Lappin-Scott. Cardiff: BioLine.

Koch, R. (1881). Methods for the study of pathogenic organisms. *Mitt Kaiserlichen Gesundheitsampte* **1**, 1–48.

Koch, R. (1884). The aetiology of tuberculosis. *Mitt Kaiserlichen Gesundheitsampte* **2**, 1–88.

Kreft, J. U., Picioreanu, C., Wimpenny, J. & van Loosdrecht, M. (1999). Individual-based modelling of biofilms: why? In *Biofilms, the Good, the Bad and the Ugly*, pp. 257–262. Edited by J. Wimpenny, P. Gilbert, J. Walker, M. Brading & R. Bayston. Cardiff: BioLine.

Larson, H. E. & Welch, A. (1993). *In vitro* and *in vivo* characterization of resistance to colonization with *Clostridium difficile*. *J Med Microbiol* **38**, 103–108.

Lawrence, J. R., Delaquis, P. J., Korber, D. R. & Caldwell, D. E. (1987). Behaviour of *Pseudomonas fluorescens* within the hydrodynamic boundary layers of surface microenvironments. *Microbiol Ecol* **14**, 1–14.

Lawrence, J. R., Korber, D. R. & Caldwell, D. E. (1992). Behavioral-analysis of *Vibrio parahaemolyticus* variants in high-viscosity and low-viscosity microenvironments by use of digital image-processing. *J Bacteriol* **174**, 5732–5739.

Lawrence, J. R., Korber, D. R., Wolfaardt, G. M. & Caldwell, D. E. (1996). Surface colonisation strategies of biofilm forming bacteria. *Adv Microb Ecol* **14**, 1–75.

Levine, J., Ellis, C. J., Furne, J. K., Springfield, J. & Levitt, M. D. (1998). Fecal hydrogen sulfide production in ulcerative colitis. *Am J Gastroenterol* **93**, 83–87.

Macfarlane, S., McBain, A. J. & Macfarlane, G. T. (1997). Consequences of biofilm and sessile growth in the large intestine. *Adv Dent Res* **11**, 59–68.

Macfarlane, G. T., Cummings, J. H. & Allison, C. (1986). Protein-degradation by human intestinal bacteria. *J Gen Microbiol* **132**, 1647–1656.

McKenney, D., Brown, K. E. & Allison, D. G. (1995). Influence of *Pseudomonas aeruginosa* exoproducts on virulence factor production in *Burkholderia cepacia*: evidence of interspecies communication. *J Bacteriol* **177**, 6989–6992.

Mackie, R. I., Sghir, A. & Gaskins, H. R. (1999). Developmental microbiology of the gastrointestinal tract. *Am J Clin Nutr* **69**, S1035–S1045.

Madsen, T. & Aamand, J. (1992). Anaerobic transformation and toxicity of trichlorophenols in a stable enrichment culture. *Appl Environ Microbiol* **58**, 557–561.

Margulis, L. & Guerrero, R. (1991). Two plus three equals one: individuals emerge from bacterial communities. In *Gaia 2. Emergence: the New Science of Becoming*, pp. 60–67. New York: Lindisfarne.

Marsh, P. D. & Bradshaw, D. J. (1998). Dental plaque: community spirit in action. In *Microbial Pathogenesis: Current and Emerging Issues*, pp. 41–53. Edited by D. J. LeBlanc, M. S. Lanz & L. M. Switalski. Indianapolis: University of Indiana.

Marsh, P. D. & Bradshaw, D. J. (1999). Microbial community aspects of dental plaque. In *Dental Plaque Revisited: Oral Biofilms in Health and Disease*, pp. 237–253. Edited by H. N. Newman & M. Wilson. Cardiff: BioLine.

Marshall, K. C. (1985). Mechanisms of bacterial adhesion at solid-water interfaces. In *Bacterial Adhesion: Mechanisms and Physiological Significance*, pp. 133–162. Edited by D. C. Savage & M. Fletcher. New York: Plenum.

Marshall, K. C. (1992). Biofilms, an overview of bacterial adhesion, activity and control at surfaces. *ASM News* **58**, 202–207.

Maynard Smith, J. (1982). *Evolution and the Theory of Games*. Cambridge: Cambridge University Press.

Mayr, E. (1993). What was the evolutionary synthesis? *Trends Ecol Evol* **8**, 31–34.

Monod, J. (1949). The growth of bacterial cultures. *Annu Rev Microbiol* **3**, 371–394.

Pines, O. & Gutnick, D. (1986). Role for emulsan in growth of *Acinetobacter calcoaceticus* rag-1 on crude-oil. *Appl Environ Microbiol* **51**, 661–663.

Roberton, A. M. (1993). Roles of endogenous substances and bacteria in colorectal-cancer. *Mutat Res* **290**, 71–78.

Rubinovitz, P. C., Passador, L. & Rosenberg, E. (1982). Emulsan production by *Acinetobacter calcoaceticus* in the presence of chloramphenicol. *J Bacteriol* **152**, 126–132.

Savage, D. S. (1977). Microbial ecology of the gastrointestinal tract. *Annu Rev Microbiol* **31**, 107–133.

Schwemmler, W. (1989). *Symbiogenesis: a Macro-mechanism of Evolution*. Berlin: Walter de Gruyter.

Shabtai, Y. & Gutnick, D. L. (1985). Exocellular esterase and emulsan release from the cell surface of *Acinetobacter calcoaceticus* RAG-I. *J Bacteriol* **161**, 1176–1181.

Sissons, C. H., Wong, L. & Cutress, T. W. (1996). Inhibition by ethanol of the growth of biofilm and dispersed microcosm dental plaques. *Arch Oral Biol* **41**, 27–34.

Stewart, P. S. (1993). A model of biofilm detachment. *Biotechnol Bioeng* **41**, 111–117.

Stoodley, P., Lewandowski, Z., Boyle, J. D. & Lappin-Scott, H. M. (1998). Oscillation characteristics of biofilm streamers in turbulent flowing water as related to drag and pressure drop. *Biotechnol Bioeng* **57**, 536–544.

Stoodley, P., Lewandowski, Z., Boyle, J. D. & Lappin-Scott, H. M. (1999a). The formation of migratory ripples in a mixed species bacterial biofilm growing in turbulent flow. *Environ Microbiol* **1**, 447–455.

Stoodley, P., Lewandowski, Z., Boyle, J. D. & Lappin-Scott, H. M. (1999b). Structural deformation of bacterial biofilms caused by short-term fluctuations in fluid shear: an *in situ* investigation of biofilm rheology. *Biotechnol Bioeng* **65**, 83–92.

Stuart, B. L. (1998). Multi-drug resistance – a sign of the times. *N Engl J Med* **338**, 1376–1378.

Tannock, G. W. (1994). The acquisition of the normal microflora of the gastrointestinal tract. In *Human Health: the Contribution of Microorganisms*, pp. 1–16. Edited by S. A. W. Gibson. London: Springer.

Vroom, J. M., de Grauw, K. J., Gerritsen, H. C., Bradshaw, D. J., Marsh, P. D., Watson, G. K., Birmingham, J. J. & Allison, C. (1999). Depth penetration and detection of pH gradients in biofilms by two-photon excitation microscopy. *Appl Environ Microbiol* **65**, 3502–3511.

Willcock, L., Holah, J., Allison, D. G. & Gilbert, P. (1997). Population dynamics and steady-state biofilms: effects of growth environment upon dispersal. In *Biofilms: Community Interactions and Control*, pp. 23–31. Edited by J. Wimpenny, P. Handley, P. Gilbert, H. Lappin-Scott & M. Jones. Cardiff: BioLine.

Wimpenny, J. (1995). How thick must a biofilm be for gradients to determine its physiology. In *The Life and Death of Biofilm*, pp. 109–112. Edited by J. Wimpenny, P. Handley, P. Gilbert & H. Lappin-Scott. Cardiff: BioLine.

Wimpenny, J. W. T. & Colasanti, R. A. (1997). Unifying hypothesis for the structure of microbial biofilms based on cellular automaton models. *FEMS Microbiol Ecol* **22**, 1–16.

Xun, L. Y., Mah, R. A. & Boone, D. R. (1990). Isolation and characterization of disaggregatase from *Methanosarcina mazei*. *Appl Environ Microbiol* **56**, 3693–3698.

Zambrano, M. & Kolter, R. (1995). Changes in bacterial cell properties in going from exponential growth to stationary phase. In *Microbial Quality Assurance: a Guide Towards Relevance and Reproducibility of Inocula*, pp. 21–30. Edited by M. R. W. Brown & P. Gilbert. Boca Raton, FL: CRC Press.

Zambrano, M., Siegele, D. A., Almiron, M., Tormo, A. & Kolter, R. (1993). Microbial competition: *Escherichia coli* mutants that take over stationary phase culture. *Science* **259**, 1757–1760.

Biodegradation by biofilm communities

Gideon M. Wolfaardt,[1] Darren R. Korber,[2]
Subramanian Karthikeyan[2] and Douglas E. Caldwell[2]

[1] Department of Microbiology, University of Stellenbosch, Private Bag X1, 7602 Stellenbosch, South Africa
[2] Department of Applied Microbiology & Food Science, University of Saskatchewan, Saskatoon, SK, Canada S7N 5A8

INTRODUCTION

It has been shown that microbial communities contribute extensively to the attenuation, mineralization and transport of both organic and inorganic contaminants in the environment. The development of biofilms by microbial communities is often a key factor contributing to the overall efficiency of these processes (Rothemund *et al.*, 1996). For instance, bacterial biofilms are able to accumulate metals through various mechanisms (Marques *et al.*, 1991; Sillitoe *et al.*, 1994). Liehr *et al.* (1994) showed that biofilms formed by algae could concentrate metals at levels more than four orders of magnitude higher than those in the surrounding water.

The potential of bioremediation as an alternative to physical and chemical remediation strategies has resulted in a significant amount of research effort on degradative biofilms. Although much emphasis has been placed on the degradation of xenobiotic compounds, the knowledge gained through these studies has also contributed to an improved understanding of processes involved in the degradation of naturally occurring molecules as well as nutrient cycling in general. Tank & Webster (1998) suggested that competition for nutrients might regulate heterotrophic microbial processes in natural streams. In their study, they found that nutrient immobilization by leaves partially inhibited other heterotrophic processes, as evidenced by low microbial respiration, fungal biomass and extracellular enzyme activity among wood biofilms in the presence of leaf litter. Lawrence *et al.* (1998) stressed the applicability of knowledge gained through the study of naturally occurring attenuation mechanisms to remediating contaminated environments. Clearly, the study of degradative biofilms is of both fundamental and applied interest.

SGM symposium 59: Community structure and co-operation in biofilms. Editors D. Allison, P. Gilbert, H. Lappin-Scott, M. Wilson. Cambridge University Press. ISBN 0 521 79302 5 ©SGM 2000

PURE CULTURES AND BIOFILMS WITH KNOWN SPECIES COMPOSITION

The study of biofilms formed by micro-organisms grown in pure culture continues to significantly contribute to efforts aimed at understanding the interactions of adherent cells with their physical and chemical environment. Such approaches are especially valuable when processes such as pure culture fermentations are of interest, or when there is a need to evaluate a specific chemical or physical control on the behaviour of selected strains. The study of pure culture biofilms also permits specific phenomena, such as stratification of respiratory activity by individual species, to be investigated (Huang *et al.*, 1998; Villaverde & Fernandez-Polanco, 1999). In addition, there is often a need to study the interactions between two organisms, for example to evaluate syntrophic co-operation or antagonism/inhibition of one by the other through the production of bacteriocin. S. Karthikeyan, G. M. Wolfaardt, D. R. Korber & D. E. Caldwell (unpublished results) used continuous flow chambers in combination with confocal scanning laser microscopy to study the association between the benzoate-degrading *Pseudomonas* strain BD1 and the benzoate-sensitive *Pseudomonas* strain BS2. When they plated the effluent from these flow cells on agar containing 0.10–0.25% (w/v) benzoic acid, BS2 formed satellite colonies near the BD1 (primary) colonies (Karthikeyan *et al.*, 1999). Even when an additional labile carbon source such as glucose was added to the medium, strain BS2 could only form colonies in the vicinity of BD1. In contrast, BS2 formed colonies at lower benzoate concentrations (0.015–0.05%) independent of the benzoate-degrading isolate BD1. Strain BS2 was also able to grow on a minimal salts medium supplemented with glucose as sole carbon source in the absence of BD1. Based on these observations, it was suggested that BS2 was not dependent on BD1 for growth factors or the supply of benzoate-degradation intermediates, and that the association between the two organisms could be explained in terms of the protective role by the primary colonies. This protection enabled BS2 to grow at benzoate concentrations that were otherwise inhibitory. Cross-streak assays further indicated that the primary colonies were able to support growth of unrelated organisms, suggesting toxicity reduction as a general mechanism for the satellite effect.

Of particular interest is to evaluate whether the interactions observed in a diffusion-dominated environment, such as the satellite effect on semi-solid media, can also be established and maintained in flowing environments dominated by advective flow. Intuitively, it seems unlikely, unless specific mechanisms exist to preserve protective associations such as toxicity reduction by lowering the concentrations of inhibitory compounds. Muralidharan *et al.* (1997), who studied hydrogen transfer between the two hyperthermophilic micro-organisms *Thermotoga maritima* (an anaerobic heterotroph) and *Methanococcus jannaschii* (a hydrogenotrophic methanogen), provided evidence for co-operation in aqueous environments. It was demonstrated that

Fig. 1. Greyscale micrograph of biofilm formation by a binary culture of the benzoate-degrading *Pseudomonas* strain BD1 (darker cells) and the green fluorescent protein labelled benzoate-sensitive *Pseudomonas* strain BS2 (bright cells). BS2 required the presence of BD1 for biofilm development at benzoic acid concentrations >0.1 % (w/v).

the numbers of *T. maritima* increased 10-fold when co-cultured with *M. jannaschii*. Furthermore, because of this association, *M. jannaschii* was able to grow in the absence of externally supplied H_2 and CO_2. However, the co-culture could not be established if the two organisms were physically separated by a dialysis membrane, suggesting that close juxtapositioning and the maintenance of a controlled microenvironment were required between the two organisms. Additional support for this contention is the study by S. Karthikeyan, G. M. Wolfaardt, D. R. Korber & D. E. Caldwell (unpublished results), which showed that biofilm development by the benzoate-sensitive strain BS2 was increased ~10-fold in the presence of the benzoate-degrading strain BD1 (Fig. 1). A possible mechanism employed by interacting organisms in flowing environments is the establishment of a diffusion-dominated microenvironment through the production of extracellular polymeric substance (EPS). Various groups attributed the establishment of a stable microenvironment to microbial EPS production. In what they referred to as 'a heterogeneous organization of cells', Boult *et al.* (1997) suggested that the diffusion properties and presence of charged reactive groups of EPS allowed the establishment and maintenance of steady state conditions. This 'organization of cells' is mostly associated with solid surfaces. However, relatively extensive areas that are void of cells are often observed between cell clusters and the substratum (Fig. 2). These voids are usually filled with EPS.

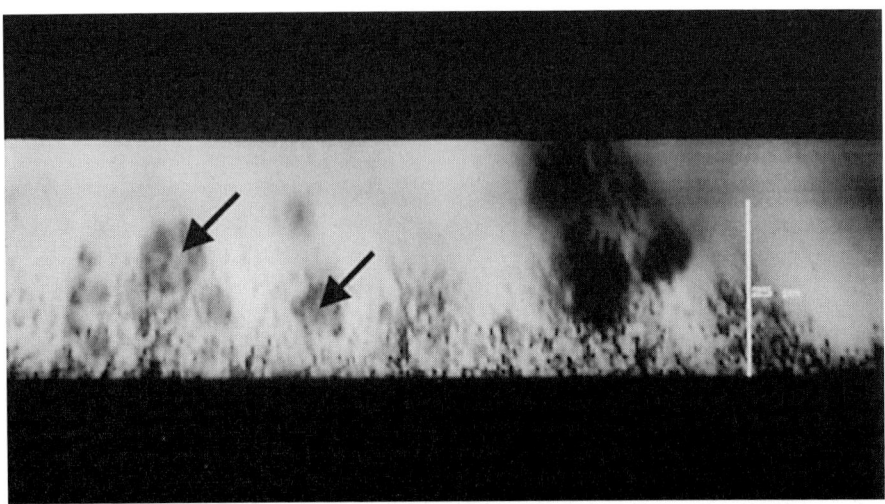

Fig. 2. *xz* section showing cell clusters that are positioned at some distance away from the attachment surface (indicated). Although biofilm formation is a surface-associated phenomenon, the substratum often contains relatively low cell densities. EPS plays a role in maintaining this biofilm architecture.

Decho (1999) reviewed the role of EPS in the chemical communication within biofilms. Notably, the author discussed the inherent hindered diffusion and sorption properties of EPS that may create a diffusion barrier capable of localizing chemicals released by the cells. It was further pointed out that the high cell densities in biofilms most likely facilitate the formation of autoinducer–receptor complexes – a phenomenon that requires high extracellular concentrations of the autoinducer. Wingender *et al.* (1999) pointed out that a common feature among aggregated forms of microbial existence, such as biofilms, is that the cells are embedded in an EPS matrix. In addition to its possible role in the localization of chemical signals as discussed above, this matrix also contributes to the functioning of biofilms in various other ways, including the attenuation of metals (Lawrence *et al.*, 1998) and the accumulation of organic substrates (Wolfaardt *et al.*, 1995).

H. Nel, L. M. Dicks & G. M. Wolfaardt (unpublished results) studied inhibition of a *Staphylococcus* strain by a bacteriocin-producing *Lactobacillus* species (Fig. 3). In addition to offering a promising alternative to the use of antibiotics and biocides for the control of pathogenic or spoilage bacteria, studying the regulation of bacteriocin production in nature poses interesting ecological questions. While it is unlikely that bacteriocin production by one organism leads to the killing of target organisms in natural environments (it is more likely to lead to the inhibition or partial suppression of

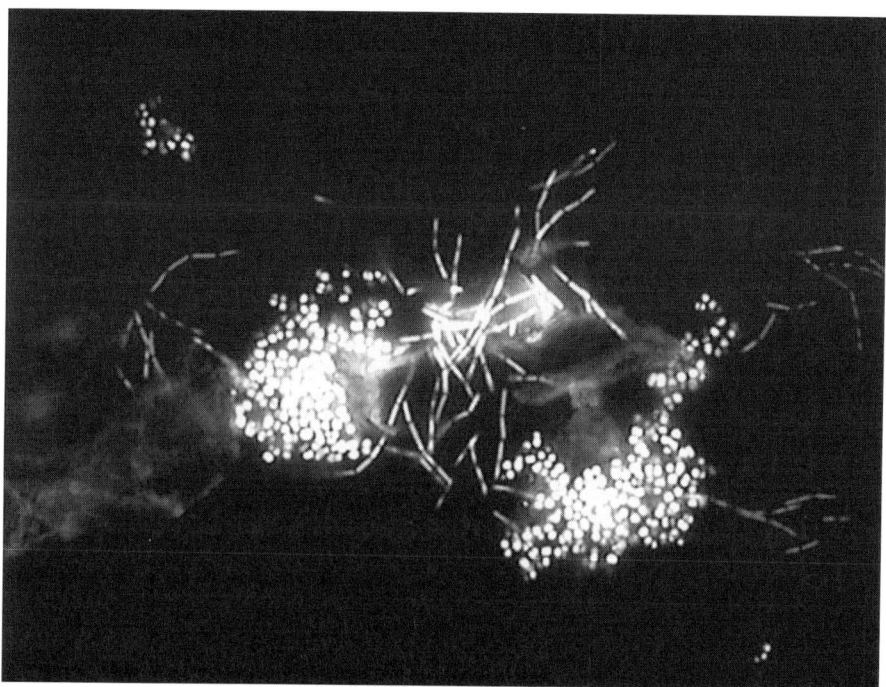

Fig. 3. Co-cultivation of binary cultures in continuous flow chambers used to study interactions between two cell lines. Obvious differences in cell morphology, such as this bacteriocin-producing *Lactobacillus* (rod-shaped cells) and a potential target *Staphylococcus* sp , facilitate easy distinction between the organisms. However, it is often difficult or impossible to distinguish between two species based on cell morphology and thus markers are ideally required for identifying each cell line. Fluorescently labelled antibodies and other recent developments, such as green fluorescent protein (see Fig. 1) and the *in situ* visualization of gene expression, offer powerful techniques in this regard.

target organisms), it is also unlikely that bacteriocins are constitutively produced uncontrolled if no mechanism or suitable environment existed to maintain concentrations high enough to benefit the producer organism. Relatively few studies have evaluated the occurrence of inhibition between members in degradative biofilm communities, and if these interactions exist, their effects on overall degradative efficiency. Therefore, because of a limited amount of information, this can at best be a subject of speculation. However, when viewed from an ecological perspective, it can be argued that even though certain members of a community are inhibited, the production of bacteriocins is one form of self-regulation that contributes to overall community stability. Degradation rates of contaminants may thus be impeded if the community members inhibited by the bacteriocin are primary degraders. An example would be low-level contamination, often observed in groundwaters, where the inhibition of opportunistic organisms may be an internal control for preservation of the community. Indeed, in oligotrophic environments, strict control of microbial activity is required to prevent complete exhaustion of available nutrients and subsequent collapse of the community. Thus partial inhibition between different members of a natural community in order to maintain metabolic activity that is in balance with the flux of nutrients, including contaminants, may result in slower biodegradation rates. This also provides the basis for addition of nutrients to oligotrophic contaminated sites. The controls on bacteriocin production in natural settings, and especially in biofilms, are not well understood. A better understanding of these interactions and subsequent manipulations of cell–cell inhibition within degradative biofilms will clearly contribute to improved biodegradation strategies.

Even though binary associations usually occur in the environment in the presence of other micro-organisms, studies conducted in the absence of the other organisms are often a critical initial step in defining associations as other species may interfere in various ways, including protection, competition and predation. Predators including amoebae, flagellates and ciliates are usually present in biofilms when soil or water from the environment is used as inoculum. They can have a significant impact on biofilm structure and species composition, including degradative biofilms. Møller *et al.* (1997) observed that seemingly stable degradative biofilms, maintained over extended periods in flow cells, can be reduced to isolated microcolonies within less than 24 h following a sudden change in the predator–prey numbers. Similar observations were made during studies to evaluate the integration of a variety of yeasts into a predominant bacterial biofilm community (G. M. Wolfaardt & A. Botha, unpublished data). Even though predators were observed from day 1, the biofilm structure remained relatively stable for 12 days until a sudden increase in the number of predators was observed which resulted in a significant reduction in biofilm biomass in less than 12 h. There was no obvious explanation for this phenomenon, although it was possible that the number of a specific sub-population of the prey organisms, or their extracellular excretions, reached a threshold minimum required to initiate a sudden increase in predator numbers. J. R. Lawrence, G. M. Wolfaardt & R. A. Snyder (unpublished data) previously demonstrated a relationship between predation and the composition of EPS produced by a degradative biofilm community. Furthermore, selective feeding has been observed in biofilms degrading labile substrates as well as chlorinated aromatic compounds, making it unrealistic to use data obtained in one predator–prey association to predict the potential effect of predation on degradation efficiency in another. Therefore, when the objective is to evaluate the effect of specific controls on biofilm behaviour, it may be necessary to eliminate variables such as predation. Fig. 4 shows the effect of predation by protistan feeders on the structure of a degradative biofilm.

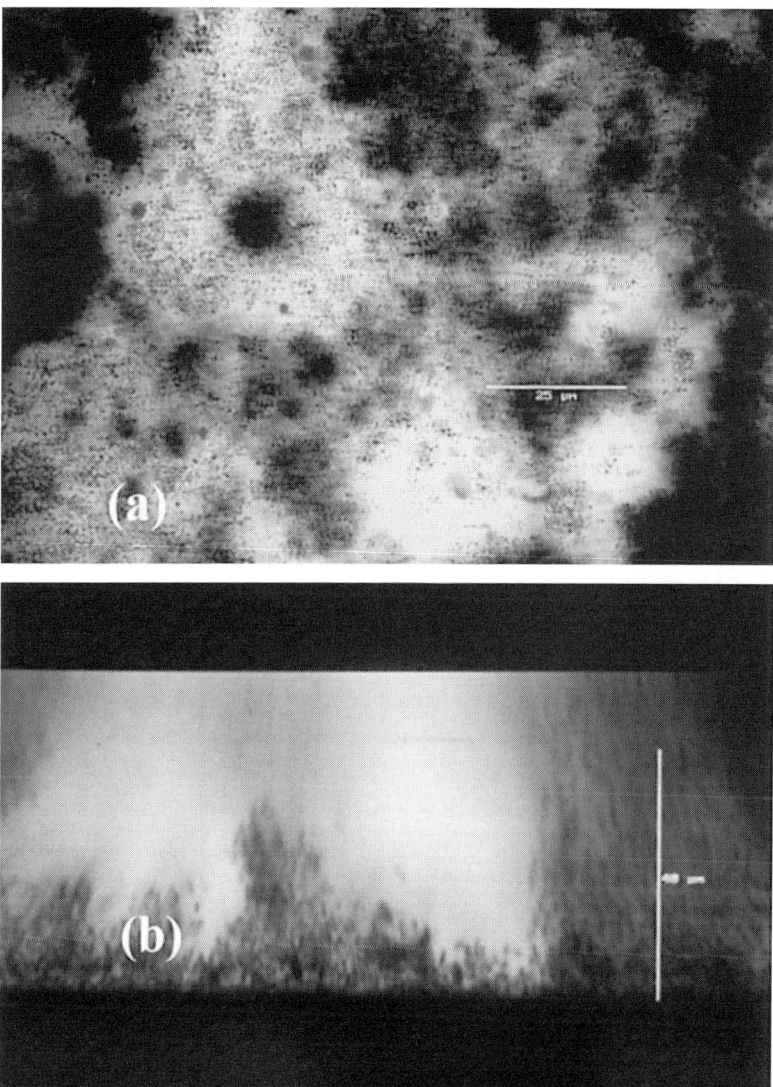

Fig. 4. Effect of predation on biofilm structure. (a) Figure showing how selective feeding on a diclofop-degrading biofilm (Wolfaardt *et al.*, 1994b), in this case by flagellates, results in extensive voids and a highly irregular biofilm structure. (b) *xz* section through such a void area. Note the 'pillar' in the centre that resisted predation. Observations of the feeding behaviour of these predators often reveal preferential feeding in certain areas of a biofilm.

Elvers *et al.* (1998) discussed earlier work indicating that microbial interactions are usually classified based on the effect of the interaction on each participating population in a binary association. These authors listed six types of interactions that can occur between binary populations, namely neutralism, mutualism, commensalism, amensalism, prey–predator relationships and competition. Elvers *et al.* (1998) further suggested that an increased number of species within a mixed community results in higher numbers of interactions. It thus seems reasonable that the cultivation of pure or binary culture would be preferred to evaluate the effect of physical (for example, temperature) or chemical (for example, nutrient type) parameters on a selected member of a degradative community under certain conditions. Indeed, to evaluate toluene degradation in a biological trickling filter, Pedersen *et al.* (1997) isolated various degraders from a multispecies toluene-degrading biofilm, and selected a *Pseudomonas putida* strain as a representative of the toluene-degrading population. They found that the selected *P. putida*, which was also genetically well characterized, played a relatively minor part (~11%) in the total toluene degradation by the mixed community.

HETEROGENEOUS BIOFILM COMMUNITIES

The range of conditions under which a microbial species can survive is defined, and thus restricted, by its limited genetic and physiological capabilities (Caldwell, 1993). One way in which individual species can extend their habitat range, and thereby improve their own overall reproductive success, is to form associations with other organisms. It is widely accepted that the establishment and maintenance of these associations are facilitated through spatial organization and the development of favourable microenvironments in biofilms. Earlier work focusing on the syntrophic relationships in anaerobic digesters, including those describing interspecies hydrogen transfer, demonstrated the advantage of microbial interactions to participating members of multispecies aggregates (see, for example, Wilson *et al.*, 1985; Macleod *et al.*, 1990). It has been suggested that similar spatial organization and interspecies associations present in aggregates are also typically found in biofilms. For example, the layered structure of the aggregates in an anaerobic sludge digester described by Macleod *et al.* (1990), where the central core contains obligate anaerobes with facultative anaerobes and aerobes positioned in layers around the centre, also fits the biofilm model. When aggregates from an anaerobic system and microcolonies formed by the same microbial community grown as biofilms are compared, striking similarities are often observed (Fig. 5). In the example shown in Fig. 5, both the aggregates and microcolonies consist of densely packed clusters connected by long filaments. Using 16S rRNA oligonucleotide probes specific for the acetoclastic methanogens *Methanosaeta concilii* and *Methanosarcina barkeri*, Rocheleau *et al.* (1999) studied the spatial organization of the organisms in granules from an upflow anaerobic sludge blanket bioreactor that treated food production wastewater. These authors observed distinct layered structures. Because of the inherent difficulty in immobilizing

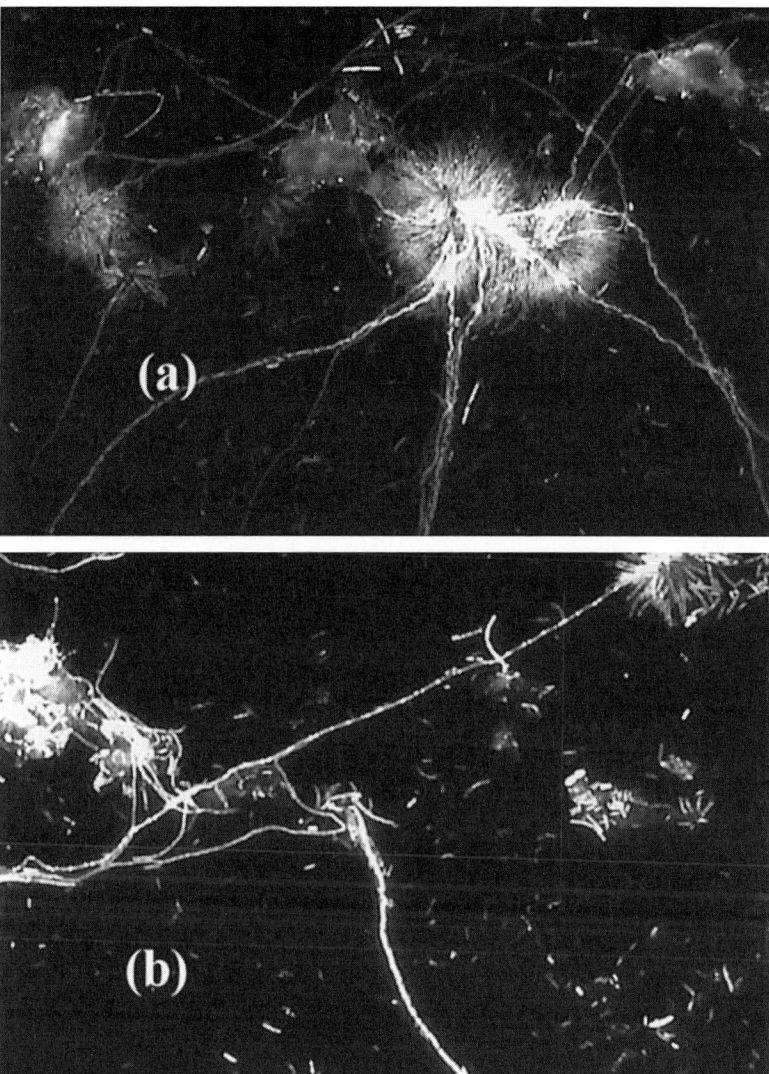

Fig. 5. Micrographs showing the resemblance between aggregates (formed in suspension) (a) and microcolonies (grown as biofilms) (b) formed by an anaerobic sludge community. Biofilm studies offer a useful substitute to aggregates because of the inherent difficulty in performing temporal analyses on aggregates.

granules for microscopy while maintaining aggregate structure and gradients such as oxygen concentration profiles, biofilm studies have been suggested as a substitute.

Another area of biofilm research that has contributed to our understanding of co-operation within heterogeneous biofilm communities is the study of oral biofilms. These biofilms are characterized by a wide range of interactions that occur between species to modify their microenvironment into a more favourable one (Sjollema *et al.*, 1990). Furthermore, within this microenvironment the different members show increased resistance to antimicrobial agents (Wilson *et al.*, 1998), a property generally attributed to biofilms, and one which is important for biofilms involved in the degradation of toxic compounds.

Semple *et al.* (1999) indicated that it is the exception rather than the rule to find a single micro-organism capable of completely degrading a mixture of xenobiotics, or even a single pollutant, under environmental conditions. Furthermore, biofilm formation is often associated with degradative communities, and in some cases is even a prerequisite. For this reason, Mirpuri *et al.* (1997) stated that for superior bioreactor design it is necessary to investigate changes in cellular activity that occur during biofilm development under conditions similar to reactor operation before predictive models for bioreactor systems can be developed. In their study of a biofilm consortium consisting of at least nine bacterial members, Wolfaardt *et al.* (1994a) found that together these organisms could mineralize 65% of ^{14}C-labelled diclofop-methyl to ^{14}CO$_2$. In contrast, when grown in pure cultures, none of the nine bacterial isolates could degrade diclofop. When a labile exogenous carbon source was added to the pure culture incubations, two of the bacteria could then mineralize diclofop to CO$_2$. The authors suggested that this induction of diclofop degradation by the presence of a labile substrate could either indicate that diclofop was used as a secondary substrate where the micro-organisms obtained most of their energy and carbon from another (more labile) nutrient, or that it could be an example of co-metabolism – the degradation of a non-growth-supporting substrate in the obligate presence of a growth-supporting substrate. In either case, the labile substrate, or a component of it, was likely responsible for the induction of enzymes involved in the degradation of diclofop. It is possible that such an induction, caused by the intermediate metabolites produced by one set of organisms, also occurs between interacting members of a degradative community.

Both the review by Semple *et al.* (1999) and the study by Wolfaardt *et al.* (1994a) indicated that, in addition to bacteria, algae also contribute to degradation processes. Semple *et al.* (1999) further discussed the ubiquitous distribution of algae in the environment, their key role in fixation and turnover of carbon and other elements, as well as their heterotrophic capabilities. The review also included a discussion of

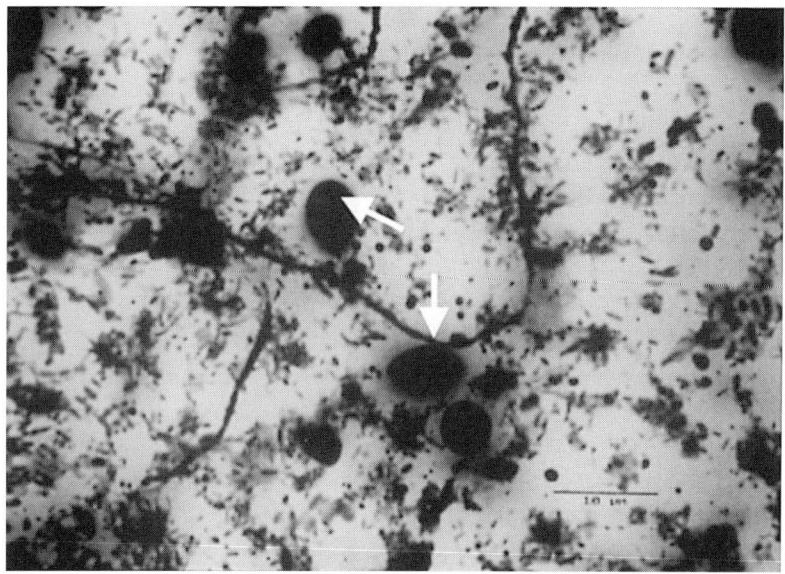

Fig. 6. Heterogeneity in degradative biofilms. This micrograph shows the diverse composition typically associated with degradative biofilms. Note the presence of cell clusters, void spaces, filaments and algal cells (the latter indicated with arrows).

accumulation and degradation of pesticides by algae. Wolfaardt *et al*. (1994a) demonstrated that an alga of the genus *Chlorococcum* improved the degradative efficiency of the diclofop-degrading biofilm community, as indicated by conversion of ^{14}C-labelled diclofop to $^{14}CO_2$, from 65% to 76%. The alga was found to be present as part of the biofilm (Fig. 6) whenever cultivation occurred in the light.

SPATIAL ORGANIZATION, STRATIFICATION AND THE ROLE OF EPS

The close proximity of cells within biofilms results in closer metabolic associations than between free-living cells. This has led to the contention that cell–cell interactions such as quorum sensing may be an important factor in biofilm physiology (McLean *et al*., 1997). Davies *et al*. (1998) demonstrated that the biofilms formed by *Pseudomonas aeruginosa* in the absence of an intercellular signal molecule differ substantially from those formed in its presence, in terms of biofilm structure as well as resistance to antimicrobial compounds. Such results strongly suggest that biofilm formation facilitates a higher degree of organization than what is possible in the planktonic phase. Geesey & Costerton (1986) suggested that many microbial processes which pure cultures cannot carry out are made possible through consortial activities between different species. Among these is the utilization, through interspecies interactions, of recalcitrant compounds that are resistant to degradation by pure cultures. In other words, micro-organisms that share a common habitat, such as those in a biofilm, maximize their metabolic capabilities through co-operative interactions. In addition to improving their own competitive advantage, interacting cells contribute to the overall maintenance of population or community integrity and stability. Wolfaardt *et al.* (1994b) described spatial positioning between members of a heterogeneous biofilm community and concluded that close associations existed between the nine bacterial members of this community when cultivated with chlorinated aromatic compounds as the sole carbon and energy source. A phenomenon observed in these biofilms was the spatial arrangement of different consortia members around a single member of the consortium. An explanation for such behaviour is that these foci represented primary degraders that carried out the critical initial step in the utilization of recalcitrant compounds. Similar behaviour was also observed in reservoirs containing river water, demonstrating some degree of similarity between biofilms degrading xenobiotic compounds and those degrading naturally occurring substrates. Figs 2, 4, 5 and 6 show clustering that is typically observed in degradative biofilms with a heterogeneous species composition. When these biofilms are cultivated in flow cells and regularly observed over extended periods, it is often found that individual clusters are maintained for weeks.

Improved degradative efficiency among microbial communities is often attributed to adaptation (see, for example, McCuster *et al.*, 1988; Holben *et al.*, 1992; Ascon-Cabrera & Lebeault, 1993). Although a number of explanations for such acclimation have been proposed, the underlying mechanisms (especially the genetic factors involved) have received relatively little attention (Madsen, 1991). Nevertheless, Wyndham *et al.* (1994) demonstrated the importance of transposable elements in the lateral movement of catabolic genes and further described the significance of genetic rearrangement during adaptation. It is clear that the presence of donor as well as recipient cells is a prerequisite for gene transfer to occur. In theory, immobilization of potential donors and recipients in biofilms should facilitate the transfer of genetic information. This, together with significant advances made recently in the field of biofilm research, as well as molecular techniques in general, justify more research in this field. Furthermore, the spatial and temporal stability of degradative biofilms (Wolfaardt *et al.*, 1994b) render them suitable for studies to address the potential problem of gene transfer between genetically modified organisms and indigenous microflora.

In oligotrophic environments, it can be expected that cells at the base of biofilms (closest to the substratum) would be least active and those at or near the biofilm–liquid interface most active. Indeed, physical measurements in degradative biofilms (Arcangeli & Arvin, 1992) and mathematical simulations (Skowlund, 1990) indicated such layering within biofilms. In contrast, the opposite may be observed when biofilms are exposed to toxic

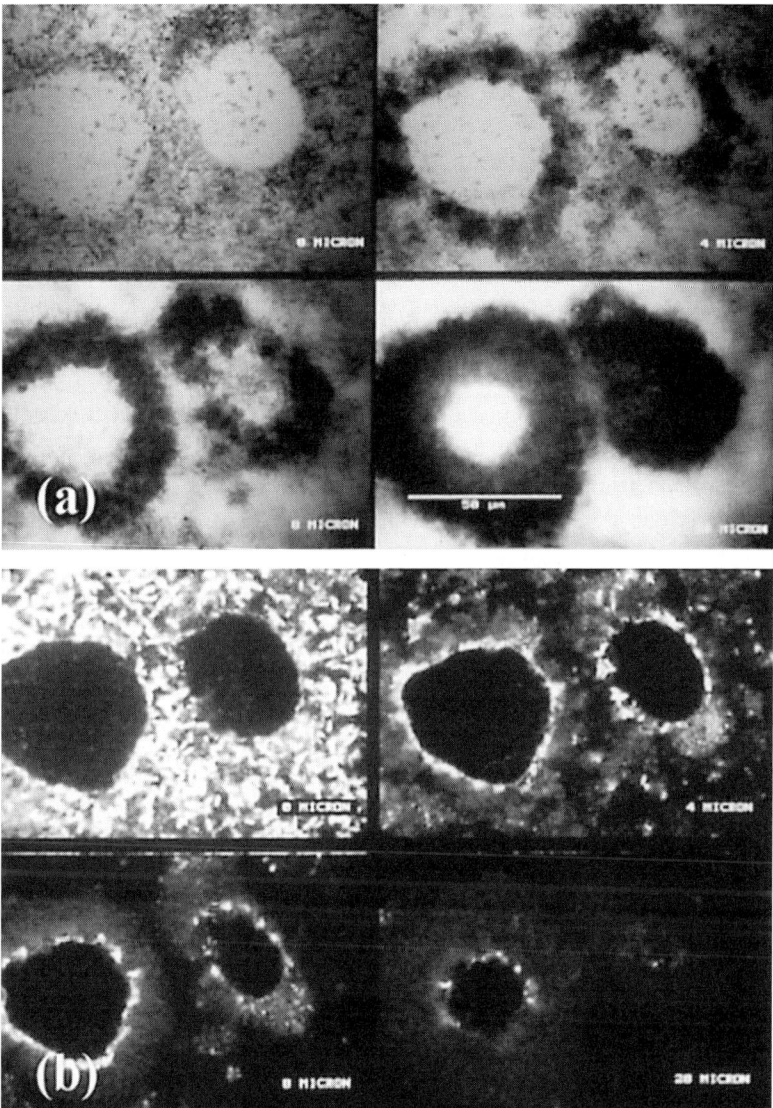

Fig. 7. Optical thin sections through clusters in a degradative biofilm community. (a) Negatively stained sections collected at 0, 4, 8 and 20 μm from the base of the biofilm showing hollow mounds formed by densely packed cells (dark against a light background). (b) Sections from the same locations and depths shown in (a) illustrating the accumulation of the herbicide diclofop-methyl (light against a dark background) by EPS. For additional details, see Wolfaardt et al. (1998).

compounds. Using a combination of cryosectioning and differential staining with 5-cyano-2,3-ditolyl tetrazolium chloride, Villaverde & Fernandez-Polanco (1999) described such stratification in biofilms where respiratory activity was highest near the substratum when volatile organic compounds were provided as energy and carbon sources. Stratification in degradative biofilms does not occur only in terms of respiratory activity; there are usually also discernible layers of gene expression or strata consisting of distinct populations. Huang *et al.* (1998) demonstrated that the expression of alkaline phosphatase was spatially and temporally variable in response to phosphate starvation, while Rothemund *et al.* (1996) described three different layers in a 2,4-dichloro-phenoxyacetic acid (2,4-D) degrading biofilm. Although the overall composition of the biofilm was characterized by a dominance of organisms of the β-subclass of *Proteobacteria*, the bottom layer consisted primarily of amoebae covered with bacterial cells representing a variety of groups. The next layer contained mostly bacteria of the β-subclass and some fungi, while a higher bacterial diversity and some amoebae were found in the surface layer. Wolfaardt *et al.* (1998) used fluorescent lectins to characterize biofilm EPS *in situ* and found that the chemical composition at the bottom layers of a diclofop-methyl-degrading biofilm differed substantially from the zones closer to the biofilm–liquid interface. These authors also found that accumulation of diclofop and other chlorinated aromatic molecules occurred primarily at the bottom of the biofilm (Fig. 7).

In conclusion, there is ample evidence that microbial interactions are important for the functioning of microbial communities, especially when challenged with recalcitrant substrates. Biofilm formation has been identified as an important mechanism to improve interactions – among micro-organisms, and also between micro-organisms and their physical and chemical environment. Two key properties of degradative biofilms are: (1) the spatial organization of cells and (2) the establishment of a stable microenvironment through the production of EPS. These characteristics promote the assemblage of a larger and more diverse genetic pool in a confined microniche, thereby expanding the range of substrates that can be degraded.

REFERENCES

Arcangeli, J. P. & Arvin, E. (1992). Toluene biodegradation and biofilm growth in an aerobic fixed-film reactor. *Appl Microbiol Biotechnol* **37**, 510–517.

Ascon-Cabrera, M. & Lebeault, J. M. (1993). Selection of xenobiotic-degrading microorganisms in a biphasic aqueous-organic system. *Appl Environ Microbiol* **59**, 1717–1724.

Boult, S., Johnson, N. & Curtis, C. (1997). Recognition of a biofilm at the sediment-water interface of an acid mine drainage-contaminated stream, and its role in controlling iron flux. *Hydrol Process* **11**, 391–399.

Caldwell, D. E. (1993). The microstat: steady-state microenvironments for subculture of steady-state consortia, communities and microecosystems. In *Trends in Microbial Ecology*, pp. 123–128. Edited by R. Guerrero & C. Pedros-Alio. Barcelona: Spanish Society for Microbiology.

Davies, D. G., Parsek, M. R., Pearson, J. P., Iglewski, B. H., Costerton, J. W. & Greenberg, E. P. (1998). The involvement of cell-to-cell signals in the development of a bacterial biofilm. *Science* **280**, 295–298.

Decho, A. W. (1999). Chemical communication within microbial biofilms: chemotaxis and quorum sensing in bacterial cells. In *Microbial Extracellular Polymeric Substances*, pp. 155–169. Edited by J. Wingender, T. R. Neu & H.-C. Flemming. Heidelberg: Springer.

Elvers, K. T., Leeming, K., Moore, C. P. & Lappin-Scott, H. M. (1998). Bacterial-fungal biofilms in flowing water photo-processing tanks. *J Appl Microbiol* **84**, 607–618.

Geesey, G. G. & Costerton, J. W. (1986). The microphysiology of consortia within adherent bacterial populations. In *Perspectives in Microbial Ecology*, pp. 238–242. Edited by F. Megusar & M. Gantar. Yugoslavia: Mladinski Knjiga.

Holben, W. E., Shroeter, B. M., Calabrese, V. G. M., Olsen, R. H., Kukor, J. K., Biederbeck, V. O., Smith, A. E. & Tiedje, J. M. (1992). Gene probe analysis of soil microbial populations selected by amendment with 2,4-dichlorophenoxyacetic acid. *Appl Environ Microbiol* **58**, 3941–3948.

Huang, C.-T., Xu, K. D., McFeters, G. A. & Stewart, P. S. (1998). Spatial patterns of alkaline phosphatase expression within bacterial colonies and biofilms in response to phosphate starvation. *Appl Environ Microbiol* **64**, 1526–1531.

Karthikeyan, S., Wolfaardt, G. M., Korber, D. R. & Caldwell, D. E. (1999). Identification of synergistic interactions among microorganisms in biofilms by digital image analysis. *Int Microbiol* **2**, 241–250.

Lawrence, J. R., Swerhone, G. D. W. & Kwong, Y. T. J. (1998). Natural attenuation of aqueous metal contamination by an algal mat. *Can J Microbiol* **44**, 825–832.

Liehr, S. K., Chen, H. J. & Lin, S. H. (1994). Metal removal by algal biofilms. *Water Sci Technol* **30**, 59–60.

McCuster, V. W., Skipper, H. D., Zublena, J. P. & Gooden, D. T. (1988). Biodegradation of carbamothioates in butylate-history soils. *Weed Sci* **36**, 818–823.

McLean, R. J. C., Whiteley, M., Stickler, D. J. & Fuqua, W. C. (1997). Evidence of autoinducer activity in naturally occurring biofilms. *FEMS Microbiol Lett* **154**, 259–263.

Macleod, F. A., Guiot, S. R. & Costerton, J. W. (1990). Layered structure of bacterial aggregates produced in an upflow anaerobic sludge bed and filter reactor. *Appl Environ Microbiol* **56**, 1598–1607.

Madsen, E. L. (1991). Determining *in situ* biodegradation. *Environ Sci Technol* **25**, 1663–1673.

Marques, A. M., Roca, X., Simon-Pujol, M. D., Fuste, M. C. & Congregado, F. (1991). Uranium accumulation by *Pseudomonas* sp. EPS-5028. *Appl Microbiol Biotechnol* **35**, 406–410.

Mirpuri, R., Jones, W. & Bryers, J. D. (1997). Toluene degradation kinetics for planktonic and biofilm-grown cells of *Pseudomonas putida* 54G. *Biotechnol Bioeng* **53**, 535–546.

Møller, S., Korber, D. R., Wolfaardt, G. M. & Caldwell, D. E. (1997). Impact of nutrient composition on a degradative biofilm community. *Appl Environ Microbiol* **63**, 2432–2438.

Muralidharan, V., Rinker, K. D., Hirsh, I. S., Bouwer, E. J. & Kelly, R. M. (1997). Hydrogen transfer between methanogens and fermentative heterotrophs in hyperthermophilic cocultures. *Biotechnol Bioeng* **56**, 268–278.

Pedersen, A. R., Møller, S., Mølin, S. & Arvin, E. (1997). Activity of toluene-degrading *Pseudomonas putida* in the early growth phase of a biofilm for waste gas treatment. *Biotechnol Bioeng* **54**, 131–141.

Rocheleau, S., Greer, C. W., Lawrence, J. R., Cantin, C., Laramee, L. & Guiot, S. R. (1999). Differentiation of *Methanosaeta concilii* and *Methanosarcina barkeri* in anaerobic mesophilic granular sludge by fluorescent in situ hybridization and confocal scanning laser microscopy. *Appl Environ Microbiol* **65**, 2222–2229.

Rothemund, C., Amann, R., Klugbauer, S., Manz, W., Bieber, C., Schleifer, K.-H. & Wilderer, P. (1996). Microflora of 2,4-dichlorophenoxyacetic acid degrading biofilms on gas permeable membranes. *Syst Appl Microbiol* **19**, 608–615.

Semple, K. T., Cain, R. B. & Schmidt, S. (1999). Biodegradation of aromatic compounds by microalgae. *FEMS Microbiol Lett* **170**, 291–300.

Sillitoe, R. H., Folk, R. L. & Saric, N. (1994). Bacteria as mediators of copper sulfide enrichment during weathering. *Science* **272**, 1153–1155.

Sjollema, J., van der Mei, H. C., Uyen, H. M. & Busscher, H. J. (1990). Direct observations of cooperative effects in oral streptococcal adhesion to glass by analysis of the spatial arrangement of adherent bacteria. *FEMS Microbiol Lett* **69**, 263–270.

Skowlund, C. T. C. (1990). Effect of biofilm growth on steady-state biofilm models. *Biotechnol Bioeng* **35**, 502–510.

Tank, J. L. & Webster, J. R. (1998). Interaction of substrate and nutrient availability on wood biofilm processes in streams. *Ecology* **79**, 2168–2179.

Villaverde, S. & Fernandez-Polanco, F. (1999). Spatial distribution of respiratory activity in *Pseudomonas putida* 54G biofilms degrading volatile organic compounds (VOC). *Appl Microbiol Biotechnol* **51**, 382–387.

Wilson, J. T., McNabb, J. F., Cochran, J. W., Wang, T. H., Tomson, M. B. & Bedient, P. B. (1985). Influence of microbial adaptation on the fate of organic pollutants in ground water. *Environ Toxicol Chem* **4**, 721–726.

Wilson, M., Patel, H. & Noar, J. H. (1998). Effect of chlorhexidine on multi-species biofilms. *Curr Microbiol* **36**, 13–18.

Wingender, J., Neu, T. R. & Flemming, H.-C. (1999). What are bacterial extracellular polymeric substances? In *Microbial Extracellular Polymeric Substances*, pp. 1–19. Edited by J. Wingender, T. R. Neu & H.-C. Flemming. Heidelberg: Springer.

Wolfaardt, G. M., Lawrence, J. R., Robarts, R. D. & Caldwell, D. E. (1994a). The role of interactions, sessile growth and nutrient amendment on the degradative efficiency of a bacterial consortium. *Can J Microbiol* **40**, 331–340.

Wolfaardt, G. M., Lawrence, J. R., Robarts, R. D., Caldwell, S. E. & Caldwell, D. E. (1994b). Multicellular organization in a degradative biofilm community. *Appl Environ Microbiol* **60**, 434–446.

Wolfaardt, G. M., Lawrence, J. R., Robarts, R. D. & Caldwell, D. E. (1995). Bioaccumulation of the herbicide diclofop in extracellular polymers and its utilization by a biofilm community during starvation. *Appl Environ Microbiol* **61**, 152–157.

Wolfaardt, G. M., Lawrence, J. R., Robarts, R. D. & Caldwell, D. E. (1998). In situ characterization of biofilm exopolymers involved in the accumulation of chlorinated organics. *Microb Ecol* **35**, 213–233.

Wyndham, R. C., Nakatsu, C., Peel, M., Cashore, A., Ng, J. & Szilagyi, F. (1994). Distribution of the catabolic transposon Tn5271 in a groundwater bioremediation system. *Appl Environ Microbiol* **60**, 86–93.

Biofilms and prosthetic devices

Roger Bayston

Biomaterials-Related Infection Group, University of Nottingham, Division of Microbiology, Clinical Sciences Building, City Hospital, Nottingham NG5 1PB, UK

THE USE OF PROSTHETIC DEVICES IN MEDICINE

Prosthetic devices are used to repair or replace a structure or function which has been damaged or is absent as a result of disease, trauma (including surgery) or congenital defect. The limits on their use are considerations of mechanical properties or function, and biocompatibility, and these are progressively being overcome as new biomaterials are developed. However, most prosthetic devices in common use are made from a narrow range of biomaterials, such as silicone elastomer, polyurethanes, fabricated polytetrafluoroethylene (Teflon) or polyethylene terephthalates (Dacron), titanium, stainless steel, ceramics and composite materials containing carbon or glass fibres.

Most prosthetic devices are used in either children or the elderly, and the proportion of the population comprising these two age groups is increasing in most developed countries. It is likely, therefore, that the absolute numbers of prosthetic devices used each year will rise rapidly. Examples of such devices are large joint (hip, knee, shoulder, etc.) replacements for arthritis, spinal fixation to stabilize the spine after cancer or trauma, prosthetic heart valves, pacemakers, vascular grafts for obstructed or weakened major arteries, voice prostheses for those whose larynx has been removed because of cancer or trauma, central venous catheters for parenteral nutrition or anti-cancer drug administration, catheters for peritoneal dialysis for those with renal failure, and shunts to control hydrocephalus, or pathological accumulation of fluid in the brain.

SGM symposium 59: Community structure and co-operation in biofilms. Editors D. Allison, P. Gilbert, H. Lappin-Scott, M. Wilson. Cambridge University Press. ISBN 0 521 79302 5 ©SGM 2000.

Incidence and resource implications of device-related infection

While biodeterioration or mechanical dysfunction are important complications, the most feared is infection, which occurs at widely varying rates between devices. Hip replacement infection occurs after about 1.5–5% of operations, depending on the age and preoperative condition of the patient and the cause of the arthritis (and the figures for knee replacement are higher), but approximately 90 000 large joint replacements are carried out each year in the UK. Estimated hospital costs for a hip replacement are between £3000 and £4000, but the estimated hospital costs of a subsequent infection are between £20 000 and £30 000, making an annual total of over £100 million. In addition, several patients awaiting joint replacement will have their operations deferred by at least several weeks while the bed is occupied by the infected patient. These costs do not include those for the patient and his or her family in terms of loss of earnings (or job), travel to hospital, social services support and rehabilitation on discharge, etc. Unfortunately, accurate figures are not available for most devices, but those for hydrocephalus shunts show that approximately 4000 are used annually in the UK, mainly but not exclusively in young children. While in adults the infection rate is about 5%, in this group it is about 15–20% (Pople *et al.*, 1992). Similar calculations can therefore be made for these and other devices, and it becomes clear that infection in prosthetic devices is extremely costly in terms of resources. At least as important are the trauma, distress and increased disability of the patients concerned. The outcome in a hip replacement infection may be permanent disability; in hydrocephalus shunts it may be diminished intellectual capacity; and in vascular grafts it may be death from disruption of a major artery (in about 50%) or lower limb amputation in some survivors (Pitsch & Lawrence, 1994).

Categorization of prosthetic devices

Some devices are totally implanted in the body, while others are partially implanted, and yet others are not surgically implanted. It is therefore useful to categorize devices according to this criterion as it has important bearing on subsequent considerations of risk. Category 1 devices are totally implanted in the tissues of the body and are generally intended to remain for the life of the patient. Category 2 devices are partially implanted, and are often intended to remain for long periods but not indefinitely. Category 3 devices are not implanted in the tissues at all, though they may require surgery to create an access aperture. Examples of devices and their categories are shown in Table 1.

Period of risk and causative organisms

In Category 1 devices, the period of risk is overwhelmingly at the time of surgical placement of the device, and organisms from the skin or mucous membranes of the

Table 1. Categorization of prosthetic devices according to their degree of implantation in the body

Category	Device	Site of implantation
Category 1	Large joint replacement	Hip, knee, etc.
	Prosthetic heart valve	Heart
	Intraocular lens	Eye
	Cardiac pacemaker	Heart, tissues
	Vascular graft	Major artery
	Hydrocephalus shunt	Brain, tissues
	Ascites shunt	Peritoneal cavity
	Spinal fixation device	Bones of spine
Category 2	Central venous catheter	Veins, heart
	CAPD* catheter	Abdomen
	External ventricular drain	Brain to exterior
Category 3	Urinary catheter	Bladder to exterior
	Voice prosthesis	Trachea to oesophagus

Note: *CAPD, continuous ambulatory peritoneal dialysis.

patient and sometimes the staff gain access to the device. This is minimized by preoperative skin preparation using antiseptics, but it is impossible to sterilize the deeper layers of the skin and the surgical wound usually becomes contaminated during the operation (Raahave *et al.*, 1986; Bayston & Lari, 1974). Most of the infections of devices in this category are therefore caused by *Staphylococcus epidermidis* and other coagulase-negative staphylococci (CoNS), *Staphylococcus aureus* and coryneforms, though the anaerobic *Propionibacterium acnes* has also been reported (Tunney *et al.*, 1999; Whitcup *et al.*, 1991). However, in most there is also a well-defined though low incidence of 'late' infection, caused by haematogenous spread of bacteria from the respiratory, alimentary or genito-urinary tracts, and these infections are usually caused by Gram-negative bacteria such as *Escherichia coli*, *Klebsiella pneumoniae* and *Haemophilus influenzae*, or Gram-positive cocci such as *Enterococcus* species. Category 2 devices require a minor surgical procedure to implant them, but the great majority of infections are due to organisms introduced during the use of the device. Again, they may be derived from the patient's skin at the site where the device exits the body, but they may also get into the lumen of the catheter from the patient's fingers or those of attendants. Nowadays, contaminated commercial intravenous fluids are only very occasionally responsible. The range of organisms is therefore wide, including skin flora as well as bacteria of enteric origin such as *E. coli*, and those from the environment such as *Pseudomonas aeruginosa*, *Sphingobacterium* species or *Stenotrophomonas* species (Sefcick *et al.*, 1999). *Candida albicans* is also sometimes involved, originating on skin or mucous membranes of patients who have often received long courses of antibiotics. In Category 1 and 2 device infections, despite the mixed flora of the site from which infecting organisms originate, it is extremely rare to find more than one clonal isolate. The reasons for this are unknown, but are probably related to relative avidity of microbial glycoprotein-binding proteins, susceptibility to complement or phagocytosis, and possibly competition between organisms which reach the device in a viable state. However, in Category 3 devices which are exposed constantly to a wide range of organisms, mixed infections occur which in some respects resemble oral biofilms. Indwelling urinary catheters, which are in place for long periods, become colonized by a variety of Gram-positive and Gram-negative bacteria of enteric or environmental origin, and the resulting biofilm also contains crystalline deposits of ammonium magnesium phosphate and hydroxyapatite resulting from urease production by many of the colonizing bacteria (Morris *et al.*, 1997). Bacteria most often found are *Proteus mirabilis*, *Morganella morganii* and *Klebsiella pneumoniae*.

It may be seen that prosthetic devices falling into different categories have different periods of risk for infection and may have widely differing biofilm populations. These considerations are important in prevention planning.

PATHOGENESIS OF DEVICE-RELATED INFECTION

Biofilms in prosthetic-device-related infection

Most bacteria which cause device-related infections, and especially the CoNS which predominate in Category 1 and 2 devices, are harmless to healthy people and form part of their normal flora. With the introduction of biomaterials and implantable devices, such bacteria were presented with an opportunity to colonize the new environment of the implant.

We are now familiar with the concept that, contrary to conventional medical microbiological teaching, most inter-relationships between bacterium and host involve not planktonic cells as in the laboratory but sessile ones, attached to the host surfaces. Such attached cells form biofilms, which are defined as functional communities of micro-organisms associated with a surface or interface (Fig. 1).

Microbial adhesion to devices

The essential primary event in medical biofilm formation is therefore adhesion, though this rarely takes place between bacterium and naked biomaterial. Within seconds of a biomaterial being implanted into the body, a cascade of plasma-derived glycoproteins begins to interact with its surface (Rosengren *et al.*, 1996), eventually leading to a steady state conditioning film. It is to components of this film that micro-organisms must adhere. The conditioning film might include fibronectin, laminin, collagen,

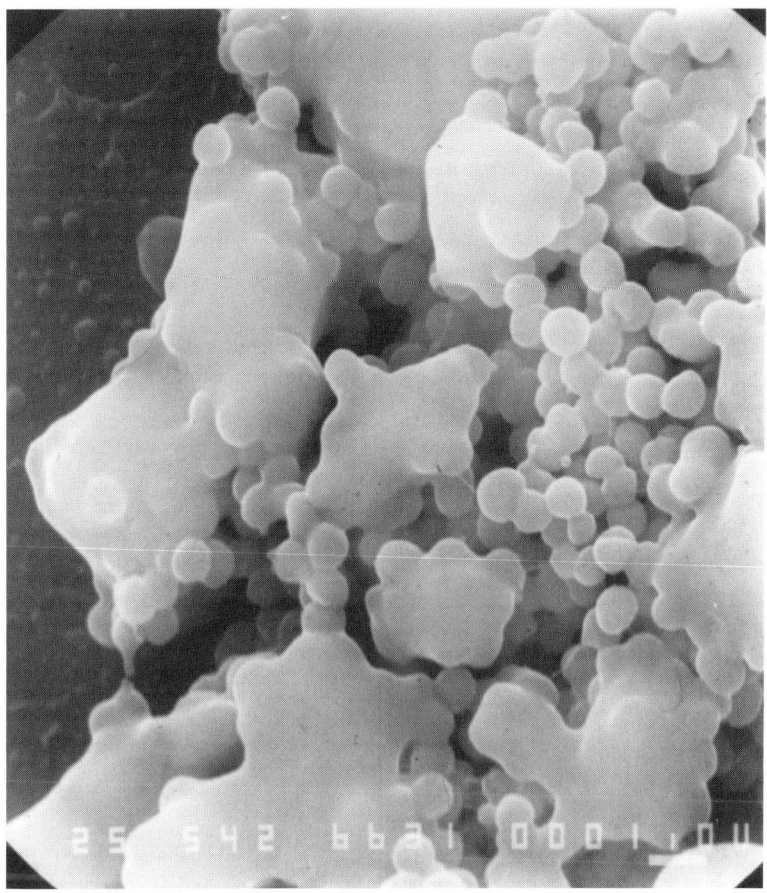

Fig. 1. Scanning electron micrograph of *Staphylococcus epidermidis* in a colonized central venous catheter removed from a patient. The catheter was fixed in absolute ethanol and dehydrated using a critical point dryer before sputtercoating with gold. The image shows the inner surface of the catheter, with possible conditioning film (top right-hand corner), and a biofilm partly covered with condensed exopolymer, in which many individual cocci can be seen. Bar, 1 μm.

albumin, immunoglobulins, mucins, haemoglobin, platelets and red blood cell fragments, depending on the site of implantation (Sullam *et al.*, 1996; Baumgartner & Cooper, 1996). Bacteria such as *S. aureus* and CoNS which are involved in device-related infections usually express binding proteins for fibronectin, collagen and other conditioning film components which are involved in adherence of the organisms (Vercellotti *et al.*, 1985). Activated platelets bound to a tissue surface can also act as binding sites for these organisms. Conditioning films are also important when the use of modified polymers to reduce the risk of infection is contemplated, as they are capable of obliterating the modified surface as well as negating the protective effect of most antimicrobial coatings.

Development of biofilm

After adhesion has taken place, those organisms which are not cleared by phagocytosis begin to proliferate and form plaques which develop into microcolonies. The bacteria in these microcolonies are usually capable of producing exopolymers which are involved in adherence of the organisms to one another, so facilitating biofilm construction. A great deal of investigation has been carried out on the exopolymer produced by CoNS, and particularly *S. epidermidis*, and several chemical identities have been proposed (Peters *et al.*, 1987; Tojo *et al.*, 1988), including teichoic acid (Hussain *et al.*, 1992). However, it now appears that the major exopolymer involved in cell–cell adhesion in CoNS is polysaccharide intercellular adhesin or PIA. This consists of a major neutral component, PIA 1, and a minor component, PIA 2 (Mack *et al.*, 1996). The genetic control of PIA production has now been elucidated by Heilmann *et al.* (1996): the *icaABC* operon encoding extracellular PIA expression includes an *N*-acetylglucosaminyltransferase and a helical porin peptide in its translation products. However, it seems likely that other polymers, including an anionic glycosaminoglycan-like substance, are present in the CoNS biofilm matrix (Bayston & Rodgers, 1990). In addition, as with Gram-negative bacteria, it is considered likely that cell–cell signalling takes place within the biofilm community, influencing exopolymer production and possibly dehiscence of cells from the biofilm (Allison *et al.*, 1999).

The biofilm phenotype and its clinical implications

The development of a mature biofilm structure at the surface of an implanted device has important clinical implications. The bacteria in the biofilm multiply very infrequently as they are deprived of essential nutrients such as iron, carbon and nitrogen sources by the host defences (e.g. transferrin and other iron-binding proteins present in host tissues mean that the bacteria must compete for this essential mineral). There is also a considerable oxygen/carbon dioxide gradient within the biofilm. This leads to cell differentiation, with some cells probably sited in the deeper layers entering a 'dormant' state and becoming auxotrophic for substances such as haem, menadione, thiamin and CO_2 (Proctor & von Humboldt, 1998). When cultured on nutrient agar plates, these organisms grow very slowly and produce very small colonies unless supplied with their complete growth requirements. The finding of these 'small colony variants' (SCV) and their larger parents often leads to the reporting of a mixed culture, indicating contamination rather than infection and confounding the diagnostic process. More importantly, SCV are considered to be largely responsible for the two major clinical characteristics of device-related infections: insusceptibility to antibiotic treatment and chronicity.

The role of SCV in antibiotic insusceptibility and treatment failure. SCV reproduce at a very slow rate, and almost all their metabolic functions are decreased. Antibiotics kill or inhibit the growth of bacteria by binding to specific target sites such as membrane-bound enzymes involved in cell wall construction (β-lactam agents such as penicillin or cephalosporins), or ribosomal subunits involved in protein synthesis (aminoglycosides such as gentamicin). If cell growth and division and protein synthesis are taking place extremely slowly, the antibiotics will have little inhibitory or lethal action on the bacterial cell. Also, most antibiotics must first enter the bacterial cell in order to reach their target site, and in the case of aminoglycosides this is dependent on membrane proton motive force generated in turn by the bacterial electron transport system (Mates *et al.*, 1982). As this is also greatly diminished in SCV, aminoglycosides are unable to enter the cell. Indeed, aminoglycoside insusceptibility is a common feature of SCV of many bacteria, and has been exploited *in vitro* to select them from cultures of clinical material. An additional factor may, in the case of some antibiotics, be exclusion by the charged components of the exopolymer layer, though many other factors are involved (Nichols *et al.*, 1989; Nichols, 1991).

Aminoglycosides are commonly used in the treatment of device-related infections. When such treatment is contemplated, the bacterial isolate from the clinical sample is tested to determine the minimum inhibitory concentration (MIC) of antibiotics, that is, the lowest amount which will inhibit growth on culture medium. If this is several times lower than the known toxic plasma level, or the concentration achievable in the patient's tissues, then the agent may be chosen for treatment of the infection. However, planktonic bacterial cultures are used universally in laboratory testing, and the results of MIC tests do not predict clinical efficacy because of the biofilm-associated factors set out above. For the same isolate, the planktonic MIC might be $1 \text{ mg } l^{-1}$, whereas the minimum biofilm eradication concentration (MBEC) is likely to exceed $100 \text{ mg } l^{-1}$, for an antibiotic which is toxic to humans at a concentration of $50 \text{ mg } l^{-1}$ (Ceri *et al.*, 1999). Treatment therefore either fails, or appears to succeed only for the infection to reappear after treatment is stopped. The device must then be surgically removed in order to eradicate the infection. The resources consumed by this lengthy process, and the distress caused to the patient, are considerable. A test system is needed which uses biofilms in order to determine the MBEC rather than the irrelevant MIC. This would allow a decision to be made either to attempt antibiotic treatment using agents shown to have a chance of success in eradicating the biofilm, or to resort immediately to surgical removal. Though several such systems have been proposed (Vergières & Blaser, 1992; Zimmerli *et al.*, 1994; Hodgson *et al.*, 1995; Houlihan *et al.*, 1997), problems of the need for relatively rapid (i.e. <48 h) results and for easy integration of the technology into a routine laboratory environment remain, with one promising exception (Ceri *et al.*, 1999).

The role of SCV in chronic device-related infection. As part of a general reduction in metabolic activity, SCV do not produce virulence factors such as alpha-toxin, coagulase, haemolysins or other products which are associated with stimulation of cytokine release and immune response (Proctor *et al.*, 1996). The inflammatory reaction to a device-related infection is therefore usually weak, and insufficient to eradicate the bacteria in the biofilm. The structure of the biofilm and its exopolymer matrix also protect most of the constituent bacteria against phagocytosis (Takeoka *et al.*, 1998). However, another factor is probably more important in the instigation and perpetuation of chronic infection in this setting. Staphylococcal SCV are known to be able to survive inside phagocytic cells such as endothelial cells or macrophages (Proctor *et al.*, 1996), and may do so by escaping from the phagosome or by inhibiting fusion with the lysosome, as is the case with some other known intracellular survivors (Roy, 1999). SCV of *S. aureus*, which normally kills endothelial cells using alpha-toxin, do not produce this toxin, are not lethal and therefore are capable of intracellular survival. Supply of menadione and haem to these intracellular survivors *in vitro* restored alpha-toxin production and precipitated a significant reduction in intracellular survival of the bacteria as well as in survival of the endothelial cells (von Eiff *et al.*, 1997). Bacteria surviving inside host cells are protected from other predatory phagocytes and from most antibiotics which cannot penetrate the host cells. Such bacterial strategies explain how device-related infection, and some other biofilm-related infections such as osteomyelitis, can persist for years, causing slow but relentless tissue damage but little inflammatory or effective immune response. SCV revert to fully competent parent forms at a low frequency, and these virulence-factor-producing cells are responsible for the low-grade tissue damage.

STRATEGIES FOR TREATMENT AND PREVENTION

Treatment of device-related infection

Approaches to treatment of infection in individual cases should ideally be based on results of tests for MBEC, but this information is unlikely to be available to the majority of institutions for some time. In most cases, therefore, treatment includes surgical removal of the device. However, attempts have been made to counter the effects of the biofilm phenotype. Khoury *et al.* (1992) found that when a DC electric current of between 15 and 64 $\mu A\ cm^{-2}$ was passed through a biofilm in the presence of an antibiotic *in vitro*, the biofilm was eradicated. There was little antibiotic effect without the current. Further work has been carried out on this phenomenon, termed the 'bioelectric effect' (Benson *et al.*, 1994; Costerton *et al.*, 1994; Wellman *et al.*, 1996), and though the mechanism of action is still unknown, it is possible that the membrane permeability of SCV in the biofilm is increased. Using an ultrasound technique termed the 'bioacoustic effect', Rediske *et al.* (1999) showed that low-frequency ultrasound applied to biofilms on implants in rabbits, while having no effect alone, significantly reduced the biofilm in the presence of gentamicin, but only at a frequency which also caused skin damage. The mechanism is unknown.

Several workers have observed that salicylic acid inhibits exopolymer production in *S. epidermidis* (Teichberg *et al.*, 1993), but only at concentrations far in excess of those achievable with normal dosing in humans. However, this line of investigation is being pursued in the hope that exopolymer synthesis might be inhibited, leading to disruption of the biofilm.

Another approach results from a reappraisal of the intracellular nature of staphylococcal infection, and targeting of intracellular bacteria by antibiotics which are known to enter host cells. An *in vitro* study using human endothelial cells showed that relatively few antibiotics penetrated cells and were able to kill intracellular *S. aureus* (Darouiche & Hamill, 1994). Those that did so were lipophilic, such as rifampicin and ciprofloxacin, and the hydrophilic drugs, such as cephalosporins and vancomycin, which are more commonly used to treat such infections, did not. A clinical trial of two lipophilic agents, rifampicin and ciprofloxacin, in cases of infection following prosthetic hip replacement has been successful in avoiding surgical removal of the infected device (Zimmerli *et al.*, 1998).

Prevention of device-related infection

Prophylactic antibiotics are commonly given at the time of surgical implantation of most devices, though evidence for their efficacy is often not forthcoming. Within Category 1 devices, the low infection rate in large joint replacements is mainly due to the use of prophylactic antibiotics, though changes in surgical practice and the introduction of laminar flow theatres is considered to have contributed (Gosden *et al.*, 1998). However, in the case of hydrocephalus shunts, another Category 1 device, prophylactic antibiotics cannot be shown to be of benefit. In Category 2 devices, the period of risk makes the administration of prophylactic antibiotics unreasonable. Modification of biopolymers has therefore been investigated as a possible way to reduce the risk of biomaterials-related infection. Various chemical or physical modifications intended to change the surface free energy, charge, surface roughness, etc. have been devised but none has yet been of clinical benefit. Conditioning films are able to obliterate many modified surfaces, though their presence in itself may reduce bacterial adhesion (Brydon *et al.*, 1996; Busscher *et al.*, 1997). Coatings attached to catheter surfaces, which are either antimicrobial themselves or are used as ligands to bind antimicrobials to the surface, have been extensively investigated. The ligands consist of those such as benzalkonium chloride or tridodecylmethylammonium chloride which bind anionic antibiotics such as penicillins or rifampicin and minocycline (Raad *et al.*, 1998), and those consisting of a hydrogel coating in which a water-soluble antibiotic is dissolved, or a hydrophilic coating in which hydrophobic antibiotics are dispersed (Rushton *et al.*, 1989). Metal coatings, particularly silver, have been used extensively, but while *in vitro* results have usually been encouraging, animal studies and clinical trials have not supported these findings (Kampf *et al.*, 1998; Riley *et al.*, 1995). The reasons for this have been set out by Schierholz *et al.* (1998). In the case of silver and of other types of coating, two main problems are encountered. One is the interaction with glycoproteins in the conditioning film, but another is that the antimicrobial agent is rapidly eluted by tissue fluids, urine or blood, depending on the site of implantation. Unless 'antimicrobial' biomaterials are tested *in vitro* using systems which include a conditioning film and under fluid flow conditions appropriate to their intended application, clinically predictive data will not be obtained. Another process, that of impregnation by expanding the polymer matrix and inserting molecules of physico-chemically compatible antimicrobials, has been tested in these conditions, and has yielded promising results (Bayston & Lambert, 1997), including long-lasting antimicrobial activity which has been shown to be protective for over 50 days.

REFERENCES

Allison, D. G., Heys, S. J. D., Willcock, L., Holah, J. & Gilbert, P. (1999). Cellular detachment and dispersal from bacterial biofilms: a role for quorum sensing? In *Biofilms: the Good, the Bad, and the Ugly*, pp. 279–286. Edited by J. Wimpenny, P. Gilbert, J. Walker, M. Brading & R. Bayston. Cardiff: BioLine.

Baumgartner, J. N. & Cooper, S. L. (1996). Bacterial adhesion on polyurethane surfaces conditioned with thrombus components. *ASAIO J* **42**, M476–M479.

Bayston, R. & Lambert, E. (1997). Duration of protective activity of cerebrospinal fluid shunt catheters impregnated with antimicrobial agents to prevent bacterial catheter-related infection. *J Neurosurg* **87**, 247–251.

Bayston, R. & Lari, J. (1974). A study of the sources of infection in colonised shunts. *Dev Med Child Neurol* **16** (suppl. 32), 16–22.

Bayston, R. & Rodgers, J. (1990). Production of extracellular slime by *Staphylococcus epidermidis* during stationary phase of growth: its association with adherence to implantable devices. *J Clin Pathol* **43**, 866–870.

Benson, D. E., Grissom, C. B., Burns, G. L. & Hohammad, S. F. (1994). Magnetic field enhancement of antibiotic activity in biofilm forming *Pseudomonas aeruginosa*. *ASAIO J* **40**, M371–M376.

Brydon, H. L., Bayston, R., Hayward, R. & Harkness, W. (1996). Reduced bacterial adhesion to hydrocephalus shunt catheters mediated by cerebrospinal fluid proteins. *J Neurol Neurosurg Psychiatry* **60**, 671–675.

Busscher, H. J., Geertsema-Doornbusch, G. I. & van der Mei, H. C. (1997). Adhesion to silicone rubber of yeasts and bacteria isolated from voice prostheses: influence of salivary conditioning films. *J Biomed Mater Res* **34**, 201–210.

Ceri, H., Olson, M. E., Stremick, C., Read, R. R., Morck, D. & Buret, A. (1999). The Calgary Biofilm Device: new technology for rapid determination of antibiotic susceptibilities of bacterial biofilms. *J Clin Microbiol* **37**, 1771–1776.

Costerton, J. W., Ellis, B., Lam, K., Johnson, F. & Khoury, A. E. (1994). Mechanism of electrical enhancement of efficacy of antibiotics in killing biofilm bacteria. *Antimicrob Agents Chemother* **38**, 2803–2809.

Darouiche, R. O. & Hamill, R. J. (1994). Antibiotic penetration of and bactericidal activity within endothelial cells. *Antimicrob Agents Chemother* **38**, 1059–1064.

Gosden, P. E., MacGowan, A. P. & Bannister, G. C. (1998). Importance of air quality and related factors in the prevention of infection in orthopaedic implant surgery. *J Hosp Infect* **39**, 173–180.

Heilmann, C., Schweitzer, O., Gerke, C., Vanittanakom, N., Mack, D. & Götz, F. (1996). Molecular basis of intercellular adhesion in biofilm-forming *Staphylococcus epidermidis*. *Mol Microbiol* **20**, 1083–1091.

Hodgson, A. E., Nelson, S. M., Brown, M. R. W. & Gilbert, P. (1995). A simple in vitro model for growth control of bacterial biofilms. *J Appl Bacteriol* **79**, 87–93.

Houlihan, H. H., Mercier, R.-C. & Rybak, M. J. (1997). Pharmacodynamics of vancomycin alone and in combination with gentamicin at various dosing intervals against methicillin-resistant *Staphylococcus aureus*-infected fibrin-platelet clots in an in vitro infection model. *Antimicrob Agents Chemother* **41**, 2497–2501.

Hussain, M., Hastings, J. G. M. & White, P. J. (1992). Comparison of cell-wall teichoic acid with high-molecular-weight extracellular slime material from *Staphylococcus epidermidis*. *J Med Microbiol* **37**, 368–375.

Kampf, G., Dietze, B., Große-Siestrup, C., Wendt, C. & Martiny, H. (1998). Microbicidal activity of a new silver-containing polymer, SPI-ARGENT II. *Antimicrob Agents Chemother* **42**, 2440–2442.

Khoury, A. E., Lam, K., Ellis, B. & Costerton, J. W. (1992). Prevention and control of bacterial infections associated with medical devices. *ASAIO J* **38**, M174–M178.

Mack, D., Fischer, W., Krokotsch, A., Leopold, K., Hartmann, R., Egge, H. & Lauks, R. (1996). The intercellular adhesin involved in biofilm accumulation of *Staphylococcus epidermidis* is a linear β-1,6-linked glucosaminoglycan: purification and structural analysis. *J Bacteriol* **178**, 175–183.

Mates, S. M., Eisenberg, E. S., Mandel, L. J., Patel, L., Kaback, H. R. & Miller, M. H. (1982). Membrane potential and gentamicin uptake by *Staphylococcus aureus*. *Proc Natl Acad Sci USA* **79**, 6693–6697.

Morris, N. S., Stickler, D. J. & Winters, C. (1997). Which indwelling urethral catheters resist encrustation by *Proteus mirabilis* biofilms? *Br J Urol* **80**, 58–63.

Nichols, W. W. (1991). Biofilms, antibiotics and penetration. *Rev Med Microbiol* **2**, 177–181.

Nichols, W. W., Evans, M. J., Slack, M. P. E. & Walmsley, H. L. (1989). The penetration of antibiotics into aggregates of mucoid and nonmucoid *Pseudomonas aeruginosa*. *J Gen Microbiol* **135**, 1291–1303.

Peters, G., Schumacher-Perdreau, F., Jansen, B., Bey, M. & Pulverer, G. (1987). Biology of *Staphylococcus epidermidis* extracellular slime. *Zentbl Bakteriol Mikrobiol Hyg Abt 1 Suppl* **16**, 15–32.

Pitsch, R. J. & Lawrence, P. F. (1994). Natural history of graft infections. In *Vascular Graft Infections*, pp. 31–42. Edited by T. J. Bunt. Armonk: Futura.

Pople, I. K., Bayston, R. & Hayward, R. D. (1992). Infection of cerebrospinal fluid shunts in infants: a study of etiological factors. *J Neurosurg* **77**, 29–36.

Proctor, R. A. & von Humboldt, A. (1998). Bacterial energetics and antimicrobial resistance. *Drug Resistance Updates* **1**, 227–235.

Proctor, R. A., Vesga, O., Otten, M. F., Koo, S.-P., Yeaman, M. R., Sahl, H.-G. & Bayer,

A. S. (1996). *Staphylococcus aureus* small colony variants cause persistent and resistant infections. *Chemotherapy* **42** (suppl. 2), 47–52.

Raad, I. I., Darouiche, R. O., Hachem, R., Abi-Said, D., Safar, H., Darnule, T., Mansouri, M. & Morck, D. (1998). Antimicrobial durability and rare ultrastructural colonization of indwelling central catheters coated with minocycline and rifampin. *Crit Care Med* **26**, 219–224.

Raahave, D., Friis-Møller, A., Bjerre-Jepsen, K., Thiis-Knudsen, J. & Rasmussen, L. B. (1986). The infective dose of aerobic and anaerobic bacteria in post-operative wound sepsis. *Arch Surg* **121**, 924–929.

Rediske, A. M., Roeder, B. L., Brown, M. K., Nelson, J. L., Robison, R. L., Draper, D. O., Schaalje, G. B., Robison, R. A. & Pitt, W. G. (1999). Ultrasonic enhancement of antibiotic action on *Escherichia coli* biofilms: an in vivo model. *Antimicrob Agents Chemother* **43**, 1211–1214.

Riley, D. K., Classen, D. C., Stevens, L. E. & Burke, J. P. (1995). A large randomized clinical trial of a silver-impregnated urinary catheter: lack of efficacy and staphylococcal superinfection. *Am J Med* **98**, 349–358.

Rosengren, A., Johansson, B. R., Danielsen, N., Thomsen, P. & Ericson, L. E. (1996). Immunohistochemical studies on the distribution of albumin, fibrinogen, fibronectin, IgG and collagen around PTFE and titanium implants. *Biomaterials* **17**, 1779–1786.

Roy, C. R. (1999). Trafficking of the *Legionella pneumophila* phagosome. *ASM News* **65**, 416–421.

Rushton, D. N., Brindley, G. S., Polkey, C. E. & Browning, G. V. (1989). Implant infections and antibiotic-impregnated silicone rubber coating. *J Neurol Neurosurg Psychiatry* **52**, 223–229.

Schierholz, J. M., Lucas, L. & Pulverer, G. (1998). Silver coating of medical devices – a review. *J Hosp Infect* **40**, 257–262.

Sefcick, A., Tait, R. C. & Wood, B. (1999). *Stenotrophomonas maltophilia*: an increasing problem in patients with acute leukaemia. *Leuk Lymphoma* **35**, 207–211.

Sullam, P. M., Bayer, A. S., Foss, W. M. & Cheung, A. L. (1996). Diminished platelet binding *in vitro* by *Staphylococcus aureus* is associated with reduced virulence in a rabbit model of infective endocarditis. *Infect Immun* **64**, 4915–4921.

Takeoka, K., Ichimiya, T., Yamasaki, T. & Nasu, M. (1998). The in vitro effect of macrolides on the interaction of human polymorphonuclear leukocytes with *Pseudomonas aeruginosa* in biofilm. *Chemotherapy* **44**, 190–197.

Teichberg, S., Farber, B. F., Wolff, A. G. & Roberts, B. (1993). Salicylic acid decreases extracellular biofilm production by *Staphylococcus epidermidis*: electron microscopic analysis. *J Infect Dis* **167**, 1501–1503.

Tojo, M., Yamashita, N., Goldmann, D. A. & Pier, G. B. (1988). Isolation and characterization of a polysaccharide adhesin from *Staphylococcus epidermidis*. *J Infect Dis* **157**, 713–722.

Tunney, M. M., Patrick, S., Curran, M. D., Ramage, G., Hanna, D., Noxon, J. R., Gorman, S. P., Davis, R. I. & Anderson, N. (1999). Detection of prosthetic hip infection at revision arthroplasty by immunofluorescence microscopy and PCR amplification of the bacterial 16S rRNA gene. *J Clin Microbiol* **37**, 3281–3290.

Vercellotti, M., McCarthy, J. B., Lindholm, P., Peterson, P. K., Jacob, H. S. & Furcht, L. T. (1985). Extracellular matrix proteins (fibronectin, laminin and Type IV collagen) bind and aggregate bacteria. *Am J Pathol* **120**, 13–21.

Vergières, P. & Blaser, J. (1992). Amikacin, ceftazidime, and flucloxacillin against suspended and adherent *Pseudomonas aeruginosa* and *Staphylococcus epidermidis* in an in vitro model of infection. *J Infect Dis* **165**, 281–289.

Von Eiff, C., Heilmann, C., Proctor, R. A., Woltz, C., Peters, G. & Götz, F. (1997). A site-directed *Staphylococcus aureus* hemB mutant is a small-colony variant which persists intracellularly. *J Bacteriol* **179**, 4706–4712.

Wellman, N., Fortun, S. M. & McLeod, B. R. (1996). Bacterial biofilms and the bioelectric effect. *Antimicrob Agents Chemother* **40**, 2012–2014.

Whitcup, S. M., Belfort, R., Desmet, M. D., Palestine, A. G., Nussenblatt, R. B. & Chan, C. C. (1991). Immunohistochemistry of the inflammatory response in *Propionibacterium acnes* endophthalmitis. *Arch Ophthalmol* **109**, 978–979.

Zimmerli, W., Frei, R., Widmer, A. F. & Rajacic, Z. (1994). Microbiological tests to predict treatment outcome in experimental device-related infections due to *Staphylococcus aureus*. *J Antimicrob Chemother* **33**, 959–967.

Zimmerli, W., Widmer, A. F., Blatter, M., Frei, R. & Ochsner, P. E. (1998). Role of rifampin for treatment of orthopedic implant-related staphylococcal infections – a randomised controlled trial. *J Am Med Assoc* **279**, 1537–1541.

Biofilms: problems of control

David G. Allison, Andrew J. McBain and Peter Gilbert

School of Pharmacy and Pharmaceutical Sciences, University of Manchester, Oxford Road, Manchester M13 9PL, UK

INTRODUCTION

Biofilms are notorious for their high level of resistance towards all forms of chemical treatments intended to kill or control their growth. Such resistance to antibiotics, biocides and disinfectants has been attributed to a variety of processes from which a number of dominant mechanisms have been identified. The most frequent attributions of resistance relate to the properties of the extracellular polymer matrix (glycocalyx). The diffusivity of the glycocalyx towards antimicrobial agents is only slightly less than that of water. It does not, therefore, present a significant diffusion barrier in its own right, but can restrict access of antimicrobial to the depths of the biofilm by acting as a substrate for chemically reactive agents or through non-specific binding of highly charged antimicrobial compounds. Additionally, the matrix binds extracellular enzymes, such as β-lactamases and formaldehyde lyase, which are then able to augment the small change in diffusivity through the destruction of susceptible compounds. Factors, other than the glycocalyx, responsible for the resistance properties of biofilms relate to the close proximity of cells. In a dense biofilm community, nutrients and oxygen are preferentially consumed at the periphery. In this respect, gradients of growth rate, associated with different growth-limiting nutrients, will develop across the community. During exposure to antimicrobial agents, the slower-growing cells, at the core of the biofilm and at the colonized surface, will generally out-survive the faster metabolizing, peripheral ones. Slow growth will, in addition, cause the cells to express dormant, starvation phenotypes which often overexpress non-specific defences such as shock proteins, multi-drug efflux pumps (*acrAB*) and also extracellular polymers. The final contributor to resistance relates to the expression of less-susceptible, biofilm-specific phenotypes,

SGM symposium 59: Community structure and co-operation in biofilms. Editors D. Allison, P. Gilbert, H. Lappin-Scott, M. Wilson. Cambridge University Press. ISBN 0 521 79302 5 ©SGM 2000.

possibly regulated through quorum-sensing mechanisms, and induced through the cellular attachment to surfaces. Whilst the data concerning each of these resistance mechanisms are convincing, none of the processes described can provide a complete explanation for the observed levels of resistance *in situ*. Rather, alone and in combination, they will delay the eradication of the treated population and provide a window of opportunity for the development/expression of more resistant phenotypes.

BIOFILM STRUCTURE AND PHYSIOLOGY

It is now apparent that the predominant mode of growth of bacteria in the natural world is as an attached biofilm. This mode of growth has enabled microbial populations to colonize many harsh environments, including those associated with industry and implant surgery. The metabolic by-products of microbial growth are often inimical to the growth and survival of other micro-organisms. Growth within a biofilm is likely to concentrate these products such that adoption of a biofilm mode of growth necessarily involves the development and expression of resistance/survival mechanisms for individual community members. As such, a biofilm phenotype will in many instances reflect general survival strategies that coincidentally confer a degree of resistance towards man-made chemical interventions. The need for man to control biofilm growth is demonstrated by their broad involvement in chronic infections and in infections relating to implanted medical devices (Costerton *et al.*, 1987), biofouling and biocorrosion of pipework and heat exchangers (Characklis, 1990; Little *et al.*, 1990). Microbial biofilms are also associated with increased frictional resistance to fluid flow on ship hulls and in water conduits, and with the corrosion of metallic substrata. To these immediate problems can be added those associated with the contamination and spoilage of manufactured foods (Holah *et al.*, 1994; Eginton *et al.*, 1998) and other sensitive products. In all of these areas, microbial biofilms exhibit a broad spectrum of resistance to antimicrobial treatments which renders them some 100–1000 times less susceptible than their free-living counterparts (Allison & Gilbert, 1995; Costerton *et al.*, 1987; Nickel *et al.*, 1985). Such resistance is demonstrated not only towards antibiotics and antiseptics, but also towards highly reactive chemical biocides. The latter include isothiazolones (Costerton & Lashen, 1984), quaternary ammonium compounds (Costerton & Lashen, 1984; Evans *et al.*, 1990a), halogens (Huang *et al.*, 1995) and halogen-release agents (Favero *et al.*, 1983).

From a physiological standpoint, biofilms must be considered as functional consortia of microbial cells rather than as a diverse collection of individual cells. These consortia are enveloped within, sometimes extensive, matrices of extracellular polymers (glycocalyx) and the concentrated products of their own metabolism. These not only include ions and nutrients sequestered from the environment, but also extracellular enzymes such as polysaccharases, proteases, β-lactamases, etc., that are derived from component cells but which are available to all. In the majority of natural habitats, such consortia are made up of a variety of species and genera, but in biomedical situations, particularly those associated with soft tissue infection or infections of indwelling devices, they are more likely to be mono-cultures.

The relative position of an individual cell within the biofilm community will affect the extent and nature of its access to potential nutrients and to oxygen. Thus, even within single-species biofilms, a plethora of phenotypes will be represented, the breadth of which reflects the extent of chemical heterogeneity within the film (Gilbert *et al.*, 1990). The outcome of any antimicrobial treatment directed against the biofilm community will therefore reflect the least-susceptible phenotype within the consortium. As the biofilm matures, and the extent of exopolymer deposition increases, then the magnitude of the nutrient and gaseous gradients within them will become increased and the net growth rate of the community will become further reduced. This may possibly bring about an onset of dormancy in the cells and trigger the expression of stringent response genes (Zambrano & Kolter, 1995) within the deeper lying cells. All of the above factors have been demonstrated to markedly decrease susceptibility towards most antimicrobial substances (Gilbert *et al.*, 1990; Brown *et al.*, 1990). This is perhaps not surprising when it is realized that the majority have generally been optimized for their activity against fast-growing, planktonic systems.

Future strategies for the control of microbial biofilms must therefore involve the design of antimicrobial agents that are specifically targeted towards biofilm-specific phenotypes or which are capable of perturbing the physico-chemical structure and biosynthesis of the glycocalyx. Such developments are likely to include those molecules that possess a high diffusion-reaction ratio and agents that have been specifically targeted towards slow- or non-growing cells. Such approaches have, to date, only met with limited success. The need to develop efficient, low cost hygienic surface-cleansing systems, and antibiotics which are capable of effectively treating biofilm infections, remains as urgent as ever. In order to rationalize our search for suitable antimicrobial targets, there is a need not only to develop our knowledge of biofilm physiology and of the nature and regulation of the biofilm phenotype, but also to examine the various mechanisms associated with biofilm resistance towards conventional antibiotics and biocides. In this article, we consider the current understanding of resistance gained by microbial biofilms as a result of attachment to surfaces, the close proximity of other cells and the deposition of an extracellular polymer matrix.

BIOFILM RESISTANCE TO ANTIMICROBIAL AGENTS

The failure of a microbial population to succumb to antimicrobial treatments might arise through a number of factors. These include an inherent insusceptibility to the agents employed, the acquisition of resistance by previously susceptible strains, by either genetic mutation or the transfer of genetic material from another organism, and the emergence of pre-existing but unexpressed resistant phenotypes. The extent to which such adaptation towards less-susceptible phenotypes is influenced through growth as a biofilm is currently a matter for debate and will be considered later in this chapter. From a mechanistic standpoint, our understanding of resistance development during biofilm formation has hinged upon three non-exclusive theories. The first of these relates to various facets of diffusion-limitation for the passage of drugs across the extracellular matrix (i.e. chemical gradients) and are independent of the innate susceptibility of the component cells. The second theory (physiological gradients) invokes the expression of less-susceptible phenotypes, possibly through stringent response, which reflect gradients of nutrient, growth rate and redox potential across the biofilm. The third explanation concerns the induction of attachment-specific, drug-resistant physiologies (Gilbert & Allison, 1999) associated with either the close proximity of a surface or of other cells. Whilst there is much evidence to support each of these explanations, it is unlikely that any single mechanism will account for the general observation of resistance: rather, that these mechanisms are compounded in biofilms to create insusceptibility and an environment ideally suited for the emergence of tolerant genotypes.

CHEMICAL GRADIENTS AND THE EXOPOLYMERIC MATRIX

An intuitive view of biofilm resistance is that since the cells are enveloped within an extensive 'slime' layer, then antimicrobial agents are prevented from accessing the more deeply implanted cells, and that these thereby escape exposure. This image of the glycocalyx as an impenetrable 'umbrella' is largely incorrect. Whilst the glycocalyx has been shown to comprise an ordered array of fine fibres (Costerton *et al.*, 1981) which provide a relatively thick coating to the cells and biofilm community, these are generally highly hydrated and composed of exopolysaccharide gels (EPS) (Sutherland, 1997). Other macromolecules such as nucleic acids, proteins, globular glycoproteins, lipids and ions may also be present (Sutherland, 1995) but only as minority components. The nature of the matrix polymers determines the physical properties, i.e. hydrophobicity, charge, elasticity, dissolution and deformation characteristics, of the biofilm (Sutherland, 1985; Allison, 1998). Similarly, extracellular polymer matrices that result from the biosynthetic effort of a number of different species within a biofilm community may combine to produce a gel with characteristics that are unique (Allison & Matthews, 1992). Whilst the degree of charge within the polymers will undoubtedly affect non-specific binding of antimicrobial agents, hydrated polysaccharide gels such as these will, however, have diffusivities, towards uncharged solutes, that are not too dissimilar to that of water (Stewart *et al.*, 1998). The barrier properties of the matrix polymers are significant, however, when the polymers either directly or indirectly react with or bind the agent. In such instances, replacement and dissolution of the matrix and its binding/reaction potential become important factors in resistance.

There is still some debate as to whether the biofilm matrix polymers differ from the extracellular polymers produced by the same organisms growing planktonically. In such instances, it has been widely demonstrated that the extent and nature of exopolymer production are heavily dependent upon physiological factors such as the relative availability of carbon and nitrogen (Sutherland, 1985), in addition to attachment events. Also yet to be discovered is whether the matrix polymers, which bind cells to other cells, differ in any way from the 'foot-print polymers' which cement the primary colonizers to the substratum (Allison, 1998; Sutherland, 1997).

Regardless as to the nature of the polymers, it has been shown that biosynthesis is generally up-regulated within minutes of the irreversible attachment of a cell to a surface (Allison & Sutherland, 1987; Davies *et al.*, 1993). Deposition of EPS throughout the developing microcolony then proceeds over a period of hours. Some changes in susceptibility towards antibiotics and biocides are concurrent with these activities (Das *et al.*, 1998).

Regulation of exopolymer synthesis

The synthesis of matrix polymers appears to be regulated by a variety of factors, of which surface attachment appears to be of particular importance. Thus Davies & Geesey (1995) showed exopolysaccharide (alginate) production to be de-repressed in individual *Pseudomonas aeruginosa* cells shortly after they had become attached to the inside surface of a flow-chamber. Similarly, using a perfused biofilm fermenter which allows for growth rate control of attached cell populations, Evans *et al.* (1994) showed that attached cells of *Staphylococcus epidermidis* produced substantially greater quantities of extracellular polymers than did planktonic cells growing at similar growth rates in identical media. These workers also noted a strong growth rate dependency of exopolymer production for the biofilm populations. In this fashion, slow-growing cells produced greater levels of EPS than fast-growing ones; a relationship that would provide for increased exopolymer production within the slow-growing heart of a thick microcolony/biofilm. This would alter the distribution and density of cells throughout the matrix, and confer some structural organization upon the community (Costerton *et al.*, 1994).

It is notable that, with the exception of the alginates, the vast majority of the bacterial polysaccharides are soluble to some degree in water. Biofilm structure is therefore dynamic, with solubilization of polymers occurring at the periphery being compensated by increased polymer production within the depths. Clearly, such replenishment of the

EPS will enhance the protective capability of the glycocalyx where this is mediated through non-specific binding of the treatment agent. Recently, it has also been demonstrated that in some Gram-negative organisms the up-regulation of EPS synthesis is partly under the control of signal substances such as N-acyl homoserine lactone (AHL) (Davies *et al.*, 1998). These are global regulators of transcriptional activation and are responsible for cell–cell signalling in dense communities of Gram-negative bacteria. Whilst the activities of AHL in bacteria have been largely documented as being responsive to critical cell densities in planktonic cultures, they have also been widely heralded as candidate regulators of biofilm specific physiology (Cooper *et al.*, 1995; Gambello *et al.*, 1993; Williams *et al.*, 1992; Heys *et al.*, 1997). In biofilms, signal substances such as AHL would become concentrated within the geometric centre of the microcolonies/ biofilm, thereby increasing exopolymer production. This would alter the distribution and density of cells throughout the matrix, and as before confer some degree of structural organization upon the community (Costerton *et al.*, 1994). Similar sensing mechanisms have been found in a range of Gram-positive bacteria, but these are hydrophobic cyclic peptides (Kleerebezem *et al.*, 1997) rather than lactones. Many of the candidate signalling molecules, be they from Gram-positive or Gram-negative organisms, are relatively hydrophobic and would concentrate at a solid:aqueous interface possibly acting as an attachment sensor for single cells.

Function of the glycocalyx as a diffusion barrier

The presence of a charged, hydrated exopolymer matrix surrounding individual cells and microcolonies influences profoundly the access of molecules and ions, including protons, to the cell wall and membranes. Restricted diffusion of agents from the surrounding medium may occur through a combination of ionic-interaction and molecular-sieving events. In this respect, the polymers of the extracellular matrix act as ion-exchange resins where strongly charged molecules are actively removed from solution as they pass through. It is not surprising, therefore, that many groups of workers have suggested that the glycocalyx functions as a protective shield and physically prevents the access of antimicrobial to the cell surface (Costerton *et al.*, 1987; Slack & Nichols, 1981, 1982; Suci *et al.*, 1994). Accordingly, if the antimicrobial agents are strongly charged (i.e. tobramycin) or chemically highly reactive (i.e. halogens/peroxygens), they will be quenched within the matrix during diffusion. Such universal explanations of resistance are in conflict (Gordon *et al.*, 1988; Nichols *et al.*, 1988, 1989) with *ex vivo* experiments that show the reductions in diffusion coefficient, for antibiotics such as tobramycin and cefsulodin, to be insufficient for the observed change in susceptibility. *In vivo* observations are also conflicting. In this context, Gristina *et al.* (1987) found no difference in the antibiotic susceptibility of slime-producing and non-slime-producing strains of *S. epidermidis*, but Evans *et al.* (1991) were able to demonstrate significant changes in the susceptibility, to the quinolone antibiotic ciprofloxacin, of mucoid and non-mucoid strains of *P. aeruginosa* grown as biofilms. In general, whilst the possession of a mucoid phenotype may be associated with decreases in susceptibility, reductions in the diffusion coefficient, across polymeric matrices relative to liquid media, are insufficient to account for it. Additionally, at equilibrium, which will be achieved relatively swiftly, the concentrations at the cell surface and in the surrounding bulk aqueous phase will be equivalent.

The possible function of the glycocalyx as a protective, diffusion shield has largely been investigated from the context of antibiotics and medically relevant biofilms rather than for biocidal molecules applied in industrial settings. In the former situation, biofilms are generally sparse relative to industrial situations, where they might be tens of centimetres thick (Nichols *et al.*, 1989). Whilst thickness of a biofilm will not affect the diffusion properties per se, substantial fouling will increase the time delay for equilibration and ultimately the mass flux of charged antimicrobial to the depths of the community (Nichols, 1993). Stewart (1996) reached similar conclusions, albeit using a theoretical model, for the penetration of antibiotics into microbial biofilms. On the basis of the available data, the extent of retardation of the diffusion of antibiotics is insufficient to account for the observed reductions in biofilm susceptibility.

Function of the glycocalyx as a reaction-sink

In addition to any effect on susceptibility through acting as a diffusion barrier (above), extracellular polymers and cellular materials at the periphery of a biofilm may act as substrates for chemically reactive biocides. In such a fashion, they might neutralize the treatment agent and thereby further reduce its availability and diffusion across the biofilm. Such effects are most pronounced with biocides such as iodine and iodine–polyvinylpyrollidone complexes (Favero *et al.*, 1983), and for chlorine and per-oxygens (Huang *et al.*, 1995). These are able to react in a consumptive manner with the exopolymer and cellular materials. Under such circumstances, if the rate of deposition of biofilm and biofilm-associated polymers exceeds the rate of their consumption by the treatment agent, then a long-lasting recalcitrance would be obtained (Kumon *et al.*, 1994). Alternatively, if the reaction between peripheral biofilm and treatment agent were sufficient to reduce the bulk-phase concentration to below effective levels, then the disinfection process would be halted. The latter situation is unlikely to occur in open systems or in treatment of infection where the reactive capacity of the biofilm community is small relative to the availability of agent. If, however, the inactivation of agent were mediated through a non-consumptive, catalytic reaction within the biofilm matrix then the reactive capacity becomes almost infinite and the need to 'turn-over' biofilm becomes less important in resisting chronic exposure to agent. In this respect, it has been noted that drug-inactivating enzymes, such as β-lactamases, are up-regulated in biofilm communities, and in those exposed to sub-lethal concentrations of imipenem and/or piperacillin (Lambert *et al.*, 1993; Giwercman *et al.*, 1991), and become bound within the glycocalyx. In a similar fashion, the deposition of enzymes, such as formaldehyde lyase and dehydrogenase (Sondossi *et al.*, 1985), or catalase and super-oxide dismutase, which would collectively degrade formaldehyde and peroxides, respectively, might confer resistance upon biofilm communities towards reactive bio-cides. Such enzyme-mediated losses of treatment agent would be in addition to those losses in the activity of highly charged drug molecules, such as the glycopeptides, attrib-utable to an irreversible binding to the matrix (Hoyle *et al.*, 1992). Curiously, macrolide antibiotics, which are also positively charged, but also very hydrophobic, are relatively unaffected by the presence of exopolymers (Ichimiya *et al.*, 1994). Poor penetration through anionic matrices might therefore be a phenomenon restricted to the more hydrophilic, positively charged agents.

The barrier properties associated with the exopolymeric matrix are dependent upon not only the chemical nature of the treatment agents and the binding capacity of the polymeric matrix, but also the levels of agent employed (Nichols, 1993). Equally, the distribution of biomass and its turnover (Kumon *et al.*, 1994), together with local hydrodynamic effects, profoundly affect the mass flux of agent to the core of a biofilm (DeBeer *et al.*, 1994). For antibiotics such as tobramycin and cefsulodin, such effects are likely to be minimal (Nichols *et al.*, 1988, 1989), but for positively charged anti-biotics such as the aminoglycosides, which will bind to polyanionic matrix polymers (Nichols *et al.*, 1988), they are high.

In such a fashion, the resistance of *Klebsiella pneumoniae* and *P. aeruginosa* biofilms towards monochloramine treatment has been explained as depletion of the biocide within the interior of the biofilm caused through reaction-diffusion interactions at the periphery (Huang *et al.*, 1995). Huang *et al.* grew biofilms of the two organisms together on stainless steel surfaces using a continuous-flow annular reactor. Biofilms were treated with 2 mg monochloramine l^{-1} for 2 h and stained using a fluorogenic redox indicator that could differentiate respiring from non-respiring cells. Epifluores-cent micrographs of frozen cross-sections taken at regular time intervals revealed gradients of respiratory activity within biofilms in response to monochloramine treat-ment. Cells near the biofilm–bulk fluid interface lost respiratory activity early in the treatment whereas residual respiratory activity persisted near the substratum or in the centre of small, viable cell clusters, even after 2 h treatment. A further study (Stewart *et al.*, 1998), using both oxidizing and non-oxidizing biocides, used an artificial biofilm construct of alginate-entrapped *Enterococcus faecalis* to demonstrate a similar lack of penetration and action against the entrapped cells by chlorine, glutaraldehyde, isothia-zolone and some quaternary ammonium biocides. In a real-life situation, however, the volume and reactive capacity of a biofilm would be insufficient to deplete the bulk availability of biocide, and interactive sites within the matrix polymers would become saturated with adsorbed/reacted biocide. The net effect would therefore be to delay, rather than prevent, the inhibitory process (Huang *et al.*, 1995). Provided that the exposure to biocide were brief, or that the biocide were not in a vast excess to requirement, then reaction-diffusion-limitation could allow the survival of cells at the base of the biofilm. Such cells would flourish once the biocide was removed.

PHYSIOLOGICAL GRADIENTS AND THE EXOPOLYMERIC MATRIX

The susceptibility of bacterial cells towards antibiotics, biocides and preservatives is significantly affected not only by their nutrient status and growth rate, but also by temperature, pH and prior exposure to sub-effective concentrations of antimicrobial (Brown & Williams, 1985; Brown *et al.*, 1990; Williams, 1988). Such factors, which are long-established and independent of growth in association with a surface, relate to changes in a variety of cellular components which include membrane fatty acids, phospholipids and envelope proteins, and the production of extracellular enzymes and polysaccharides. Such plasticity of form not only enables the cells to compete effectively for nutrients within hostile, oligotrophic environments, but coincidentally also alters the abundance, accessibility and physiological importance of various targets for antimicrobial action (Brown *et al.*, 1990; Gilbert *et al.*, 1990). Biofilm communities provide for a close proximity of cells that enables functional partnerships to become established through the localized concentration of enzymes and metabolic products, and which also minimizes the consequences of fluctuations in the surrounding macroenvironment. On the other hand, this proximity of cells establishes concentration gradients of gases such as oxygen and of readily utilized nutrients within the community (Costerton *et al.*, 1987). Thus cells deep within the biofilm are exposed to concentrations of substrates, hydrogen ions and also oxidation potentials that are substantially altered from those experienced by cells on the periphery and by those cells growing planktonically in the same medium. Growth rates will therefore be generally reduced within the biofilm through an imposition of nutrient deficiencies, which may or may not reflect the composition of the bulk aqueous phase. In nutrient-rich environments, oxygen and nutrients are rapidly utilized by aerobic bacteria at the biofilm:liquid interface, thereby diminishing the availability of such nutrients in the depths of the biofilm. This leads to the formation of anaerobic and anoxic zones (Marshall, 1992). In this manner, nutrient gradients will become established within thick biofilms that will generate communities that are spatially heterogeneous with respect to growth rate. At any particular time, a plethora of phenotypes would be represented within the community, reflecting the chemical heterogeneity of the biofilm. A wealth of data gathered from conventional planktonic experiments would suggest that each of these phenotypes would have a different susceptibility towards a variety of chemical antimicrobial agents or antibiotics.

Whilst the existence of physiological gradients, and hence gradients of susceptibility, within biofilm communities had long been assumed (Brown *et al.*, 1990; Gilbert *et al.*, 1990), it has been elegantly visualized by Wentland *et al.* (1996). These workers used acridine orange to differentially stain *K. pneumoniae* biofilms and showed that the regions of fastest growth were in the outer 30 μm of the biofilm, closest to the bulk liquid. Dispersed regions of slow growth, tending towards the substratum, were also noted. A major contributor towards the observed resistance of biofilms can therefore be identified as associated with physiological gradients of growth rate and nutritional status.

A distinction should be made between those effects related to the nature of the least available nutrient (nutrient limitation/depletion) and the cellular growth rate per se. Within the depths of a biofilm, growth rates will generally be suppressed relative to planktonic cells growing in the same environment. In this respect, Ashby *et al.* (1994) determined biofilm:planktonic ratios of isoeffective concentration (growth inhibition and bactericidal activity) for a wide range of antibiotics against cells that had been grown either in broth culture or as biofilms on urinary catheter discs. They noted that such ratios paralleled closely those generated between non-growing and actively growing cultures. With the exception of ciprofloxacin, those antibiotic agents that were most effective against non-growing cultures (i.e. imipenem, meropenem) were also the most active against the biofilms. Other workers have used perfused biofilm fermenters (Gilbert *et al.*, 1989) to study the effects of growth rate within biofilm populations. Such fermenters generate relatively homogeneous biofilms (approx. 40 μm thickness) that are attached to bacteria-proof, cellulose membranes through which nutrient media are perfused. These biofilms achieve a pseudo-steady state with respect to population size, where the rate of cell division is governed by the rate of perfusion of medium. Using cells grown planktonically at specified growth rates within a chemostat as comparators, then the separate contributions of growth rate and association within a biofilm to antimicrobial susceptibility can be evaluated. Using such methods, it was possible to compare the susceptibility of *S. epidermidis* (Duguid *et al.*, 1992) and *Escherichia coli* to tobramycin (Evans *et al.*, 1990b), and of *E. coli* to cetrimide (Evans *et al.*, 1990a), for cells resuspended from biofilms and those grown planktonically at the same growth rate. In these instances, whilst susceptibility was found to be markedly affected by growth rate, the two culture types possessed virtually identical susceptibilities to these agents. When intact, rather than resuspended, biofilms were examined, however, then susceptibility was decreased somewhat from that of planktonic and resuspended biofilm cells. This indicated some benefit to the cells of organization within an exopolymeric matrix.

Stewart (1994) developed a mathematical model which incorporated these concepts of metabolism-driven oxygen gradients and growth-rate-dependent killing to examine the susceptibility of *S. epidermidis* biofilms to various antibiotics. The model accurately predicted that susceptibility would thereby be reduced in thicker biofilms due to oxygen limitation. Oxygen availability within the biofilm may also have a direct influence upon the activity of some antibacterial agents (i.e. Bronopol; Shepherd *et al.*, 1988; Zabinski *et al.*, 1995). Since the extent of nutrient and gaseous gradients will increase as biofilms thicken and mature, then the effects of growth rate upon susceptibility will become increasingly marked as the biofilms increase in age (Anwar *et al.*, 1989; Anwar & Costerton, 1990). Such changes probably contribute to reports that aged biofilms are more recalcitrant to antibiotic and biocide treatment than are younger ones (Anwar *et al.*, 1989).

Whilst the contribution of reduced growth rate to a general lack of susceptibility to antibiotics within biofilm communities is likely to be profound, as for diffusion-limitation, it can not be the sole explanation of resistance. The establishment of physiological gradients within biofilm communities is dependent upon the growth and metabolism of cells at the periphery. Such cells will have growth rates and nutrient profiles that are not very dissimilar from those of planktonic cells and will deplete nutrients before they become accessible to the more deeply placed cells. The peripheral cells ought therefore to be as sensitive to biocides and antibiotics as are the planktonic ones and will quickly succumb to treatments. Failure of these cells to consume nutrients, and the release of the products of lysis from the killed cells, would dramatically increase the supply of nutrients to underlying regions of the biofilm. If growth rate were the sole moderator of resistance, then these regions would step up their metabolism and growth rate, adopt a more susceptible phenotype and die. Indeed, the growth rate of cells within this layer might even exceed that of untreated peripheral cells through mobilization of nutrients from the dead biomass. This phenomenon would occur throughout the biofilm, proceeding inwards from the outside, until the biofilm had been killed in its entirety. Should the flux of antimicrobial agent cease, then the biofilm could re-establish almost as fast as it was destroyed because of localized abundance of nutrients. In such a fashion, whilst such growth-rate-related differences in susceptibility within a biofilm might delay the onset of killing in the recesses of the film, they could not confer resistance to sustained exposures.

Neither reaction-diffusion-limitations on the access of treatment agent to a biofilm nor the existence of physiological gradients across them can provide a complete explanation of their resistance to long-term exposure to antimicrobial agents. In order for such tolerance of antimicrobial agents to develop, the biofilm populations must adapt to more resistant phenotypes either at the time of their formation or during the 'time-window' of opportunity provided by this buffering of antimicrobial action.

BIOFILM-SPECIFIC, ANTIBIOTIC-RESISTANT PHENOTYPES

Neither the generation of concentration gradients of antimicrobial nor the existence of physiological heterogeneities across biofilms can fully explain the observed levels of resistance towards chronic exposure to high levels of antibiotics and biocides. Either mechanism is capable of delaying the time of death for a small proportion of the biofilm community. These remnants of the treated population must adopt a more resistant phenotype before their expected demise if the community as a whole is to persist.

Clonal selection of resistant phenotypes

With respect to treatment with many antibiotics, then for many agents [i.e. staphylococci and fusidic acid (Chopra, 1976); pneumococci and sulphonamides (Wolf & Hotchkiss, 1963); *Enterobacteriaceae* and triclosan (McMurray *et al.*, 1998)], resistance can develop as a result of single point mutation. These often occur spontaneously with a frequency of approximately $1/10^8$ divisions. The numbers of individual cells within biofilms associated with disease will generally exceed the required number for resistant cells to be present. During antibiotic treatment of bacteraemias, such occasional cells will not only be nutrient-starved, but they will also be open to attack by the immune defences of the host. Treated within a biofilm, on the other hand, such cells will not only be protected against phagocytosis, but they will also be capable of very rapid division when bathed in the degradation products of their dead neighbours. In this latter respect, the glycocalyx will act as a nutrient reservoir for the stressed community and circumvent the normal iron restriction of growth that would be imposed within a mammalian host. The appearance of sub-lethally treated biofilms with small clusters of surviving cells is supportive of the validity of clonal selection in biofilm resistance development. Clinical data do not support this view, however, since a number of studies have demonstrated that the susceptibility of recovered, treated biofilms is not too dissimilar from that of the 'wild-type' (Nickel *et al.*, 1985; Gristina *et al.*, 1987).

Induction of multi-drug-resistant phenotypes

If resistance development is not simply clonal selection, then it is possible for the sublethal exposure experienced by the initial survivors to regulate the expression of a more resistant phenotype. It is interesting to note, in this light, that the expression of multi-drug-resistance operons such as *mar* and efflux pumps such as *acrAB* has been shown to be up-regulated by exposure to sub-effective concentrations of antibiotics such as tetracycline and chloramphenicol (George & Levy, 1983; Ma *et al.*, 1993), and to xenobiotics such as salicylates and triclosan (Alekshun & Levy, 1999). *mar* is chromosomal, variously induced and represented within a wide range of Gram-negative bacteria. Induction of operons such as *mar* during the delayed onset of the action of inducer antibiotics directed at biofilms is a tempting explanation of the long-term resistance of biofilms. The importance of *mar* would be far greater, however, if it were induced by growth as a biofilm per se, and conferred a more resistant phenotype upon the cells prior to treatment. Ciprofloxacin exposure does not induce the expression of *mar* or *acrAB* in *E. coli*, but such expression will confer limited protection against this agent. Exposure to ciprofloxacin of biofilms comprised of wild-type, constitutive and *mar*-deleted strains ought to evaluate whether or not such genes were up-regulated in unexposed biofilm communities. Maira-Litran *et al.* (2000a) perfused biofilms of such *E. coli* strains for 48 h with various concentrations of ciprofloxacin. These experiments, whilst demonstrating reduced susceptibility in the *mar* constitutive strain, showed little or no difference between wild-type and *mar*-deleted strains (Maira-Litran *et al.*, 2000a). Similar experiments, using biofilms constructed from strains in which the efflux pump *acrAB* was either deleted or constitutively expressed (Maira-Litran *et al.*, 2000b), showed the *acrAB* deletion to not significantly affect susceptibility over that of the wild-type strain. Clearly, neither *mar* nor *acrAB* is induced by sub-lethal treatment of biofilms with other than inducer substances. On the other hand, constitutive expression of *acrAB* protected the biofilm against low concentrations of ciprofloxacin, and studies conducted in continuous culture with a *lacZ* reporter gene fused to *marO* showed *mar* expression to be inversely related to specific growth rate (Maira-Litran *et al.*, 2000b). In this fashion, exposure of biofilms to sub-lethal levels of any inducer substance, of which there are a large number, will cause *mar* expression to be greatest within the depths of the biofilm where growth rates are suppressed. This factor will further enhance the likelihood of long-term survival from the treatment. Similar systems under the regulation of different inducer agents might extend this explanation of biofilm tolerance to include other treatment agents.

Attachment-specific phenotypes

Not only can bacteria sense the proximity of a surface and up-regulate production of EPS, but changes in phenotype coincidental with attachment to a surface can be associated with changes in susceptibility towards antibiotics (Ashby *et al.*, 1994) and biocides (Das *et al.*, 1998). Das *et al.* (1998) developed a novel spectrophotometric method to simultaneously monitor the growth rate of both planktonic and biofilm bacteria within microtitre plates. When performed in the presence of various concentrations of biocides, it was possible to demonstrate that the susceptibility of *P. aeruginosa* and *Staphylococcus aureus* to a range of different agents changed rapidly after cellular attachment. Such changes occurred before sufficient time had elapsed for other than microcolony formation to occur. In some instances, three- to fivefold decreases in susceptibility occurred immediately on attachment and could occur in the presence of biocide concentrations that exceeded the MIC for the planktonic population. In a similar study, Fujiwara *et al.* (1998) investigated the immediate effects of adherence on the susceptibility of *P. aeruginosa*, *Serratia marcescens* and *Proteus mirabilis*. Their results demonstrated that after 1 h incubation, during which time the bacteria were allowed to adhere to a plastic surface, the minimal bactericidal concentrations were markedly elevated. Whilst the magnitude of the changes in susceptibility associated with attachment per se is insufficient to account for the reported levels of resistance in mature biofilm communities, they will contribute towards the overall adoption of recalcitrance. Changes such as these are possibly mediated through the accumulation of AHL-like signal substances at the occluded surface (Davies *et al.*, 1998), as for the up-regulation of EPS deposition. Alternatively, attachment to surfaces might trigger a general stress response. The *rpoS*-encoded sigma factor σ^S is a master regulator in a complex network of stationary phase responsive genes in *E. coli*. There is a hierarchical link between AHL and *rpoS* expression (Latifi *et al.*, 1996) and a role for AHL has now been demonstrated in biofilm formation. It is, therefore, not inconceivable that biofilm growth leads to an early accumulation of density-dependent signals and to an early general stress response, and possibly to a more complete expression relative to that in conventional planktonic culture (Foley *et al.*, 1999).

SUMMARY AND CONCLUSIONS

There is an unambiguous association between the growth of cells within a biofilm community and the resistance to a wide variety of antimicrobial agents. In spite of a wealth of studies into the phenomena associated with this development, the mechanisms remain unclear. This is particularly the case when such resistance relates to a long-term, sustained treatment with biocides or antibiotics. Organization of the biofilm population within an exopolymeric matrix establishes a reaction-diffusion-limitation to the access of agent from its point of application to the deeper lying cells. These deeper lying cells will out-survive those on the surface and, if the bulk of the treatment agent is depleted or the exposure transient, will multiply and divide. Nutrients, as well as antimicrobial agents, will suffer from reaction-diffusion-limitation of their availability to individual cells within biofilms. This will lead to the establishment of spatial gradients of growth rate within the community structure. Different growth-limiting nutrients will also prevail at different points in the biofilm. This will provide for a plethora of phenotypes within the biofilm, each reflecting the physico-chemical microenvironment of individual cells and their proximity to neighbours. Faster-growing, more-susceptible cells will generally lie on the periphery of the biofilm with slow-growing recalcitrant cells being more deeply placed. In both instances, at the fringes of action, selection pressures will enrich the populations with the least susceptible genotype. It is possible, under such circumstances, for repeated chronic exposure to sub-lethal treatments to select for a resistant population that shows cross-resistance to other forms of antimicrobial. Whilst neither of these mechanisms can provide a complete explanation of recalcitrance, together they will delay eradication of the treated population and allow other selection and regulation events to occur.

Interestingly, these mechanisms of resistance acquisition are not dissimilar to *in vitro* 'training' experiments whereby micro-organisms that survive a low concentration of an antimicrobial agent are gradually exposed to increasing concentrations of the agent in a step-wise manner. In this fashion, cells can be 'trained' to acquire resistance (Brown *et al.*, 1969; Gilleland *et al.*, 1989). This process has often been dismissed in the past as being 'artificial' on the grounds that it could not possibly occur in nature. However, biofilms exposed to antimicrobial agents in a pulsed fashion will, assuming incomplete eradication of the cells, produce a front of killing within the biofilm that contains a mixture of live and dead cells. The live cells will subsequently be selected as being more resistant, and at the next passage of antimicrobial treatment will move closer to the biofilm periphery. Such resistance problems will be compounded by the inappropriate use of different antimicrobial agents.

More recent explanations of the resistance of biofilm communities lie with the expression of biofilm-specific phenotypes that are so different to those of planktonic cells that the agents developed against the latter fail to be effective. Whilst such phenotypes are known to be expressed, and might be regulated through quorum-sensing mechanisms and general stress response regulons, they do not appear to contribute greatly to the susceptibility pattern of individual biofilm cells. Such processes do, however, offer the possibility of novel agents that might prevent the formation of dense, polymer-encased communities and thereby circumvent the problem of killing intact, mature biofilms.

REFERENCES

Alekshun, M. N. & Levy, S. B. (1999). Alteration of the repressor activity of MarR, the negative regulator of the *Escherichia coli marRAB* locus, by multiple chemicals *in vitro*. *J Bacteriol* **181**, 4669–4672.

Allison, D. G. (1998). Exopolysaccharide production in bacterial biofilms. *Biofilm* **3**, BF98002.

Allison, D. G. & Gilbert, P. (1995). Modification by surface association of antimicrobial susceptibility of bacterial populations. *J Ind Microbiol* **15**, 311–317.

Allison, D. G. & Matthews, M. J. (1992). Effect of polysaccharide interactions on antibiotic susceptibility of *Pseudomonas aeruginosa*. *J Appl Bacteriol* **73**, 484–488.

Allison, D. G. & Sutherland, I. W. (1987). The role of exopolysaccharides in adhesion of freshwater bacteria. *J Gen Microbiol* **133**, 1319–1327.

Anwar, H. & Costerton, J. W. (1990). Enhanced activity of combination of tobramycin and piperacillin for eradication of sessile biofilm cells of *Pseudomonas aeruginosa*. *Antimicrob Agents Chemother* **34**, 1666–1671.

Anwar, H., Dasgupta, M., Lam, K. & Costerton, J. W. (1989). Tobramycin resistance of mucoid *Pseudomonas aeruginosa* biofilm grown under iron limitation. *J Antimicrob Chemother* **24**, 647–655.

Ashby, M. J., Neale, J. E., Knott, S. J. & Critchley, I. A. (1994). Effect of antibiotics on non-growing cells and biofilms of *Escherichia coli*. *J Antimicrob Chemother* **33**, 443–452.

Brown, M. R. W. & Williams, P. (1985). The influence of environment on envelope properties affecting survival of bacteria in infections. *Annu Rev Microbiol* **39**, 527–556.

Brown, M. R. W., Watkins, W. M. & Foster, J. H. (1969). Step-wise resistance to polymyxin and other agents by *Pseudomonas aeruginosa*. *J Gen Microbiol* **55**, 17–18.

Brown, M. R. W., Collier, P. J. & Gilbert, P. (1990). Influence of growth rate on the susceptibility to antimicrobial agents: modification of the cell envelope and batch and continuous culture. *Antimicrob Agents Chemother* **34**, 1623–1628.

Characklis, W. G. (1990). Microbial fouling. In *Biofilms*, pp. 523–584. Edited by W. G. Characklis & K. C. Marshall. New York: Wiley.

Chopra, I. (1976). Mechanisms of resistance to fusidic acid in *Staphylococcus aureus*. *J Gen Microbiol* **96**, 229–238.

Cooper, M., Batchelor, S. M. & Prosser, J. I. (1995). Is cell density signalling applicable to biofilms? In *The Life and Death of Biofilm*, pp. 93–97. Edited by J. Wimpenny, P. Handley, P. Gilbert & H. Lappin-Scott. Cardiff: BioLine.

Costerton, J. W. & Lashen, E. S. (1984). Influence of biofilm on the efficacy of biocides on corrosion-causing bacteria. *Mat Per* **23**, 34–37.

Costerton, J. W., Irwin, R. T. & Cheng, K. J. (1981). The bacterial glycocalyx in nature and disease. *Annu Rev Microbiol* **35**, 399–424.

Costerton, J. W., Cheng, K. J., Geesey, G. G., Ladd, T. I., Nickel, J. C., Dasgupta, M. & Marrie, T. J. (1987). Bacterial biofilms in nature and disease. *Annu Rev Microbiol* **41**, 435–464.

Costerton, J. W., Lewandowski, Z., DeBeer, D., Caldwell, D., Kober, D. & James, G. (1994). Biofilms, the customised microniche. *J Bacteriol* **176**, 2137–2142.

Das, J. R., Bhakoo, M., Jones, M. V. & Gilbert, P. (1998). Changes in the biocide susceptibility of *Staphylococcus epidermidis* and *Escherichia coli* cells associated with rapid attachment to plastic surfaces. *J Appl Microbiol* **84**, 852–858.

Davies, D. G. & Geesey, G. G. (1995). Regulation of the alginate biosynthesis gene *algC* in *Pseudomonas aeruginosa* during biofilm development in continuous culture. *Appl Environ Microbiol* **61**, 860–867.

Davies, D. G., Chakrabarty, A. M. & Geesey, G. G. (1993). Exopolysaccharide production in biofilms: substratum activation of alginate gene expression by *Pseudomonas aeruginosa*. *Appl Environ Microbiol* **59**, 1181–1186.

Davies, D. G., Parsek, M. R., Pearson, J. P., Iglewski, B. H., Costerton, J. W. & Greenberg, E. P. (1998). The involvement of cell-to-cell signals in the development of a bacterial biofilm. *Science* **280**, 295–298.

DeBeer, D., Srinivasan, R. & Stewart, P. S. (1994). Direct measurement of chlorine penetration into biofilms during disinfection. *Appl Environ Microbiol* **60**, 4339–4344.

Duguid, I. G., Evans, E., Brown, M. R. W. & Gilbert, P. (1992). Effect of biofilm culture upon the susceptibility of *Staphylococcus epidermidis* to tobramycin. *J Antimicrob Chemother* **30**, 803–810.

Eginton, P. J., Holah, J., Allison, D. G., Handley, P. S. & Gilbert, P. (1998). Changes in the strength of attachment of microorganisms to surfaces following treatment with disinfectants and cleansing agents. *Lett Appl Microbiol* **27**, 101–106.

Evans, D. J., Brown, M. R. W., Allison, D. G. & Gilbert, P. (1990a). Growth rate and resistance of Gram-negative biofilms towards Cetrimide USP. *J Antimicrob Chemother* **26**, 473–478.

Evans, D. J., Brown, M. R. W., Allison, D. G. & Gilbert, P. (1990b). Susceptibility of bacterial biofilms to tobramycin: role of specific growth rate and phase in the division cycle. *J Antimicrob Chemother* **25**, 585–591.

Evans, D. J., Brown, M. R. W., Allison, D. G. & Gilbert, P. (1991). Susceptibility of *Escherichia coli* and *Pseudomonas aeruginosa* biofilms to ciprofloxacin: effect of specific growth rate. *J Antimicrob Chemother* **27**, 177–184.

Evans, E., Brown, M. R. W. & Gilbert, P. (1994). Iron chelator, exopolysaccharide and protease production of *Staphylococcus epidermidis*: a comparative study of the effects of specific growth rate in biofilm and planktonic culture. *Microbiology* **140**, 153–157.

Favero, M. S., Bond, W. W., Peterson, N. J. & Cook, E. H. (1983). Scanning electron microscopic observations of bacteria resistant to iodophor solutions. In *Proceedings of the International Symposium on Povidone*, pp. 158–166. University of Kentucky, Lexington, USA.

Foley, I., Marsh, P., Wellington, E. M. H., Smith, A. W. & Brown, M. R. W. (1999). General stress response master regulator *rpoS* is expressed in human infection: a possible role in chronicity. *J Antimicrob Chemother* **43**, 164–165.

Fujiwara, S., Miyake, Y., Usui, T. & Suginaka, H. (1998). Effect of adherence on antimicrobial susceptibility of *Pseudomonas aeruginosa*, *Serratia marcescens* and *Proteus mirabilis*. *Hiroshima J Med Sci* **47**, 1–5.

Gambello, M. J., Kaye, S. & Inglewski, B. H. (1993). *LasR* of *Pseudomonas aeruginosa* is a transcriptional activator of the line protease gene (*apr*) and an enhancer of exotoxin A expression. *Infect Immun* **61**, 1180–1184.

George, A. M. & Levy, S. B. (1983). Amplifiable resistance to tetracycline, chloramphenicol, and other antibiotics in *Escherichia coli*: involvement of a non-plasmid-determined efflux of tetracycline. *J Bacteriol* **155**, 531–540.

Gilbert, P. & Allison, D. G. (1999). Biofilms and their resistance towards antimicrobial agents. In *Dental Plaque Revisited: Oral Biofilms in Nature and Disease*, pp. 125–143. Edited by H. N. Newman & M. Wilson. Cardiff: BioLine.

Gilbert, P., Allison, D. G., Evans, D. J., Handley, P. S. & Brown, M. R. W. (1989). Growth rate control of adherent bacterial populations. *Appl Environ Microbiol* **55**, 1308–1311.

Gilbert, P., Collier, P. J. & Brown, M. R. W. (1990). Influence of growth rate on susceptibility to antimicrobial agents: biofilms, cell cycle and dormancy. *Antimicrob Agents Chemother* **34**, 1865–1868.

Gilleland, L. B., Gilleland, H. E., Gibson, J. A. & Champlin, F. R. (1989). Adaptive resistance to aminoglycoside antibiotics in *Pseudomonas aeruginosa*. *J Med Microbiol* **29**, 41–50.

Giwercman, B., Jensen, E. T., Hoiby, N., Kharazmi, A. & Costerton, J. W. (1991). Induction of β-lactamase production in *Pseudomonas aeruginosa* biofilms. *Antimicrob Agents Chemother* **35**, 1008–1010.

Gordon, C. A., Hodges, N. A. & Marriot, C. (1988). Antibiotic interaction and diffusion through alginate and exopolysaccharide of cystic fibrosis derived *Pseudomonas aeruginosa*. *J Antimicrob Chemother* **22**, 667–674.

Gristina, A. G., Hobgood, C. D., Webb, L. X. & Myrvik, Q. N. (1987). Adhesive colonisation of biomaterials and antibiotic resistance. *Biomaterials* **8**, 423–426.

Heys, S. J. D., Gilbert, P. & Allison, D. G. (1997). Homoserine lactones and bacterial biofilms. In *Biofilms: Community Interactions and Control*, pp. 103–112. Edited by J. Wimpenny, P. Handley, P. Gilbert, H. Lappin-Scott & M. Jones. Cardiff: BioLine.

Holah, J. T., Bloomfield, S. F., Walker, A. J. & Spenceley, H. (1994). Control of biofilms in the food industry. In *Bacterial Biofilms and their Control in Medicine and Industry*, pp. 163–168. Edited by J. T. Wimpenny, W. W. Nichols, D. Stickler & H. Lappin-Scott. Cardiff: BioLine.

Hoyle, B. D., Wong, C. K. & Costerton, J. W. (1992). Disparate efficacy of tobramycin on Ca(2+)-, Mg(2+)-, and HEPES-treated *Pseudomonas aeruginosa* biofilms. *Can J Microbiol* **38**, 1214–1218.

Huang, C. T., Yu, F. P., McFeters, G. A. & Stewart, P. S. (1995). Nonuniform spatial patterns of respiratory activity within biofilms during disinfection. *Appl Environ Microbiol* **61**, 2252–2256.

Ichimiya, T., Yamaski, T. & Nasu, M. (1994). *In-vitro* effects of antimicrobial agents on *Pseudomonas aeruginosa* biofilm formation. *J Antimicrob Chemother* **34**, 331–341.

Kleerebezem, M., Quadri, L. E. N., Kuipers, O. P. & deVos, W. M. (1997). Quorum sensing by peptide pheromones and two-component signal-transduction systems in Gram-positive bacteria. *Mol Microbiol* **24**, 895–904.

Kumon, H., Tomochika, K.-I., Matunaga, T., Ogawa, M. & Ohmori, H. (1994). A sandwich cup method for the penetration assay of antimicrobial agents through *Pseudomonas* exopolysaccharides. *Microbiol Immunol* **38**, 615–619.

Lambert, P. A., Giwercman, B. & Hoiby, N. (1993). Chemotherapy of *Pseudomonas aeruginosa* in cystic fibrosis. In *Bacterial Biofilms and their Control in Medicine and Industry*, pp. 151–153. Edited by J. T. Wimpenny, W. W. Nichols, D. Stickler & H. Lappin-Scott. Cardiff: BioLine.

Latifi, A., Foglino, M., Tanaka, K., Williams, P. & Lazdunski, A. (1996). A hierarchical quorum sensing cascade to phagocytosis in *Pseudomonas aeruginosa* links the transcriptional activators LasR and RhlR (VsmR) to expression of the stationary phase sigma factor RpoS. *Mol Microbiol* **21**, 1137–1146.

Little, B. J., Wagner, P. A., Characklis, W. G. & Lee, W. (1990). Microbial corrosion. In *Biofilms*, pp. 635–670. Edited by W. G. Characklis & K. C. Marshall. New York: Wiley.

Ma, D., Cook, D. N., Alberti, M., Pong, N. G., Nikaido, H. & Hearst, J. E. (1993). Molecular cloning and characterization of *acrAB* and *acrE* genes of *Escherichia coli*. *J Bacteriol* **175**, 6299–6313.

McMurray, L. M., Oethinger, M. & Levy, S. B. (1998). Triclosan inhibits lipid synthesis. *Nature* **394**, 531–532.

Maira-Litran, T., Allison, D. G. & Gilbert, P. (2000a). An evaluation of the potential of the multiple antibiotic resistance operon (*mar*) and the multidrug efflux pump *acrAB* in the resistance of *Escherichia coli* biofilms towards ciprofloxacin. *J Antimicrob Chemother* **45**, 789–795.

Maira-Litran, T., Allison, D. G. & Gilbert, P. (2000b). Expression of the multiple antibiotic resistance operon (*mar*) during growth of *Escherichia coli* as a biofilm. *J Appl Microbiol* **88**, 243–247.

Marshall, K. C. (1992). Biofilms: an overview of bacterial adhesion, activity and control at surfaces. *ASM News* **58**, 202–207.

Nichols, W. W. (1993). Biofilm permeability to antibacterial agents. In *Bacterial Biofilms and their Control in Medicine and Industry*, pp. 141–149. Edited by J. T. Wimpenny, W. W. Nichols, D. Stickler & H. Lappin-Scott. Cardiff: BioLine.

Nichols, W. W., Dorrington, S. M., Slack, M. P. E. & Walmsley, H. L. (1988). Inhibition of tobramycin diffusion by binding to alginate. *Antimicrob Agents Chemother* **32**, 518–523.

Nichols, W. W., Evans, M. J., Slack, M. P. E. & Walmsley, H. L. (1989). The penetration of antibiotics into aggregates of mucoid and non-mucoid *Pseudomonas aeruginosa*. *J Gen Microbiol* **135**, 1291–1303.

Nickel, J. C., Ruseska, I., Wright, J. B. & Costerton, J. W. (1985). Tobramycin resistance of cells of *Pseudomonas aeruginosa* growing as a biofilm on urinary catheter material. *Antimicrob Agents Chemother* **27**, 619–624.

Shepherd, J. E., Waigh, R. D. & Gilbert, P. (1988). Antimicrobial action of 2-bromo-2-nitropropan-1,3-diol (Bronopol) against *Escherichia coli*. *Antimicrob Agents Chemother* **32**, 1693–1698.

Slack, M. P. E. & Nichols, W. W. (1981). The penetration of antibiotics through sodium alginate and through the exopolysaccharide of a mucoid strain of *Pseudomonas aeruginosa*. *Lancet* **11**, 502–503.

Slack, M. P. E. & Nichols, W. W. (1982). Antibiotic penetration through bacterial capsules and exopolysaccharides. *J Antimicrob Chemother* **10**, 368–372.

Sondossi, M., Rossmore, H. W. & Wireman, J. W. (1985). Observation of resistance and cross-resistance to formaldehyde and a formaldehyde condensate biocide in *Pseudomonas aeruginosa*. *Int Biodeterior* **21**, 105–106.

Stewart, P. S. (1994). Biofilm accumulation model that predicts antibiotic resistance of *Pseudomonas aeruginosa* biofilms. *Antimicrob Agents Chemother* **38**, 1052–1058.

Stewart, P. S. (1996). Theoretical aspects of antibiotic diffusion into microbial biofilms. *Antimicrob Agents Chemother* **40**, 2517–2522.

Stewart, P. S., Grab, L. & Diemer, J. A. (1998). Analysis of biocide transport limitation in an artificial biofilm system. *J Appl Microbiol* **85**, 495–500.

Suci, P. A., Mittelman, M. W., Yu, F. U. & Geesey, G. G. (1994). Investigation of ciprofloxacin penetration into *Pseudomonas aeruginosa* biofilms. *Antimicrob Agents Chemother* **38**, 2125–2133.

Sutherland, I. W. (1985). Biosynthesis and composition of Gram-negative bacterial extracellular and wall polysaccharides. *Annu Rev Microbiol* **39**, 243–270.

Sutherland, I. W. (1995). Biofilm specific polysaccharides: do they exist? In *Life and Death of the Biofilm*, pp. 103–107. Edited by J. Wimpenny, P. Handley, P. Gilbert & H. Lappin-Scott. Cardiff: BioLine.

Sutherland, I. W. (1997). Microbial biofilm exopolysaccharides – superglues or velcro? In *Biofilms: Community Interactions and Control*, pp. 33–39. Edited by J. Wimpenny, P. Handley, P. Gilbert, H. Lappin-Scott & M. Jones. Cardiff: BioLine.

Wentland, E. J., Stewart, P. S., Huang, C. T. & McFeters, G. A. (1996). Spatial variations in growth rate within *Klebsiella pneumoniae* colonies and biofilm. *Biotechnol Prog* **12**, 316–321.

Williams, P. (1988). Role of the cell envelope in bacterial adaption to growth *in vivo* in infections. *Biochimie* **70**, 987–1011.

Williams, P., Bainton, N. J., Swift, S., Chhabra, S. R., Winson, M. K., Stewart, G. S. A. B., Salmond, G. P. C. & Bycroft, B. W. (1992). Small molecule-mediated density dependent control of gene expression in prokaryotes: bioluminescence and the biosynthesis of carbapenem antibiotics. *FEMS Microbiol Lett* **100**, 161–168.

Wolf, B. & Hotchkiss, R. D. (1963). Genetically modified folic acid synthesising enzymes of Pneumococcus. *Biochemistry* **2**, 145–150.

Zabinski, R. A., Walker, K. J., Larsson, A. J., Moody, J. A., Kaatz, G. W. & Rotschafer, J. C. (1995). Effect of aerobic and anaerobic environments on antistaphylococcal activities of five fluoroquinolones. *Antimicrob Agents Chemother* **39**, 507–512.

Zambrano, M. M. & Kolter, R. (1995). Changes in bacterial cell properties on going from exponential growth to stationary phase. In *Microbial Quality Assurance: a Guide Towards Relevance and Reproducibility*, pp. 21–30. Edited by M. R. W. Brown & P. Gilbert. Boca Raton, FL: CRC Press.

Biofilms in the New Millennium: musings from a peak in Xanadu

J. William Costerton

Director, Center for Biofilm Engineering, Montana State University, Bozeman, MT 59717, USA

INTRODUCTION

Most biological sciences had defined their subject organisms, grossly and at the microscopic level, long before the advent of molecular biology. For this reason, sciences like botany had characterized the structure, even at the ultrastructural level, of the cells and tissues that comprise all of the forms adopted by plants, including all vegetative and reproductive manifestations of thousands of organisms. Therefore, the detailed molecular mechanisms discovered by modern plant molecular biologists fit nicely into a framework of understanding of exactly where these mechanisms operate, in terms of the structure and function of the organism concerned. We visualize the process of transcription occurring on the cytoplasmic aspect of the endoplasmic reticulum, and we can follow the resultant peptides through their various fates, including excretion in vesicles via the Golgi apparatus. The intellectual synthesis within a field like botany is essentially very satisfying, because we could previously see the central genetic control of shape and function, and now we can understand the molecular mechanisms by which this control is exercised and we can manipulate plants using this knowledge.

The relatively new science of microbiology, which has done so much to eradicate epidemic diseases and to improve our lives, is presently reeling in confusion because it has not followed the same orderly development of concepts and techniques. At a very early stage in its development, microbiology embraced two moda operandi that served it well, but damaged it in the long term. First, following Robert Koch, we removed bacterial cells from the ecosystems in which they lived, and grew them in monospecies cultures in fluids or on agar. This decision served humanity well in the conquest of

SGM symposium 59: Community structure and co-operation in biofilms. Editors D. Allison, P. Gilbert, H. Lappin-Scott, M. Wilson.
Cambridge University Press. ISBN 0 521 79302 5 ©SGM 2000.

many acute diseases, and we still use the same 150-year-old methods in such practical areas as diagnostic microbiology. Secondly, we set certain criteria (Koch's postulates) that licensed us to extrapolate from our *in vitro* single-species data to complex problems like the microbiology of the gut, in which many species operate in concert and we understand only what 'our' target organism is doing. The powerful armamentarium of classic microbiology was unleashed, generally, only when we were threatened by diseases. We developed the vaccines and antibiotics that have served us well, until recent times when the emergence of resistant strains has us profoundly worried, but we have crippled our science by three errors to which we subscribe:

(1) We still study bacteria in monospecies cultures in rich fluid media, and thus we study only one of many phenotypes.
(2) We still extrapolate from studies of the planktonic phenotype in monospecies culture in rich media to attempt to understand the behaviour of the same organism in the biofilm phenotype in a multispecies community living in a nutrient-limited ecosystem.
(3) We have amassed huge amounts of molecular data, including several complete genetic sequences, but we still do not understand that the phenotypic expression of the virulence factors with which we are preoccupied is controlled by relationships with non-pathogenic bacteria, and with environmental factors that we rarely study.

Here amongst the mountains of Xanadu we see real bacteria growing in many different phenotypes, and forming integrated multispecies communities of remarkable complexity, and we weep over the errors and omissions of the past. What will our young colleagues find when they use confocal microscopes and species-specific probes to explore the wonderful and complex biofilm communities that we didn't even know existed? We have studied the same ecosystems for 150 years, and all we have to show for it are freezers full of monospecies cultures in broth, and reams of molecular data on the planktonic phenotypes that now cause only a minority of human infections in the developed world.

BIOFILM STRUCTURE

The confocal scanning laser microscope (CSLM), which allows the examination of living and fully hydrated biological materials, has revolutionized the study of bacteria actually growing in their native environments (Lawrence *et al.*, 1991). By subtracting out-of-focus planes from the image, the confocal microscope allows us to examine both the surface and the bulk fluid components of virtually any ecosystem, and thus to confirm the conclusion of earlier studies (Geesey *et al.*, 1977) that the vast majority of bacteria grow in slime-enclosed biofilms. These valuable but laborious confocal studies have served to define the exact distribution of bacteria in a limited number of ecosystems, but they also provide a valid basis for the re-evaluation of the thousands of papers in which transmission and scanning electron microscopy have been used to study similar biofilms. The electron microscopy methods could only examine dehydrated specimens, because electrons cannot travel through water vapour. If we use our imaginations to 'rehydrate' the older images produced by electron microscopy, we can salvage information concerning the presence and number of cells on a surface, even if we cannot see the matrix material or define the complex cellular architecture of the sessile community. Both confocal and electron microscopic images will be used in this treatise, with this codicil in mind. In addition to its advantages in sophisticated optical analysis, and the fact that it can examine living biofilms and tissues, the confocal microscope can be used with a burgeoning array of physical and chemical probes. These probes, which range from oligonucleotide sequences to simple pH meters, now allow us to identify cells of various species and to measure physical and chemical parameters (e.g. dissolved oxygen) in living biofilms with a resolution of better than 5 μm. This is a wonderful time to be a microbiologist.

Bacterial adhesion

When a bacterial cell approaches a surface, which these cells do with notable avidity (Marshall, 1985), a very large number of behavioural and phenotypic changes take place. The cells adhere to the surface, in a reversible manner, and they use their pili (O'Toole & Kolter, 1998) and other surface appendages to move over the surface and to form aggregates and other formations of considerable variety. Darren Korber and his colleagues in Saskatoon have studied the mobile 'behaviours' of the recently adherent cells of several species of bacteria, and have noted that some roll and make 'windrows', while others form single-cell monolayers or discrete microcolonies. These recently adherent bacterial cells may compete for space on the surface, and Henk Busscher and his colleagues have correctly suggested that a whole array of surfactants and other bacterial weapons may come into play to determine which cells go on to form biofilms, and which interspecies alliances are formed. It would be intellectually facile to link these early arrangements of adherent cells to the complex towers and channels of mature biofilms, but, at this stage in this field, these connections simply have not been established with any certainty.

The first indication that bacterial cells undergo a profound phenotypic change following their adhesion to a surface came from the imaginative use of confocal microscopy and a reporter construct in which an indicator enzyme gene (*lacZ*) was inserted downstream from the gene (*algC*) that encodes phosphomannomutase. Davies & Geesey (1995) used this reporter construct to show that cells of *Pseudomonas aeruginosa* up-regulate their *algC* genes within a few minutes of adhesion, as one might expect because of the role of this enzyme in the production of the matrix material (alginate) that eventually cements the cells to the surface in an irreversible manner. Non-morphological studies (Yu, 1994) had previously shown that the *algD* gene, which also forms a part of the genetic sequence that controls the production of the alginate matrix of *P. aeruginosa* biofilms, is also up-regulated soon after cells of this species adhere to surfaces. Because these two genes are part of the alginate synthesis cascade, which is controlled by a sigma factor produced by *algU* (also known as *algT*), we know that the alginate synthesis system of this organism is up-regulated within minutes of adhesion. We must not extrapolate between species, but the observation that the cells of hundreds of different marine bacteria produce visible amounts of matrix material shortly after they adhere to surfaces suggests that this response to adhesion may be universal.

While the up-regulation of the matrix synthesis system following the adhesion of bacteria to a surface is expected, and the result of this up-regulation can be seen in the form of extruded exopolysaccharide, we were not prepared for the subsequent revelation that many other unrelated phenotypic changes are triggered by adhesion. Hongwei Yu has used PAGE gel techniques to separate the outer-membrane proteins (OMPs) from planktonic cells of *P. aeruginosa* grown in liquid media and in biofilms, and the OMPs of the planktonic phenotype are profoundly different from those of the biofilm phenotype (Fig. 1). Other laboratories have now used much more sophisticated techniques, including genetic library analysis, to examine the differences between the planktonic phenotype and the biofilm phenotype of *P. aeruginosa*, and it is clear that many more differences exist and that there is not a single biofilm phenotype. For example, it appears that the *rpoS* gene that controls senescence in *Pseudomonas* is up-regulated in biofilms, and it is of great interest to note that this gene is also up-regulated in cells of *P. aeruginosa* recovered directly from the lungs of cystic fibrosis patients (Foley *et al.*, 1999). The practical implications of a profound difference between the planktonic and the biofilm phenotype of a significant number of bacteria, if this proves to be the case, will take some time for full intellectual digestion, but the first logical reaction is that much work will have to be redone. Obviously the genotype of a bacterial species is constant, but phenotypes will differ in their response to environmental factors, and bacteria that produce a certain virulence factor when growing as planktonic cells in a monospecies culture may not even produce the same factor when growing in a biofilm. Cells growing in the biofilm phenotype may be resistant to antibiotics, not because the agents cannot penetrate the biofilm matrix (Stewart, 1996), but because the sessile phenotype has a profoundly different cell envelope structure, and perhaps even a different metabolic process that no longer constitutes a suitable target. We may have more success in the treatment of biofilm infections (Costerton *et al.*, 1999) when the antibiotics that we use have been selected

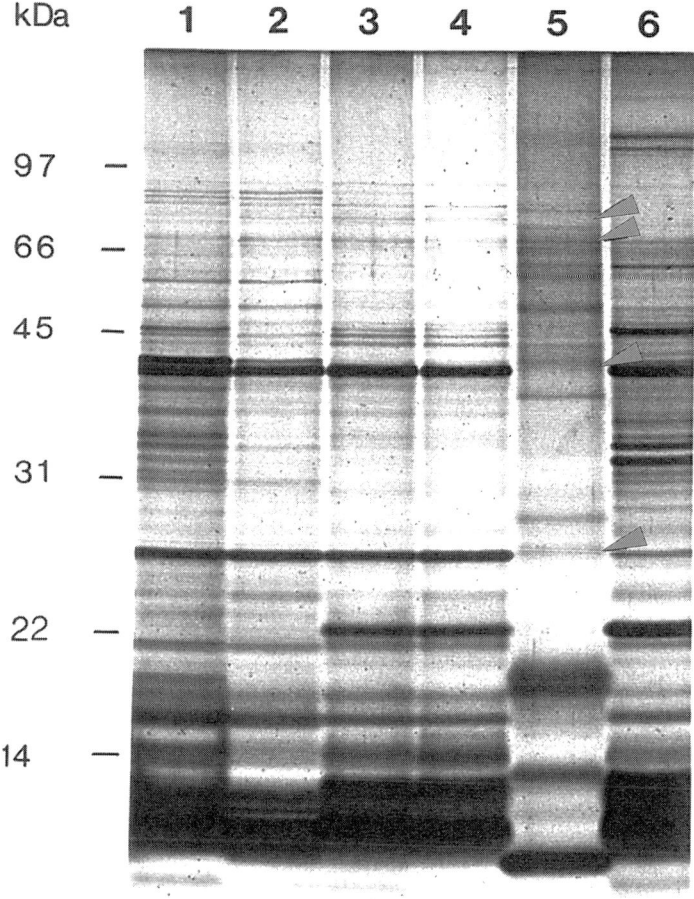

Fig. 1. PAGE gel of the outer-membrane proteins (OMPs) extracted from planktonic cells (lanes 1–4 and 6) and biofilm cells (lane 5) of *P. aeruginosa*. The planktonic OMPs seen in lane 6 are from cells grown in the same medium as the biofilm cells that yielded the OMPs seen in lane 5. Note the radical differences in the gene products of these two distinct phenotypes.

and evaluated for their ability to kill bacterial cells growing in the biofilm phenotype, and not planktonic cells growing in liquid media.

Biofilm formation

When bacterial cells have adhered to a surface, arranged themselves in species-specific patterns, assumed the biofilm phenotype, and up-regulated their matrix synthesis machinery, they have the potential to form a biofilm. Our direct examinations of living mature biofilms, by confocal microscopy, show that these sessile communities

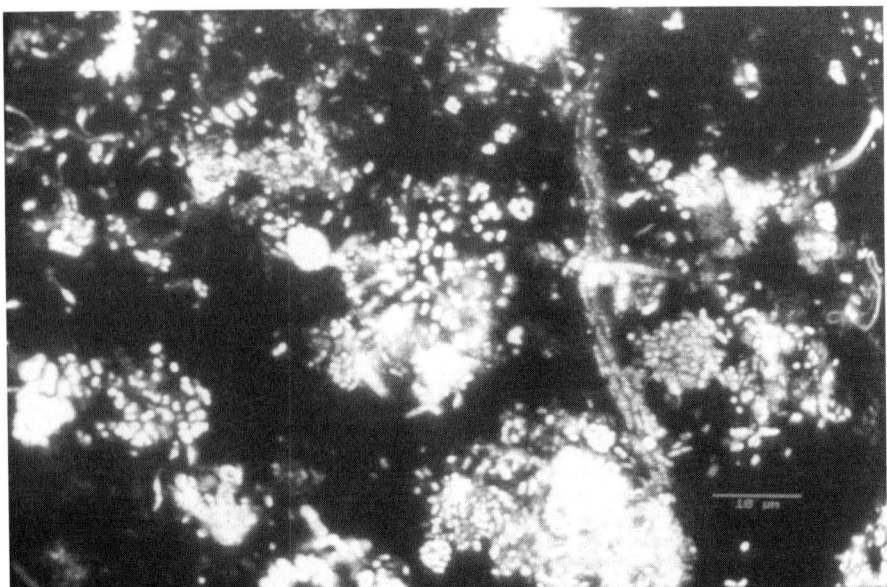

Fig. 2. Confocal scanning laser micrograph, in the x–y-axis and just above the colonized surface, of a natural multispecies biofilm formed on a polished rock surface in the Bow River over a period of 10 days. Note the discrete microcolonies, many of which contain more than one morphotype (species) of bacteria, and the open water channels that separate these basic structural units of this natural biofilm. Bar, 10 μm.

are very complex and they indicate that the newly adherent cell has a multitude of 'decisions' to make. Most of the natural biofilms examined to date are composed of discrete microcolonies, separated by open water channels (Fig. 2), and the sessile bacteria grow in the microcolonies where they comprise ±15% of the volume while the matrix material comprises ±85%. Even in monospecies biofilms, the cells are not evenly distributed within the microcolonies, and some species tend to produce a majority of cells in the apical regions of these structural units of the biofilm. Many mature biofilms are composed of microcolonies in which cells of many different species are present (Fig. 2), and we know that many of these combinations are predicated on physiological co-operativity (Costerton *et al.*, 1995). As we begin a simple-minded analysis of the early stages of biofilm formation, it is relatively easy to conceive of the spatial association of cells of two or more metabolically co-operative species to form the initial stages of a microcolony. Individual cells would grow quickly where they experienced the metabolic advantage of an interspecies association, and the co-operative cells would proliferate most exuberantly where they were in closest proximity. However, a careful computer-assisted analysis of the

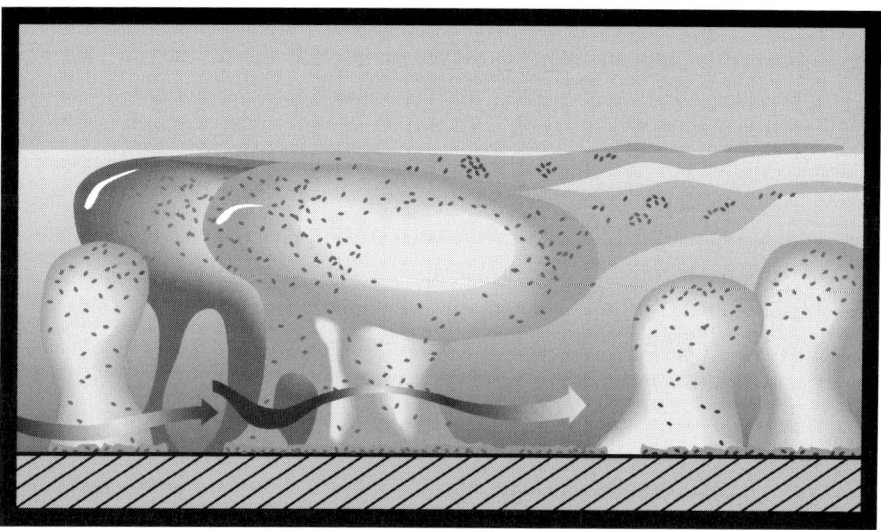

Fig. 3. Diagrammatic representation of morphological data from dozens of natural and *in vitro* biofilms, in the *x–z*-axis, showing the microcolonies and water channels that comprise these complex and highly structured communities. The sessile biofilm cells actually grow in matrix-enclosed microcolonies, of various shapes, and these microcolonies are often deformed by high shear forces to produce the streamers seen to project into the bulk fluid.

structure of monospecies and natural biofilms (Fig. 3) clearly shows that the microcolonies that comprise the biofilm are shaped like mushrooms, or stacks, and the water channels are open throughout the sessile community. This degree of structural complexity clearly precludes random growth, and dictates that we consider some sophisticated form of communication and control when considering the development of biofilms.

Biofilm structure

The first bacterial control signals were discovered in marine organisms that actually grow in biofilms in the light organs of higher animals (Fuqua *et al.*, 1994), but their manifold effects on cellular behaviour were studied using planktonic cells in monospecies fluid culture. For this reason, the actual ecological roles of these signal molecules may not have been realized, but a large and rapidly expanding literature now catalogues their manifold effects on planktonic cells (Parsek *et al.*, 1999). It became obvious that these acyl homoserine lactone (AHL) signal molecules, which control so many metabolic activities of planktonic Gram-negative cells, also control biofilm formation when mutants lacking the ability to synthesize these compounds were seen to be unable to produce structurally differentiated biofilms (Davies *et al.*, 1998). Other signal minus mutants have been able to produce structured biofilms (Stoodley *et al.*, 1999a), but it still appears that some of the signals that control the general area of quorum sensing in bacteria are also involved in controlling the development of biofilms. Ancillary support for this notion is provided by the observation that natural signal analogues (furanones), which block the activity of specific AHL signals, also block the formation of natural biofilms in the marine environment. Fig. 4 is a simple-minded cartoon that suggests that the development of a structurally differentiated multispecies biofilm requires the control of cell proliferation and matrix production by signals that allow sessile cells to control the activities of neighbouring cells of the same, or of different, species. While the use of AHL signals in quorum sensing by planktonic cells is logical and well-documented, the transitory nature of the spatial associations between planktonic cells in most ecosystems make it unlikely that these cells communicate extensively in nature where quorums are rare. However, the sessile cells in the microcolonies that comprise multispecies biofilms (Fig. 4) are in stable juxtaposition with many cells, and I predict that the next few years will see the discovery of literally hundreds of signals that serve as the hormones and pheromones that regulate these complex communities. The small distances between adjacent cells in biofilms (6–8 μm) does not preclude the possibility that electrical signals may also operate within these communities.

BIOFILMS AS MULTICELLULAR COMMUNITIES

A successful community is invariably 'greater than the sum of its parts', and the history of microbiological research illustrates this point very well. In rumen microbiology, we could see that a biofilm community formed on the surfaces of cellulosic plant materials undergoing digestion, but the rates of cellulose digestion by isolated members of this community fell far short of the rates of digestion seen in the rumen of even the most inept bovine. Direct microscopic observation of material taken from the rumen showed that cells of two bacterial species tended to coexist on the feed (Fig. 5), and, when cells of these two species were mixed and used to colonize sterile cellulose (Kudo *et al.*, 1987), the rate of digestion approached that of the natural system. The mobile cells of the *Treponema* species scavenged butyrate from the biofilm formed by the primary cellulose degrader, and reversed the feedback inhibition that was inhibiting the whole process of cellulose digestion in the monospecies biofilm. It is obvious that the structured juxtaposition of metabolically co-operative cells is as beneficial to biofilm communities as it is to the tissues and organ systems of higher multicellular organisms. Doug Caldwell rightly insists (Caldwell & Costerton, 1996) that the characteristics of a bacterial species that make it an effective partner in a metabolically co-ordinated community are much more important to its evolution than any properties that allow it to succeed as a single planktonic cell in nature. It is perhaps serendipitous that the dental and industrial biofilms that were recognized long before the word was coined

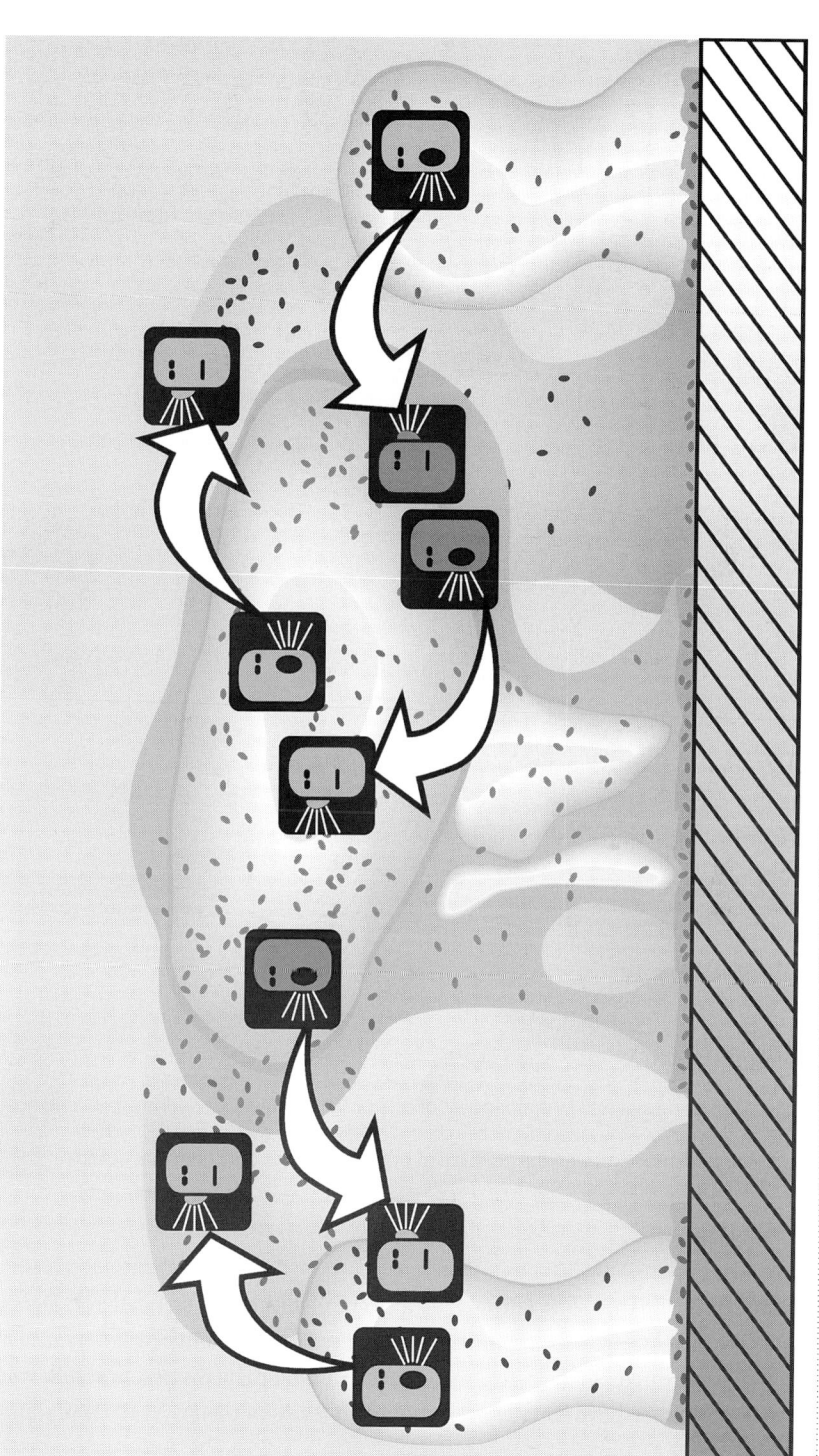

Fig. 4. Cartoon illustrating the point that interspecies and intraspecies signalling mechanisms must be operative in biofilms, in order for these sessile communities to form their elaborate structures and, especially, for the maintenance of their open water channels.

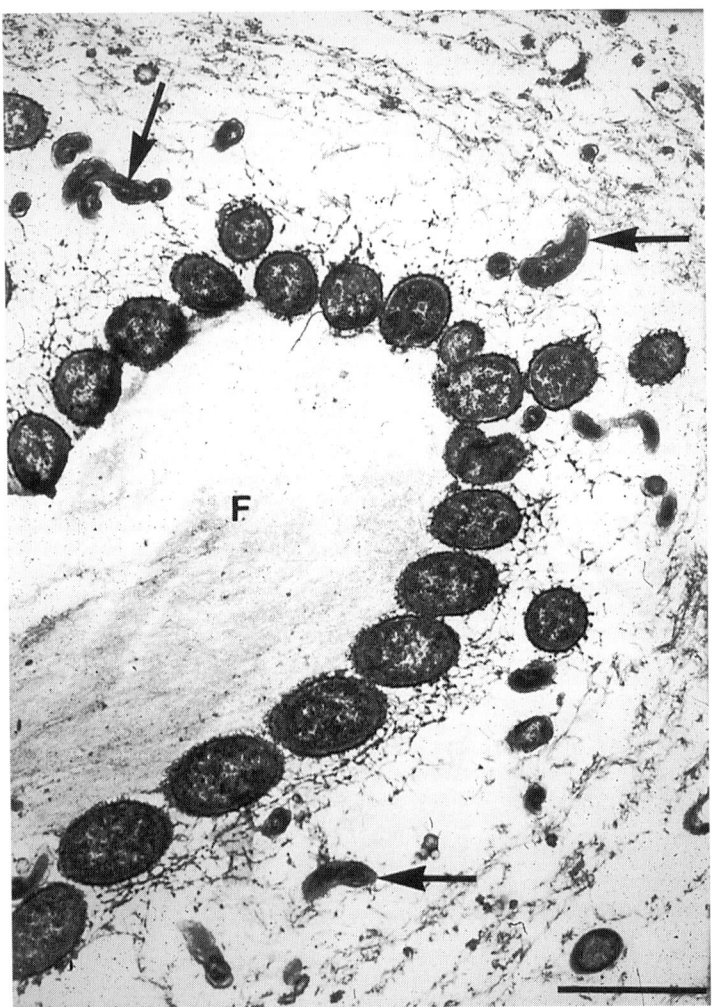

Fig. 5. Transmission electron micrograph of a stained section of a rumen specimen in which a cellulose fibre (F) is being digested by a monolayer of cells of *Fibrobacter fibrosolvens* within a fibrous biofilm. This biofilm has been invaded by mobile cells of a *Treponema* species (arrows), which accelerate this digestion by removing the products of primary digestion (butyrate) that would otherwise inhibit this process by feedback inhibition.

(Costerton *et al.*, 1978) are sufficiently thick as to be visible to the unaided eye. Dental plaque and the biofilms that form on the fixed film reactors used in sewage treatment have always been treated as integrated microbial communities by the practical folk who manage them, and they have usually been studied in terms of their overall community activities.

Biofilms as adaptive and ecologically functional communities

Biofilm communities assume a certain austere elegance when we consider their structure and function in terms of the evolutionary circumstances in which they evolved. It is certain that these assemblages of prokaryotic cells were the first multicellular communities to evolve on earth, and their basic organization certainly developed in response to the conditions operative in the primitive environment. Modern biofilms place their component bacteria in immediate juxtaposition to photosynthetic organisms and to other nutrient sources, including such autotrophic sources as oxidized minerals (e.g. elemental sulphur). Even when heterotrophic bacteria are not juxtaposed to sources of organic nutrients, the biofilm matrix has the property of trapping and concentrating organic nutrients (Costerton *et al.*, 1995), and biofilms are always the predominant form of bacterial growth in pristine mountain streams. The microcolony and channel structure of the biofilm, as illustrated in Fig. 3, represents a primitive circulatory system in terms of nutrient delivery and waste removal, and constitutes a very efficient way of supporting the growth of a large number of cells in a limited surface area. The short and stable diffusional distance between metabolically co-operative cells within mixed-species microcolonies represents a physiologically integrated structure, not dissimilar to that seen in the tissues of higher organisms, except that the co-operating elements do not all share the same genotype. In some ways, microbial biofilms presage the subsequent development of multicellular organisms whose cells are derived from the same genotype because these sessile communities were able to achieve a primitive circulation, and a measure of cellular specialization and metabolic integration, using cells of many different genotypes.

One of the most important properties of any successful community is its ability to respond to stress, and to survive in the face of adverse circumstances. In this matter, the biofilm mode of growth and the multigenomic nature of the mixed-species biofilm are both of pivotal importance. Biofilm communities are, by definition, stationary, and this stationary mode of growth reduces the hazards of a very hostile environment in which planktonic cells would be swept from one dangerous location to the next, until they perished. Cells growing in the biofilm mode of growth, in the protected biofilm phenotype, show a phenomenal resistance to chemical antagonists, including acids and antibiotics (Nickel *et al.*, 1985). They are also remarkably resistant to uptake by the amoebae that were probably amongst the first eukaryotes to challenge prokaryotic biofilms for dominance in the primitive earth. In addition to the survival values inherent in the biofilm mode of growth, and in the biofilm phenotype, the physiological and genetic heterogeneities that are characteristic of biofilms add another level of assurance of survival. Local areas within biofilms develop their own chemically distinct microenvironments, or microniches (Costerton *et al.*, 1994), and modern studies of the killing of biofilms with antibiotics and sterilants clearly show that sessile cells often survive in local 'pockets' in which the agent is less effective. An antibacterial agent must be active in a very wide range of pH values, and in an equally wide range of oxygen tensions, if it is to be able to kill all of the sessile cells even in a monospecies biofilm. If all of the microcolonies within a biofilm are not killed by an antibacterial agent, the community will recover to approximately its pre-challenge dimensions in only a few days. However, it is probably the multigenomic character of mixed-species biofilms that gives them their most effective properties in survival and in their ability to respond to changing environments. A biofilm community is usually composed of dozens of different species of bacteria, with different genomes that dictate both the metabolic potentials and the environmental susceptibilities of each individual sessile cell. In any given nutrient situation, or in any stage in the development of the biofilm, some sessile cells will be more metabolically active than others and we know that many of the cells in biofilms are often dormant or quiescent (Stewart, 1996). Each sessile cell has its own genetically determined metabolic potentials, and its own genetically determined susceptibilities to antibacterial agents, and this gives these microbial communities their remarkable resistance to adverse circumstances and their equally remarkable ability to respond to new opportunities. In any stress situation, one or two species are likely to survive, because of some property of their genome, and any new nutrient opportunity is bound to suit the genetically determined capabilities of some members of the community.

Biofilms as dynamic communities

Microbial biofilms are dynamic in both their mechanical properties and their community structure. Because of the limitations of pictures on a printed page, and of our imaginations, we imbue biofilms with a certain rigidity, even though we know that many macroscopic biofilms are relatively soft and very pliable. Recently, Paul Stoodley and some members of Zbigniew Lewandowski's group of engineers have explored the viscoelastic properties of biofilms, in the same terms that would be used to describe similar properties in any material, and biofilms have proven to be highly compliant materials (Stoodley et al., 1999b). Microcolonies that grow symmetrically (Fig. 3) in low shear laminar flows become elongated when exposed to higher shear turbulent flows, where they form filamentous streamers that oscillate in the bulk fluid and exert a measurable drag on the fluid flow (Stoodley et al., 1998). Also, microcolonies are not necessarily stationary at a given location on a colonized surface, as depicted in Fig. 3, but they move along the surface at a measurable 'creep' rate (Stoodley et al., 1999c). Even more remarkably, biofilms in high shear environments (Reynolds number of >3200) can form well-defined wave structures (Stoodley et al., 1999c) that travel downstream throughout the sessile community. Biofilms are not mechanically stable, and streamers detach into the bulk fluid when the tensile strength of their matrix is exceeded, while large pieces of biofilm are detached where the waves in biofilms 'break' on the 'shore' of the colonized area of a surface. Generally, biofilms that form at high shear rates are stronger, and less liable to break and detach when stressed, but the mechanism(s) of this alteration in tensile strength is not understood.

The community structure of biofilms is as fully dynamic as their mechanical properties. Biofilms are formed by the recruitment of planktonic cells from the bulk phase of the ecosystem (Fig. 6), and these sessile communities shed similar planktonic cells into the same bulk phase throughout their life cycle. The detachment of planktonic cells from biofilm microcolonies is an active and natural process that, at least in biofilms formed by cells of *P. aeruginosa*, is mediated by a specific lyase enzyme (Boyd & Chakrabarty, 1994) and controlled by a specific AHL signal (D. Davies, personal communication). The detachment of planktonic cells from biofilm microcolonies is a spontaneous process that occurs frequently during the growth of a biofilm, and large areas of the sessile thermal mats that colonize hot springs are sometimes denuded by mass sloughing. It has even been suggested that there may be a measure of diurnal control of biofilm detachment in some aquatic ecosystems, so that planktonic cells of certain species are shed at certain times of the day. Detachment may involve the release of a few cells from a microcolony, but spectacular detachment events have often been recorded in time-lapse studies of biofilm development. In these cases, cells at the centre of a particular microcolony will start to move in a 'seething' manner, which indicates that they have dissolved the local matrix material and developed flagella, and the mobile cells will gradually come to occupy more and more of the bulk of the microcolony. At some point in this process, the mobile cells will breach the integrity of the matrix somewhere at the edge of the microcolony, and they will then swim away, as the remainder of the microcolony dissolves and releases the remaining sessile cells as transformed planktonic swarmer cells. Often a new microcolony forms in the space left by the wholesale detachment of all of the cells of the former microcolony, and the whole biofilm community looks little changed.

In many aquatic ecosystems, like deep groundwater and the abyssal areas of the oceans, there are few nutrients to support the growth of biofilms, and huge numbers of planktonic cells exist as dormant forms as a result of the well-documented starvation survival strategy (Kjelleberg, 1993). Because planktonic bacteria are swept into virtually all aquatic ecosystems, from rich terrestrial sources and from animal excretions, the bulk phases of these systems contain an almost infinite variety of bacterial species whose genomes are intact even if their cells are not metabolically active. These genomes are the building blocks of new biofilms, they are functionally ubiquitous, and even their dormant components are readily resuscitated when conditions become favourable. For these reasons, biofilms can self-assemble, in a matter of hours, wherever nutrients become available at an uncolonized surface. Cases in point are the 'black smokers' that spew H_2S wherever volcanic activity breaches the ocean floor, and virtually identical biofilms form at these widely distributed random

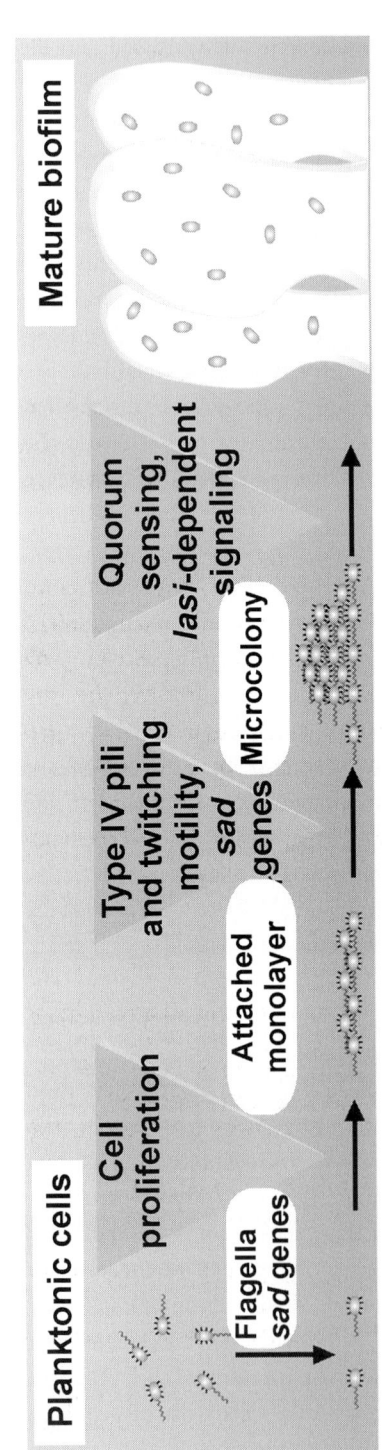
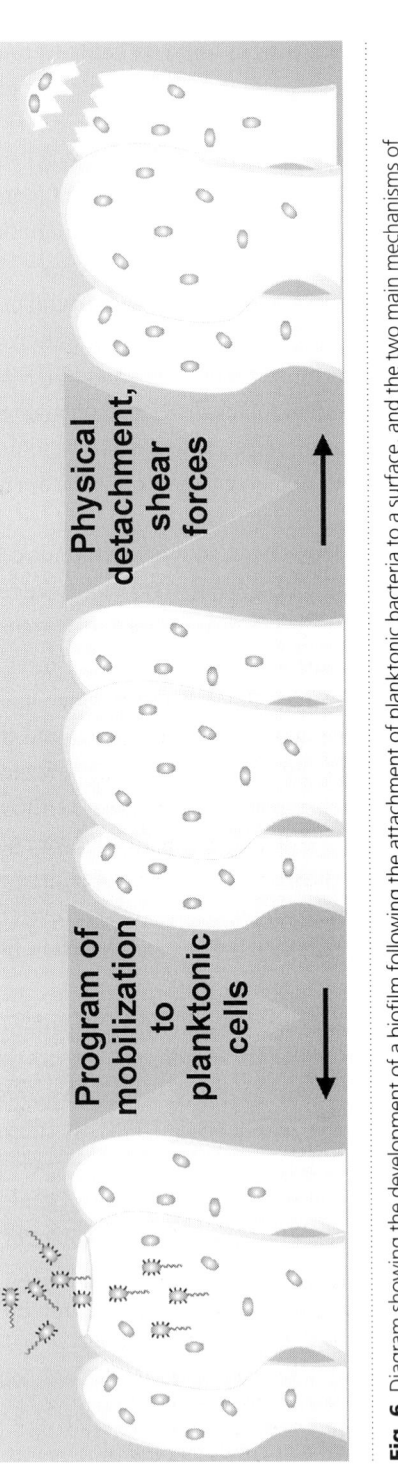

Fig. 6. Diagram showing the development of a biofilm following the attachment of planktonic bacteria to a surface, and the two main mechanisms of detachment of planktonic cells from the biofilm. Reprinted from *Science* May 21, 1999, p. 1321, with permission.

locations, from the reservoir of bacterial species in the bulk phase of the ocean and in response to the spontaneous presence of an available nutrient. The remarkable community dynamics of biofilms allow bacteria to react to colonization opportunities with amazing speed, and the detachment strategy of these sessile populations guarantees that planktonic cells of suitable species will always be readily available to colonize any suitable sites downstream from the 'mother' community. This dynamic interplay between biofilms and the planktonic cells in the bulk fluid of many ecosystems contrasts with the limited reactions of multicellular plants and animals, who usually must invoke sexual reproduction in response to similar opportunities and therefore cannot respond as quickly. This ability of microbial biofilms to self-assemble from a vast reservoir of planktonic cells of an almost infinite number of different genomes may be the most important single factor in the predominance of these remarkable multicellular communities on surfaces in even the most hostile of earth's environments.

VALETE

It should make us happy to discover new opportunities, and there should be no recriminations concerning opportunities missed. The present flood of exciting direct observations of biofilm structure and function leave us frantically searching for explanations, and rummaging through the molecular mechanisms discovered in planktonic cells to see if any of them fit these circumstances. Decades will probably pass before we understand the workings of different sessile communities sufficiently well to invoke the correct signals and enzymes to explain what we see by direct observations of living biofilms. However, we are catching up, and we are finally studying bacteria in the communities in which most of them actually live.

REFERENCES

Boyd, A. & Chakrabarty, A. M. (1994). Role of alginate lyase in cell detachment of *Pseudomonas aeruginosa*. *Appl Environ Microbiol* **60**, 2355–2359.

Caldwell, D. E. & Costerton, J. W. (1996). Are bacterial biofilms constrained to Darwin's concept of evolution through natural selection? *Microbiologia* **12**, 347–358.

Costerton, J. W., Geesey, G. G. & Cheng, K.-J. (1978). How bacteria stick. *Sci Am* **238**, 86–95.

Costerton, J. W., Lewandowski, Z., DeBeer, D., Caldwell, D., Korber, D. & James, G. (1994). Minireview: Biofilms, the customized micronich. *J Bacteriol* **176**, 2137–2142.

Costerton, J. W., Lewandowski, Z., Caldwell, D. E., Korber, D. R. & Lappin-Scott, H. M. (1995). Microbial biofilms. *Annu Rev Microbiol* **49**, 711–745.

Costerton, J. W., Stewart, P. S. & Greenberg, E. P. (1999). Bacterial biofilms: a common cause of persistent infections. *Science* **284**, 1318–1322.

Davies, D. G. & Geesey, G. G. (1995). Regulation of the alginate biosynthesis gene *algC* in *Pseudomonas aeruginosa* during biofilm development in continuous culture. *Appl Environ Microbiol* **61**, 860–867.

Davies, D. G., Parsek, M. R., Pearson, J. P., Iglewski, B. H., Costerton, J. W. & Greenberg, E. P. (1998). The involvement of cell-to-cell signals in the development of a bacterial biofilm. *Science* **280**, 295–298.

Foley, I., Marsh, P., Wellington, E. M. H., Smith, A. W. & Brown, M. R. W. (1999). General stress response regulator *rpoS* is expressed in human infection: a possible role in chronicity. *J Antimicrob Chemother* **43**, 164–165.

Fuqua, W. C., Winans, E. P. & Greenberg, E. P. (1994). Quorum sensing in bacteria: the LuxR-LuxI family of cell density-responsive transcriptional regulators. *J Bacteriol* **176**, 269–275.

Geesey, G. G., Richardson, W. T., Yeomans, H. G., Irvin, R. T. & Costerton, J. W. (1977). Microscopic examination of natural sessile bacterial populations from an alpine stream. *Can J Microbiol* **23**, 1733–1736.

Kjelleberg, S. (1993). *Starvation in Bacteria*. New York: Plenum.

Kudo, H., Cheng, K.-J. & Costerton, J. W. (1987). Interactions between *Treponema bryantii* and cellulolytic bacteria in the *in vitro* degradation of straw cellulose. *Can J Microbiol* **33**, 244–248.

Lawrence, J. R., Korber, D. R., Hoyle, B. D. & Costerton, J. W. (1991). Optical sectioning of microbial biofilms. *J Bacteriol* **173**, 6558–6567.

Marshall, K. C. (1985). Mechanisms of bacterial adhesion at solid-liquid interfaces. In *Bacterial Adhesion: Mechanisms and Physiological Significance*, pp. 133–161. Edited by D. C. Savage & K. C. Marshall. New York: Plenum.

Nickel, J. C., Ruseska, I., Wright, J. B. & Costerton, J. W. (1985). Tobramycin resistance of cells of *Pseudomonas aeruginosa* growing as a biofilm on urinary catheter material. *Antimicrob Agents Chemother* **27**, 619–624.

O'Toole, G. A. & Kolter, R. (1998). Flagellar and twitching motility are necessary for *Pseudomonas aeruginosa* biofilm development. *Mol Microbiol* **30**, 295–304.

Parsek, M. R., Val, D. L., Hanzelka, B. L., Cronan, J. E., Jr & Greenberg, E. P. (1999). Acyl homoserine-lactone quorum-sensing signal generation. *Proc Natl Acad Sci USA* **96**, 4360–4365.

Stewart, P. S. (1996). Theoretical aspects of antibiotic diffusion into microbial biofilms. *Antimicrob Agents Chemother* **40**, 2517–2522.

Stoodley, P., Lewandowski, Z., Boyle, J. & Lappin-Scott, H. M. (1998). Oscillation characteristics of biofilm streamers in flowing water as related to drag and pressure drop. *Biotechnol Bioeng* **57**, 536–544.

Stoodley, P., Jørgensen, F., Williams, P. & Lappin-Scott, H. M. (1999a). The role of hydrodynamics and AHL signalling molecules as determinants of the structure of *Pseudomonas aeruginosa* biofilms. In *Biofilms: the Good, the Bad, and the Ugly*, pp. 323–330. Edited by J. W. T. Wimpenny, P. Gilbert, J. Walker, M. Brading & R. Bayston. Cardiff: BioLine.

Stoodley, P., Lewandowski, Z., Boyle, J. D. & Lappin-Scott, H. M. (1999b). Structural deformation of bacterial biofilms caused by short term fluctuations in flow velocity: an *in situ* demonstration of biofilm viscoelasticity. *Biotechnol Bioeng* **65**, 83–92.

Stoodley, P., Lewandowski, Z., Boyle, J. D. & Lappin-Scott, H. M. (1999c). The formation of migratory ripples in a mixed species bacterial biofilm growing in turbulent flow. *Environ Microbiol* **1**, 447–457.

Yu, H. (1994). *Transcriptional and phenotypic variations of Pseudomonas aeruginosa during in vitro and in vivo biofilm development*. PhD thesis, University of Calgary.

INDEX

References to tables/figures are shown in italics